Controlled Release of Pesticides for Sustainable Agriculture

Rakhimol K. R. · Sabu Thomas ·
Tatiana Volova · Jayachandran K.
Editors

Controlled Release of Pesticides for Sustainable Agriculture

Springer

Editors
Rakhimol K. R.
International and Inter University Centre
for Nanoscience and Nanotechnology
Mahatma Gandhi University
Kottayam, Kerala, India

Tatiana Volova
Institute of Biophysics SB RAS
Siberian Federal University
Krasnoyarsk, Russia

Sabu Thomas
School of Chemical Sciences
Mahatma Gandhi University
Kottayam, Kerala, India

Jayachandran K.
School of Biosciences
Mahatma Gandhi University
Kottayam, Kerala, India

ISBN 978-3-030-23395-2 ISBN 978-3-030-23396-9 (eBook)
https://doi.org/10.1007/978-3-030-23396-9

This Springer imprint is published by the registered company Springer Nature Switzerland AG
The registered company address is: Gewerbestrasse 11, 6330 Cham, Switzerland

Preface

The rapidly increasing demand for food presents a major challenge to the scientific community. More efficient developments in agricultural and industrial sector are needed to face this issue. Novel techniques which result in high-quality products with high yield and reduced adverse effects are required. In ancient times, a major problem which scientists faced was the inefficiency of the pest control agents due to the inability of these products to reach the target and deliver optimum quantity within a short span of time. Mostly, the conventional mode of pesticide application resulted in wastage of 80% of applied pesticides through leaching, degradation, and evaporation. This intern resulted in reduced performance and environment pollution. This led the scientists to think about delivering chemicals to the target in a controlled manner. With the use of novel technologies such as nanotechnology and polymer technology, scientists could develop controlled release pesticides.

In this book, we discuss different aspects of controlled release pesticides. Chapter 1 of this book gives a general introduction about the conventional pesticide application and its adverse effects on health and environment. Chapter 2 discusses the methods for pesticide analysis and their role in the agriculture fields. Chapter 3 deals with the chemical nature and toxicology behind different chemical pesticides and insecticides of today's use. In Chap. 4, we give an overview of controlled release pesticides and their bioengineering aspects for the sustainable crop production. In Chap. 5, we explain different methods for synthesizing controlled release pesticides.

Types of slow and controlled release pesticides and their characteristics are discussed in Chap. 6. Types of pesticides based on different parameters and their physical as well as chemical characteristics are included in this chapter. Nanotechnology has played a major role in the field of controlled release technology. Nanotechnological aspects in pest control are included in Chap. 7. To release pesticides in a controlled manner, encapsulation in a porous carrier is needed. Polymers play a major role as pesticide carrier. Polymer technology for pesticide release is incorporated in Chap. 8. To detect the release of pesticides from carriers, novel sensors were developed by many scientists. Chapter 9 provides a detailed data on fabrication and working of such sensors. Chapters 10 and 11 are dealing with controlled release of herbicides and plant hormones to tune the crop production.

Kottayam, India Rakhimol K. R.
Kottayam, India Sabu Thomas
Krasnoyarsk, Russia Tatiana Volova
Kottayam, India Jayachandran K.

Contents

Chapter 1
Conventional Methods of Pesticide Application in Agricultural Field and Fate of the Pesticides in the Environment and Human Health

V. Dhananjayan, S. Jayakumar and B. Ravichandran

Abstract The increasing growth of world population required raising overall food production. In order to meet the requirement, food supply factors of increasing yields, expanding agricultural area and increasing the productivity of crop and animal agricultural are necessary. As a consequence, application of pesticide for protection of crop involves wide variety of chemicals with various applicable techniques. The pesticides are very commonly used in agriculture sector with several methods of applications in recent years. However, in developing countries, the conventional method of application of pesticide is the major practice. The knowledge and awareness of application play a major role in fate of pesticides in the environment and human health. Several studies have reported the handling practices and subsequent exposure of farmers to pesticides and pesticide accumulation. In addition to the preparation and application of pesticides, there are several important problems related to pesticide use that should be understood by every applicator. Generally, in the conventional way, application of pesticides did not include any regular inspections for safety procedure and follow-up inspections of problem areas. Some of the methods, however, may result in problems such as over application, not reaching the pest habitat while using a minimum amount of pesticide, and depositing unsafe residues. Proper techniques of application not only aid in effectiveness but also ensure workers' safety, public protection and protection of the environment. Although newer pesticides are capable of degrading fast, the adverse impact of these chemicals with already existing persistent pesticides on environment including humans is reported in recent years. The application of pesticide is not merely the operation of sprayer or duster. It has to be coupled with a thorough knowledge of the pest problem. The application techniques ideally should

V. Dhananjayan (✉) · B. Ravichandran
Industrial Hygiene and Toxicology Division, ICMR—Regional Occupational Health Centre (Southern), NIOH, ICMR Complex, Bangalore 562110, India
e-mail: Dhananjayan_v@yahoo.com

S. Jayakumar
Department of Zoology and Wildlife Biology, A.V.C. College (Autonomous), Mannampandal, Mayiladuthurai 609305, Tamil Nadu, India

Rakhimol K. R. et al. (eds.), *Controlled Release of Pesticides for Sustainable Agriculture*, https://doi.org/10.1007/978-3-030-23396-9_1

be target oriented so that safety to the non-targets and the environment is ensured. Therefore, environmentally responsive controlled release of pesticide through advance technologies is needed and recommended. This chapter deals with the different types of pesticide application, their distribution and occurrence in the environment and human health-related issues.

Keywords Pesticide application · Fate of pesticides · Health effects

1.1 Introduction

Agriculture is the most important enterprise and it is the major key to economic development in many developed countries. The agriculture remains traditional and labour intensive; little or no external inputs are used. Pest management practice under these conditions is a built-in process in the overall crop production system rather than a separate well-defined activity [1]. The most important concern associated with agricultural production is the problem of pests (weeds and insects). An important step for raising productivity and increasing the gross output of agricultural products is the elimination of losses of the harvest due to pests, plant diseases and weeds. The farmers use pesticides to control weeds, microbes, insects, rodents and other higher organisms to protect their crops and preserve food materials [2–4]. In the initial stages of agricultural practice, the cultivation together with crop and animal husbandry was largely eco-friendly, although the yields were not as high as of the present day. During 1874, the German chemist Othmar Zeidler had first synthesized dichlorodiphenyltrichloroethane (DDT) compound, but failed to realize its value as an insecticide. Similarly, a series of chemical pesticides were synthesized to protect crop plants against insect pests. While the immediate benefits of these pesticides were indeed quite impressive, their long-term harmful effects on environment and non-target organisms came to be understood only after much damage to ecosystems had already been done. After the description of Carson [5] on deleterious effects of DDT and other chlorinated pesticides in her book on 'Silent Spring', there has been a growing campaign against the use of chemical pesticides in agriculture. As an alternative to the organochlorine pesticides (OCPs), organophosphates (OPs) and carbamate pesticides are largely used in agriculture field. The introduction of these synthetic insecticides—organophosphate (OP) in the 1960s, carbamates in 1970s and pyrethroids in 1980s—and herbicides and fungicides in the 1970s–1980s contributed greatly to pest control and agricultural output. Ideally, a pesticide must be lethal to the targeted pests, but not to non-target species, including man. Unfortunately, this is not the case, so the controversy of use and abuse of pesticides has surfaced. The rampant use of these chemicals, under the adage, "if little is good, a lot more will be better" has played havoc with humans and other life forms.

During the past three decades or so, a significant amount of Integrated Pest Management (IPM) or IPM-related research has been conducted in many parts of the world including in the developing countries [4, 6]. However, the use of chemical pesticides continue for a long time in small quantities, particularly in highly populated developing countries like India and China [7]. Agriculture in the twenty-first century faces multiple challenges: it has to produce more food and fibre to feed a growing population with a smaller rural labour forces and more feedstocks for a potentially huge bio-energy market, contribute to overall development in many agriculture-dependent developing countries, adopt more efficient and sustainable production methods and adapt to climate changes [8]. There is limited scope for significantly expanding agricultural land after constraints and trade-offs are considered. The incorporation of new lands into production is likely to come with important social and ecological constraints and costs. Therefore, the initiation of exploitative agriculture without a proper understanding of the various consequences of every change introduced into traditional agriculture, and without first building up a proper scientific and training base to sustain, it may only lead us, in the long run, into an era of agricultural disaster rather than one of agricultural prosperity.

The conventional methods of pesticide applications are the major practice in developing countries. Persistence and accumulation nature of pesticides on various stages of food chain and environment affects human. Humans are exposed to pesticides through occupational or environmental exposure. The occupational exposures occur among rural workers, greenhouse workers, workers in pesticide manufacturing [2]. The group that receive the greatest exposure are those who involved in mixing, loading, transporting and applying pesticides. Furthermore, the lack of training and equipment to safely handle pesticides increases the health risk [9]. Gangemi et al. [10] summarized the most recent findings on the association between occupational pesticide exposure and the development of chronic diseases. As regards to environmental exposure, this mainly affects the general population through the consumption of polluted food and drinking contaminated water or using pesticides in the home or living close to sprayed fields. Thus, risk assessment and prevention from pesticide exposure are not a simple process, particularly when we consider several limitations, such as differences in the time and the levels of exposure, the class of pesticides (chemical structure and toxicity), mixtures or cocktails used, and the geographical and climate features of the areas where pesticides are applied. For these reasons, it has been suggested that pesticide safety should be tested before and after marketing, in order to evaluate whether the estimated or predicted individual exposure on a given task is suitable [11].

Pesticide applicators utilize many methods in the application of pesticides. Most of these methods have been developed as the result of years of studying the most effective methods. Some of the methods, however, may result in problems such as over application, not reaching the pest habitat while using a minimum amount of pesticide, and depositing unsafe residues. Proper techniques of application not only aid in effectiveness but also ensure workers' safety, public protection and protection of the environment [2–4, 12]. Competence in pest control includes basic knowledge of pests and pest problems, the ability to choose the right pesticides and equipments, and knowledge of proper methods of application. The proper technique of application probably plays a greater part in achieving the target job. Therefore, the present chapter was aimed at exploring the existing conventional method of pesticide application and the consequence of pesticide in man and environment. A brief continuation of the types of equipment and techniques of application used in agriculture practices, maintenance of equipment, knowledge of pesticide applicator, fate of pesticides in the environment, health of effects associated with pesticides and various modern techniques in controlled release of pesticides in agriculture field are presented in the following section.

1.2 Importance of Pesticide Application

Agriculture was developed to produce crops and livestock for human consumption. As the human population increases, the amount of food produced is very important. Unfortunately, there are other organisms out there that want to consume the crops that are meant for humans. It is estimated that nearly 37% of all crops produced in the USA each year are destroyed by agricultural pests, which results in an economic loss of around $122 billion a year. Due to this high loss in food production, pesticides are often used to combat the problem. There are many different types of pesticides on the market today, but the most common are herbicides and insecticides, which kill or manage unwanted plants and insects. The damage caused by agricultural pests is a global problem, and over the past half-century, the amount of pesticides used have increased severalfold. Over the years, the widespread use of pesticides had several benefits and also have caused damage to the environment and human health [12–15]. The benefits of pesticides include increased food production, increased profits for farmers and the prevention of diseases. Although pests consume or harm a large portion of agricultural crops, without the use of pesticides, it is likely that they would consume a higher percentage. Due to the use of pesticides, it is possible to combat pests and produce larger quantities of food. By producing more crops, farmers are also able to increase profits by having more produce to sell. Pesticides also increase farm profits by helping the farmer save money on labour costs. Usage of pesticides reduces the amount of time required to manually remove weeds and pests from fields. In addition to saving crops and livestock, pesticides also have direct benefits to human health. It is estimated that since 1945, the use of pesticides has prevented the deaths of around seven million people by killing pests that carry or transmit diseases. Malaria is transmitted by infected mosquitoes, known as deadly diseases to human that have decreased in prevalence due to the use of pesticides. Other diseases were also minimized due to the use of pesticides include the bubonic plague transmitted by both fleas and body lice [13]. Similary several diseases are being curtailed by application of pesticides.

1.2.1 Benefits of Pesticides

The primary benefits are the consequences of the pesticides' effects—the direct gains expected from their use. For example, the effect of killing caterpillars feeding on the crop brings the primary benefit of higher yields and better quality. The main effects result in primary benefits ranging from protection of recreational turf to saved human lives. The secondary benefits are the less immediate or less obvious benefits that result from the primary benefits. They may be subtle, less intuitively obvious or of longer term. It follows that for secondary benefits it is therefore more difficult to establish cause and effect, but nevertheless they can be powerful justifications for pesticide use. For example, the higher cabbage yield might bring additional revenue that could be put towards children's education or medical care, leading to a healthier, better educated population. There are various secondary benefits identified, ranging from fitter people to conserved biodiversity.

1.2.2 Improving Productivity

Tremendous benefits have been derived from the use of pesticides in forestry, public health and the domestic sphere—and, of course, in agriculture, a sector upon which the countries like Indian economy is largely dependent. Food grain production stood at a mere 50 million tons in 1948–49, had increased almost fourfold to 198 million tons by the end of 1996–97 from an estimated 169 million hectares of permanently cropped land. This result has been achieved by the use of high-yield varieties of seeds, advanced irrigation technologies and agricultural chemicals (Employment Information: Indian Labour Statistics, 1994). Similarly, outputs and productivity have increased dramatically in most countries, for example wheat yields in the UK and corn yields in the USA. Increases in productivity have been due to several factors including use of fertilizers, better varieties and use of machineries. These pesticide are used to control various pests and diseaes carriers, such as mosquitoes, ticks, rats and mice [16]. Therefore, pesticides have been an integral part of the process by reducing losses from the weeds, diseases and insect pests that can markedly reduce the amount of harvestable produce. There are many different types of pesticides; each is meant to be effective against specific pests. Some examples are given in Table 1.1.

Table 1.1 Type of pesticides used and its target function

S. no.	Pesticide	Target species/function
1	Acaricide	Mites, ticks
2	Algaecide	Algae
3	Anticoagulant	Rodents
4	Attractant	Attracts insects or birds
5	Avicide	Birds
6	Bactericide	Bacteria
7	Defoliant	Plant leaves
8	Desiccant	Disrupts water balance in arthropods
9	Fungicide	Fungi
10	Growth regulator	Regulates insect and plant growth
11	Herbicide	Weeds
12	Insecticide	Insects
13	Miticide	Mites
14	Molluscicide	Snails, slugs
15	Nematicide	Nematodes
16	Piscicide	Fish
17	Predacide	Vertebrate predators
18	Repellent	Repels vertebrates or arthropods
19	Rodenticide	Rodents
20	Silvicide	Woody vegetation

1.3 Types of Application

Pesticide application plays an important role in pest management. Proper technique of application of pesticide and the equipment used for applying pesticide are vital to the success of pest control operations. The application of pesticide is not merely the operation of sprayer or duster. It has to be coupled with a thorough knowledge of the pest problem [2, 4, 17]. The use of pesticides involves not only of knowledge on application equipment, but also of pest management as well. The main purpose of pesticide application technique is to cover the target with maximum efficiency and minimum efforts to keep the pest under control as well as minimum contamination of non-targets. All pesticides are poisonous substances, and they can cause harm to all living things. Therefore, their use must be very judicious. The application techniques ideally should be target oriented so that safety to the non-targets and the environment is ensured. Therefore, proper selection of application equipment, knowledge of pest behaviour and proper dispersal methods are vital [15, 17–19]. The complete knowledge of pest problem is important to define the target, i.e. location of the pest (on foliage, under the leaves, at root zone, etc.). The most susceptible stage of the pest for control measures will help to decide the time of application. The requirement of coverage and spray droplet size depends upon the mobility and size of the pest. The mode of action of pesticide, its relative toxicity and other physicochemical properties help to decide the handling precautions, agitation requirement, etc. Further, the complete knowledge of the equipment is necessary to develop desired skill of operation, to select and, to estimate the number and type of equipment needed to treat the crop in minimum time and to optimize use of the equipment.

The objective of the application of pesticide is to keep the pest under check. The pest population has to be kept suppressed to minimum biological activities to avoid economic loss of crop yields. Thorough killing of pest or eradication of pest is neither practical nor necessary. The objective of pesticide application besides keeping the pest population under check should also be to avoid pollution and damage to the non-targets. The success of pest control operations by pesticide application greatly depends on the following factors:

1. Quality of pesticide.
2. Timing of application.
3. Quality of application and coverage.

Different types of pesticides are used for controlling various pests. For example, insecticides are applied against insect pests, fungicides against crop diseases, herbicides against weeds, etc., in order to protect the crop losses. But it is essential that besides choosing an appropriate pesticide for application it has to be a quality product, i.e. proper quantity of pesticide active ingredient (*a.i*) must ensure that the quantity is maintained in production and marketing of pesticide formulations. The application of pesticide is very successful when applied at the most susceptible stage of the pest. If the timing of pesticide application is carefully considered and followed, the results will be good and economy. Therefore, for large area treatment, careful selection of equipment becomes necessary so that within the available time the area could be treated. Even though good quality pesticide is used, an optimum timing for the application of pesticide is also adopted; unless the pesticide is applied properly, it will not yield good results. Therefore, the quality of application of pesticides is very important in pest control operations. The following points can be ensured during pest control operation.

1. Proper dosage should be applied evenly.
2. The toxicant should reach the target.
3. Proper droplet size.
4. Proper density of droplet on the target.

The dosage recommendation is generally indicated for acre or hectare, e.g. kg/ha or L/ha or g *ai*/ha. It should be properly understood, and the exact quantities of the formulated pesticide should be applied.

Pesticides are dispersed by different methods like spraying, dusting, etc. For spraying of pesticides, different types of nozzles such as hydraulic, air blast, centrifugal and heat energy type are used. Water is a common carrier of pesticides, but air or oils or soap solutions are also used as carriers. Selection of proper droplet is an important consideration. The shape, size and surface of the target vary greatly. For spraying against flying insects, the hydraulic nozzles will not be effective. Hence fine size spray particles are needed to remain airborne for longer time. However, for weed control operation, usually the requirement is drift-free application or coarse spray droplets. An adequate number of spray droplets should be deposited necessarily. For fungicide application, the number of droplets deposited per unit area should be more, and maybe for translocated herbicide application, it can be less in number. It may need fewer numbers of droplets to be deposited in case of highly mobile (crawling) insect pests. The pesticides are formulated in liquid form, dust powder or granule forms such that it makes possible to apply small quantities of pesticides over large area. Some of the pesticides are applied as low as few grams *a.i.* per hectare. Therefore, the adoption of proper application technique is vital for uniform depositing of pesticide.

1.4 Equipments Used for Applications

1.4.1 Spraying Techniques

The liquid formulations of pesticide are either diluted (with water, oil) or directly applied in small drops to the crop by different types of sprayers [20]. Usually, the formulations and wettable powder are diluted suitably with water which is a common carrier of pesticides. In some cases, however, oil is used as diluent or carrier of pesticides. The important factors for spray volume consideration are depends upon the spray type and coverage, total target area, size of spray droplet and number of spray droplets. It is obvious that if the spray droplets are in coarse size, then the spray volume required will be larger than the small size spray droplets. Also if the thorough coverage (e.g. both the sides of leaves) is necessary, then the spray volume requirement has to be more.

On the basis of the volume of spray mix, the technique of spraying is classified as:

1. High-volume spraying (300–500 L/ha).
2. Low-volume spraying (50–150 L/ha).
3. Ultra-low-volume spraying (<5 L/ha).

The range of volume of spray mix in each of the above case is arbitrary. Usually, for field crop spraying, the spray volume ranges are taken as guide. There is distinct advantage in the case of lower volume application over the high-volume application. The higher the volume to be applied, the more the time, the more the labour and the more the cost of application due to labour cost. However, the lower volume applications are concentrated spraying of pesticide which should also be considered properly.

1.4.2 Dusters

Dust formulations are pesticides in which the active ingredient is mixed with a substance which is usually light in weight (talc or clay). However, heavier weighted carriers are also essential for the proper distribution in certain situations. Many lightweight dusts are not acceptable in many areas since they may be a hazard to the applicator or vicinity in which the application is made [17, 21, 22]. Ultimate drift of dispersed dust must be of considerable concern to the applicator. Precautionary measures in the use of dust equipment consist of the following:

- Because dusts are usually exerted under some degrees of pressure, all working parts of the equipment should be carefully examined.
- All literature provided by the manufacturer should be carefully read.
- Goggles and respirators should be included in the safety equipment.
- The operator should be acquainted with operations and limitations of each piece of equipment.
- Complete removal of dust remaining in the hopper and any extensions of the duster should be accomplished after the completion of job.

1.4.2.1 Hand-Operated Dusters

Getz Applicator A simple device in which a spring is surrounded by a rubber sleeve sealed at one end by a filler plug and the other by a small release tube. Contracting the two ends compresses the spring and releases a limited quantity of dust through the tube orifice. Releasing the grip on the spring forces the sleeve to resume its original position. The capacity of this device is very small (six ounces), and usually, it is only used for crack and crevice treatment. It is one of the most versatile equipment used for all field crops. Small void areas can be expertly treated with this device [17].

Flexible Bulb This is similar to the Getz applicator but does not depend upon a spring to re-establish its flexibility. Instead, by the nature of its construction, it resumes its original shape and upon compressing releases dust through a small orifice. Depending upon the weight of the dust used, it may hold slightly more than the Getz applicator. Its uses are identical as for the Getz applicator. It is highly applicable for any powdered insecticides and being used effectively in home gardens [17].

Plunger Tube Dusters It is simple in construction and consists of a dust chamber, a cylinder with a piston or plunger, a rod and a handle. Retraction of the plunger allows air and dust to mix in the cylinder and upon depression of the cylinder, repeats the above process. This device is suitable for crack, crevice or void applications. Its capacity is usually greater than both of the previously mentioned dusters. It is highly useful for small-scale use in kitchen garden and in household (TNAU Aagritech Portal, http://agritech.tnau.ac.in/index.html).

Hand Crank Dusters These dusters have a hopper for the dust which varies in capacity from 1 to 10 lb. At the bottom of the hopper is an orifice which is adjustable in size and which allows the dust to fall into an airstream created by fins attached to a shaft. The shaft in turn is attached to a crank which is driven by hand power. A tube of varying length leads from the hopper to the final area of dispersion. This unit is typically used exteriorly or occasionally under the sub-areas of a house (TNAU Aagritech Portal, http://agritech.tnau.ac.in/index.html).

Foot Pump Dusters This is really a duplicate of the plunger-type duster, but in the place of the fixed release tube a hose is substituted and a place to set the operator's foot is provided. The plunger propels the dust through the tube, usually into a rodent burrow. In most cases, this style of duster is used in exterior areas (TNAU Aagritech Portal, http://agritech.tnau.ac.in/index.html).

Power Dusters As their name implies, the following equipment either uses electric motors, gasoline engines or compressed air as the power to run the mechanism which propels the dust. Where the hand-type units are usually for small applications, power dusters are obviously used where large quantities of dust are to be dispersed. In the structural pest control industry, this type of duster is usually restricted for use in large enclosed areas such as attics or sub-areas. However, where necessary and practical, exterior application is in order. Drift hazard and time of application are definite limiting factors in their use (TNAU Aagritech Portal, http://agritech.tnau.ac.in/index.html).

Electric Motor-powered Dusters This equipment, as indicated by the name, uses an electric motor which powers the propeller that creates the air to force the dust through an applicator tube. The hoppers for these dusters usually hold from 5 to 15 lb. One of the distinct limiting factors is the necessity for the availability of electricity which, in some instances, could limit the application area [17].

Gasoline Motor-powered Dusters The most common example of this is the knapsack or backpack duster. However, there are also wheel-mounted units for large-scale applications. In this type of duster, gasoline-operated engines accomplish the same effect as outlined above for the electric motor. Particularly, with the knapsack or backpack, there is an extremely large area of mobility because the power unit is self-sustaining. Also, many of these units can be used with minor changes for both dust and liquid applications. The hopper in the case of a dust applicator has a capacity of 5–25 lb [17].

Air Pressure Dusters This type of duster is a modification of the fire extinguisher so as to enable the air pressure built up within the tank to expel dust through a hose. Its limitations are that the quantity of dust is small (1–2 lb) and additional equipment must accompany the unit to build up air pressure. A truck-mounted air compressor or gasoline facilities could also be used. Once the unit has been filled with dust and pressured with air, it is self-sustained until either the dust or the air is dissipated. Except for the above limitations, it is a very flexible piece of equipment for the treatment of cracks, voids, sub-areas and attics (TNAU Aagritech Portal, http://agritech.tnau.ac.in/index.html).

Bellows Duster It has a pair of bellows made of leather, rubber or plastic. The bellows can work with a handle just like a Blacksmith does. The dust is placed either in the bellows or in a separate container made of wood, metal or plastic attached to one end of the bellows. The air current that is created runs through the container and drives the dust out through an opening.

Hand Rotary Duster They are also called crank dusters and fan-type dusters. They may be shoulder mounted, back or belly mounted. Basically, a rotary duster consists of a blower complete with gearbox and a hopper with a capacity of about 4–5 kg of dust. The duster is operated by rotating a crank, and the motion is transmitted through the gear to the blower. The air current produced by the blower draws the dust from the hopper and discharges out through the delivery tube which may have one or two nozzles. It is used for dusting field crops, vegetables and small trees and bushes in orchards. The efficiency of these dusters is 1–1.5 ha/day (TNAU Aagritech Portal, http://agritech.tnau.ac.in/index.html).

1.4.3 Sprayers

There are many ways to classify equipment in this category. The present system is based on the method by which the pressure is originated to force the liquid from any designated holding container [20–22].

1.4.3.1 Hand-Operated Sprayers

Flit Gun This is probably one of the oldest types of sprayers used in the industry. Its basic principle of operation is that a plunger is situated above a small supply container. As the plunger is compressed, air is diverted through a small hole situated above a syphon tube from the supply container. When the air passes through the tube and siphons, the pesticide liquid from the tank reaches the nozzle tip. Because of tremendous advancements in equipment, this type of sprayer is seldom used today in pest control operation.

Small Hydraulic Sprayer This sprayer is the adaptation of the equipment that used in service stations to spray oil in a pin stream to springs and other lubricated parts of an automobile. It consists of a small supply container, rarely exceeding a quart. Inserted into this container is a tube enclosed plunger which is connected to a hand-pulled trigger. By exerting pressure on the trigger, the plunger forces liquid through a tube to an adjustable nozzle. The nozzle can be adjusted from a pinpoint sprayer to cone shape. Either oil base or stable emulsions can be used in this equipment. More often than not, this sprayer is used to apply a residual insecticide rather than a knock-down formulation. One distinct disadvantage is the small capacity of this applicator. This type of application has few uses for spraying the yard and sub-area (TNAU Aagritech Portal, http://agritech.tnau.ac.in/index.html).

Compressed Air Sprayers This is probably the most commonly used sprayer in our industry. It is composed of a tank (usually holding from one-half to 3 gal of liquid), a pump to compress air and a discharge hose with a valve to control the discharge through a nozzle. An airtight tank, preferably stainless steel, is filled approximately to three-fourths of its capacity with a pesticide, and its operation is relatively simple. The remaining space is utilized for the compressed air to be generated by the hand-powered plunger-type cylinder within the tank. A check valve is located at the bottom of the cylinder to allow the air to enter the tank but closes to prevent the liquid from entering the cylinder. A tube within the tank is located so that its source originates near the bottom of the tank but closes to prevent the liquid from entering the cylinder. Somewhere prior to the nozzle is a hand-controlled shut-off valve. There is usually a pressure valve where the pipe emerges from the tank and the hose connection. The sprayer should not be used with pressure exceeding 50 psi nor less than 25 psi. The nozzle can be either a multi-purpose type (pin stream to fan or hollow cone) or a fixed pattern. This type of application is usually confined to inside work and sometimes outside in monthly service calls. Most likely the greatest use of this type of application is in cockroach clean-out calls (TNAU Aagritech Portal, http://agritech.tnau.ac.in/index.html).

1.4.3.2 Electric or Gasoline-Operated Sprayers

Powered Spraying For the most part, power spray usually refers to the typical 50–100 gal spray rigs owned by most structural pest control firms. Most will have some form of agitation, either jet agitation or blades mounted inside the tank. When spraying either a yard or sub-area, a moderately low pressure with a relatively high volume will aid in a safe even distribution. Where grass or weeds are thick, it may be necessary to adjust the pressure upwards to be sure and drive the pesticide down into the soil or turf. However, the spray pattern should be kept coarse to ensure wetness. Low pressure, high volume is particularly valuable when spraying a dry, dusty sub-area. High pressure tends to ball up the top layer of dust or soft soil and move it around in a sweeping motion, without ever getting the soaking action that is desirable. Needless to say, some servicemen will use higher pressure to speed up a job. A conscientious applicator will move about with a low-pressure, high-volume type of application and reach all areas. This power can be transmitted to the pump by belts and pulleys, chains and sprockets, power take-off assemblies or direct drive. In all cases, the liquid is ejected by the action of a pump through hoses or wands and finally through a nozzle or groups of nozzles. In this industry, the most commonly used type of pumps is the centrifugal, gear and piston pumps. The selection of a sprayer must obviously be governed by the magnitude of the job. Various types of formulations also play an important part in deciding the tank, pump, hose, regulator or nozzle to be used. The following section indicates the various ramifications of this selection by pointing out the various components of some of the systems now in use (TNAU Aagritech Portal, http://agritech.tnau.ac.in/index.html).

Tanks They should be of a capacity commensurate with the job. This seems, in most cases, to be of the 50–100 gal capacity. There are tanks available as large as 500 gal. The tank should be of stainless steel, aluminium, fibreglass, plastic or steel which is coated interiorly with a protective lining or is galvanized. The filling opening should be large, equipped with a strainer and be of a size large enough to enable easy access for repair of the lining or any mechanical device within the tank. A secure hatch should cover the filling opening to avoid spillage of the formulation while in use. In order for the tank to be adaptable to all types of formulations, it should have a mechanical agitator or a properly placed return of the overflow from the regulator. A drain should be located at the lowest point of the tank and, preferably, should be easily accessible. The inlet to the pump should be of adequate capacity to supply the needs of the pump. An exterior plastic or glass liquid level gauge should be available to check the actual or remaining quantity of pesticide in the tank. All of the newer tanks and liners have been developed to overcome the serious problem of corrosion.

Pumps A pump is the most important part of the spraying system. It is imperative that it is chosen to satisfy the widest range of applications unless its use is for a single purpose. In most instances, the pump comes as an integral part of a complete unit in which the engineering requirements have been satisfied. In other words, it has been specifically designed and manufactured by a company in this field. As one can imagine, there are many styles of pumps, some specifically for high gallonage delivery with little pressure, while others are styled for both small and large delivered quantities and high pressure. The most common types of pumps are discussed in the following section.

1. **Piston Pump**: It is one of the most common pumps in existence, which has the ability to produce large volumes at high pressure—for example, 55 gal per minute at 800 psi. However, it can also be regulated to deliver 2 gal per minute at 150 psi. The only feature which makes this type of pump undesirable is its pulsating action on hoses and regulators which causes them to wear at a rate faster than that with other types of pumps.
2. **Centrifugal Pumps**: It is one among the best pumps noted for its ability to deliver high volumes of liquid at low pressure. These are constructed to handle corrosive and abrasive materials.
3. **Roller Pumps**: This is a popular pump in the industry because of its wide variety of uses. It has either nylon or rubber rollers and can produce a wide range of volumes and pressures.
4. **Internal and External Gear Pumps**: These are the high-pressure pumps which produce limited volumes. Although all pumps diminish in efficiency from wear, this is more critical in this specific model.
5. **Diaphragm Pump**: It is one of the low-volume pumps and the main restriction is related to using a material which will not affect the material of the diaphragm.
6. **Flexible Impeder Pump**: It is a pump restricted to low pressure and limited volume.
7. **Vane Pump**: This pump requires materials which afford some lubricating properties.

Hoses The object of any hose is to convey a liquid from the power source to the target. The material of which hoses are made varies considerably. Originally, most hoses were constructed of natural rubber, but today, synthetic rubber, plastic or sometimes both are commonly used. As with so many other aspects of equipment in this industry, it is important to know what is expected of the hose relative to performance.

Nozzles There are many types of nozzles, of which the primary concern is that the specific type provides the pattern desired. Generally, the pattern is confined to a solid/pin stream, to a fan or to a hollow/solid cone nozzle. Many of the above combinations are available in adjustable brass, but can also be made of stainless steel, aluminium and sometimes wearable parts of plastic. Except for nozzles used on small equipment, the gallonage delivered by a specific nozzle, regardless of design, should be clearly known by the applicator. The capacity of the pump, the pressure on the liquid, the friction loss and size of the hose, and the size of the orifice in the nozzle will govern the ultimate gallonage delivered from the nozzle.

Strainers These are screens made of various materials, preferably stainless steel, which usually are 50 mesh or coarser. They are located at the filling opening, suction line to the pump and the nozzle tip. Their presence is to prevent any foreign substance access to the spraying system. Routine cleanings are imperative not only for the sprayers to function properly, but also extremely important if the calibrated delivery system is functioning accurately.

Valves There are many types of valves with some being extremely sophisticated. Their purpose is to shut off the flow of the liquid. Consequently, they are strategically located to be most effective in the advent of an emergency or in the normal function of the shutting off of the supply of liquid pesticides.

Pressure Regulators As the name implies, this unit controls the pressure of the liquid being delivered to the nozzle. As with many other parts, there are numerous types of regulators. Spring tension which is controlled by a hand-adjusted screw mechanism exerts pressure on the liquid as it flows through the regulator on its way to the nozzle. They can be operated from zero pressure to as high as 800 psi. The indicator/gauge for pressure is located adjacent to or is a component part of the regulator and, as indicated, records the pressure as pounds per square inch. In addition to establishing the pressure, a provision exists for the excess or overflow to bypass the regulator and return to the tank. In many sprayers, this return line is located near the bottom of the tank and can be either the primary or secondary measure in the agitation of the spray material in the tank.

Agitators This is a means by which the contents of a spray tank are mixed and agitated. The object is to keep the pesticide in continuous suspension so that it results in an even distribution of the material. In some formulations (wettable powders), mechanical is the only type of agitation that maintains suspension. In this method, a set of paddles is attached to a horizontally located shaft at the lower portion of the tank. Exteriorly, the shaft is connected to the power source. Bypass agitators are a frequently used method of agitation. This technique utilizes the overflow from the regulator to stir the contents of a tank.

Gas Generating Sprayers There are essentially two types of gas generated sprayers, floating piston type and the very common aerosol spray-cans. In the former, a cylinder with a floating piston having nitrogen gas is in one end of the cylinder and the pesticide, which is pumped under pressure of 300–1000 lb/in.^{2}, at the opposite end. This compresses the piston against the nitrogen and forces the insecticide through a high-pressure hose into a very fine orifice nozzle. Although it can be used independently for a limited time, it depends upon a pump situated reasonably close to the job for reloading. Aerosol can dispersal of pesticides, although not new, has been very popular both with the public and with pest control operators. Their range in capacity, from 6 oz to 30 gal, has made them very popular. The small sizes have push type release nozzles, while those with a capacity of 5 lb and up have a mechanical shut-off nozzle. They also come particularly in the sizes of less than 5 lb, in disposable type containers. Their construction is very simple, consisting of a container in which designated amounts of pesticides and freon are placed. As long as the temperature in the vicinity of the area, the material to be dispersed, is warm, the freon will propel the pesticide out of the orifice.

Farmworker applying liquid insecticide using hand-operated backpack sprayer (photo by S. Jayakumar)

Farmworker applying liquid insecticide using power-operated sprayer (photo by S. Jayakumar)

Farmworker applying liquid insecticide using backpack sprayer with round nozzle (photo by S. Jayakumar)

Farmer applying granular pesticide without using personal protective equipment (PPE) (photo by S. Jayakumar)

Granular Application As the name implies, the applicator used to apply granular formulations of pesticides in field. Essentially, these are small hopper-type units with an orifice for the material to drop onto the crank-operated spinning disc which throws the material over a 15- to 20-foot swath. Also, the same equipment is available in gasoline or power take-off drives which employ the same above principle only on a larger scale. Granules are ready for immediate application. Granular applicators, except in the case of weed control, are rarely used in pest control process (TNAU Aagritech Portal, http://agritech.tnau.ac.in/index.html).

Brush Application Brush application quite often is recommended on the label for use in areas where a wet application is desired; however, care is to be exercised in preventing the movement of a pesticide to non-target areas. This technique is usually confined to application along baseboards, window sills and door thresholds.

Fumigant Injection The only fumigant injection method which might be used by a structural pesticide applicator, other than those described in the section on fumigation, is local spot injection for control of dry wood termites and wood boring beetles. This technique utilizes a small hand-held CO_2 pressurized application with a nozzle designed to penetrate into either the small emergence holes created by the insects or specially drilled holes which open into the galleries.

Spot Treatment Spot treatment is an application to limited areas on which insects are likely to occur, but which will not be in contact with food or utensils and will not ordinarily be in contact with workers. These areas may occur on floors, walls and bases or outsides of equipment. For this purpose, a 'spot' shall not exceed two square feet.

Crack and Crevice Treatment Crack and crevice treatment is the application of small amounts of insecticides into cracks and crevices in which insects hide or through which they may enter into a building. Such small openings commonly occur in expansion joints, between different elements of construction, and between equipment and floors. These openings may lead to voids such as hollow walls, equipment legs and bases, conduits, motor housings or switch boxes.

Ultra-Low-Volume Application (ULV) Ultra-low-volume application is the spraying of undiluted pesticides in small volume, usually at rates of 1/2 gal per acre or even less. This term when used in structural applications has a somewhat different meaning. It is sometimes referred to as ultra-low dosage (ULD). The principle of ULV application is the dispensing of a low volume of insecticides over relatively larger areas. This technique is accomplished by breaking the insecticides down into tiny particles prior to applications.

Soil Injection This technique employs a long hollow, pointed probe connected to a spray rig and a shut-off valve at the top of the injector. It is mostly used in placing termiticides deep into the soil around exterior foundation.

Sub-Slab Injection This technique is used for treating soil beneath slabs for the control of subterranean termites. It is one of the most useful tools developed for termite control. The sub-slab injector can be adapted to both 1/2 and 3/4-in holes drilled in slab. The inability to know exactly the direction of the flow beneath the slab is considered as the disadvantage of this technique. The application method depends on the nature and habits of the target pests, characteristics of the target sites and properties of the pesticide formulations. One must consider the suitability of the application equipment, cost and efficiency of alternative methods.

Wiper Applicator It could be used to wipe a non-selective herbicide to selectively kill individual weeds. The wiper's wetness must be less than dripping and handled carefully to avoid accidentally treating desired plants.

1.5 Maintenance of Equipment

Plant protection machines in general are not well maintained regularly either in godowns/depots where they are stored or in the field where they are used. Life of a machine depends entirely on its care and maintenance. Even though machines are made with high standards of skill and workmanship, they can easily be ruined due to improper care and maintenance. Good and constant performance from machines can be obtained only when they are used and serviced periodically. The purpose of maintaining a machine is for increasing the useful life of the machine and to be available in working order whenever put to use. The maintenance of a machine involves proper care, operation, servicing, repair and keeping it in good working order.

1.5.1 Maintenance

Normal maintenance jobs include cleaning the equipments and applying necessary lubricating oils and greases to the rubbing and moving parts. If this normal maintenance is neglected, the machine gets rusted and moving parts wear out quickly resulting in loss of efficiency, frequent replacement of spare parts and finally uneconomical working. Besides the normal maintenance as above, special care has to be taken for maintaining the plant protection equipments. The pesticide formulations are chemically aggressive on metals/other materials. The cleaning and washing of the chemical tanks, discharge lines, nozzles, etc., are to be done regularly after the day's spraying work is completed; otherwise the residues of chemicals used for spraying act on the parts and cause corrosion and deterioration of materials. If this aspect of thorough cleaning is not done on the plant protection machine, even though it is made of high standard materials, it will not serve its normal life and would lead to premature condemnation.

1.5.2 Maintenance of Hand-Operated Equipments

1. Cleaning the chemical tanks, hoses, valves and nozzles, etc., and flushing sufficiently to avoid pesticide residue which is corrosive.
2. Cleaning the machine equally well from outside also as it is contaminated due to leakage and spilling of pesticides.
3. Lubricating suitably the pump parts like piston, cylinder, valves and other rotating, sliding, moving parts.
4. Storage of machine in dry place duly protected from sun and rain.

1.5.3 Maintenance of Power-Operated Equipments

All the above maintenance jobs apply to power equipments also. But the engines have to be taken care of specially. All engines need fuel, air and proper system of ignition. Thus, in petrol engine, clean petrol, clean air and healthy ignition (spark plug and magnets) are essential. Besides those, the engine also needs perfect lubrication. In two-stroke petrol engine, care must be taken to mix lubricating oil and petrol in exact ratio as recommended by engine manufacturer. Similarly, in four-stroke petrol engine, the lubricating oil should be kept in sufficient quantity by observing the level gauge. The air cleaner should be cleaned occasionally. The spark plugs should be also cleaned, carbon removed and proper electrode gap should be maintained. The two-stroke petrol engines used in low-volume spraying should invariably be in good order; otherwise, the pesticide spraying will not be efficient. Sufficient care should be taken at the depots to clean, oil and check equipment periodically when they are stored, whenever machines are sent out to work, and when returned from fieldwork. This minimum care to inspect the equipment, clean and flush and keep it duly oiled, would go a long way in improving the availability of good working sprayers and dusters and also prolonging their useful life [21].

1.5.4 Repairs and Replacements

The plant protection equipment is often found requiring frequent repairs and replacements in nature. Hand-operated equipment generally needs minor repairs such as replacement of plunger washers, springs and nozzle, and these repairs could as well be attended by the operators themselves with little training and experience. It is essential to supply them the necessary spare parts and tools well in time for repairing. In the case of power-operated sprayers, the engine repairs are classified into minor and major respires.

1. **Minor Repairs**: Spark plug cleaning and adjustment, air cleaner, carburettor cleaning, fuel cock and lines cleaning and starter repairs are considered as minor retirements; it can be attended by the operators themselves with little experience and training.
2. **Major Repairs**: These repairs include replacement of parts like piston, rings, liners, crankshaft, bearings, valves, etc. These repairs have to be carried out systematically in well-equipped workshops by the competent and trained mechanics. Untrained personnel should not be allowed to handle such major repairs.

1.5.5 Suggestions on Maintenance

To improve the maintenance of sprayers and dusters, the following suggestions are made:

1. Plant protection equipment manufacturers, their dealers, state agricultural engineering workshops and extension officers need better coordination and cooperation to reduce the number of sick equipment.

2. The field operating staff needs to be given the orientation training on proper maintenance, repairs, operations and calibration of equipment on periodic basis.
3. An adequate number of mechanics and supervisory staffs have to be posted for checking the maintenance and effectiveness of equipment.
4. A district-wise service station could be established, which certainly facilitates the users to repair their power-operated equipment within their zone.

1.5.6 Safety Systems

Pesticide Containment Pad If one often stores, handles, mixes and loads pesticides, or cleans equipment at the same location, one must have to install a pesticide containment pad (as per EPA to determine when a containment pad is required). These pads are designed to contain spills, leaks, overflows and wastewater for reuse by the applicator or for disposal by a commercial waste management contractor. They make it easier to clean up spills and help to prevent environmental contamination.

Impervious Containment Pad Generally, the containment pad must be made of impermeable material. It should be concave or have curbs and berms/walls high enough to hold the largest amount of spill, leak or equipment wash water likely to occur at the site. It must also have a system to remove and recover spilled, leaked or released material by either an automatic pump system or a manually operated pump. Smaller, portable pads and lightweight trays made of heavy-duty plastic may be used when mixing and loading at the application site.

1.5.7 Sprayer Components

Tank A tank is necessary to contain the spray mix. Choose one made of, or coated with, a material that does not corrode and that can be cleaned easily. Cleaning prevents accumulations of corrosion and dirt that clog screens and nozzles, increasing wear on the equipment. Large tanks require an opening in the bottom to aid in cleaning and draining. A large top opening is useful for filling, cleaning and inspecting the tank. The opening must have a watertight cover to prevent spills. A tank agitation system/device is useful for most sprayable formulations, especially for wettable powders or dry flowables. Constant mixing of a pesticide and liquid carrier produces a uniform spray mixture (suspension) and results in an even application of the chemical.

Exposure to sunlight and corrosive chemicals can shorten the life of polyethylene tanks.

Three common signs of wear and potential tank failure are as follows:

- **Scratches** are on the surface could be noted.
- **Crazing** is a network of fine lines/cracks that may look like a patchwork, but often cannot be seen with a visual inspection. Crazing can be seen when using one of the testing methods explained below. Crazing occurs within the tank wall and can be a sign of deterioration of the plastic, which may lead to cracks. Tanks that show signs of crazing will still hold liquids, but the integrity of the tank is questionable. For this reason, caution should be used when putting any hazardous substance in tanks that show crazing.
- **Cracks** extend through the plastic wall and can be visually seen and felt. Cracks may run parallel or at right angles to each other.

Pump A pump agitates the spray mixture and produces a steady flow to the nozzles. Pump parts must resist corrosion and abrasion, especially when wettable powders or similar formulations are used. Pump will never be operated at speeds or pressures above those recommended by the manufacturer. Pumps depend on the spray liquid for lubrication and to prevent overheating.

Nozzle Maintenance In general, the correct screen size for each nozzle could be selected. Cleaning of nozzle tips could be carefully done with a soft brush, but not wire or a knife tip, which may damage the nozzles.

Sprayer Cleanup Spray equipment should be cleaned in the field after the spray job has been completed. Some pesticide labels provide specific information on cleaning spray equipment, and one should consult the label for guidelines. The users do not clean spray equipment in areas where rinse water will contaminate water supplies, streams or injure susceptible plants, wildlife and human.

Flushing spray equipment with water may be sufficient to remove potentially harmful amounts of many pesticides. However, certain groups of pesticides may require special attention. Thorough clean-out procedures are critically important when switching applications between crops to help avoid significant crop injury. As a rule, a sprayer that has been used to apply 2,4-D or other growth regulator-type herbicides should not be used to treat susceptible plants. A triple rinse—water, then ammonia, then water again—minimizes the risk of injury from dicamba and 2,4-D. Inconsistent applications result in control failures or injury to turf or landscape plants. Things to watch for when using this type of application are as follows:

- Pump pressure is set correctly.
- Consistent, accurate walking speed.
- Nozzle is correct for desired flow rate.
- No partially blocked nozzle openings.
- Hose is not kinked.
- Strainer screen is not clogged.

1.6 Knowledge of Pesticide Applicator

Pesticide usage in agricultural practices has become a crucial part of present-day farming and takes part in a foremost role in increasing the productivity of agricultural product. However, the indiscriminate and widespread use of pesticides stands for one of the major environmental and public health problems across the world [23, 24]. Pesticides are very important risk factors on human life not only effects on health as a result of misuse or accident, but also via leave a lasting harmful chemicals into the environment. The improper usage of pesticides is a significant sources of air, water and soil contamination, which can lead to destruction of non-target species [25, 26]. The most persistent and toxic nature of pesticide residues in agricultural products jeopardizes both the environment and human health. In global scale, it was estimated that approximately 200,000–300,000 people die every year due to the pesticide exposure [27, 28]. It has also been estimated that only about 0.1% of pesticides used reach the target organisms, and all other remaining portions of pesticides applied were entered into the surrounding environment.

At a global scale, a large number of casual and temporary workers are engaged in agriculture sector. Unpaid family members carry out agricultural work as unrecognized farm labour or support small-scale family farming. In recent years, there has been an increase in the use of pesticides in developing countries, and they now account for about 20% of the world's expenditure on pesticides [29]. Researchers have concluded that farmworkers in developing countries will continue to use pesticides in increasing quantities because of the lack of alternatives to pesticides and ignorance of the sustainability of pesticide use. Extensive use of such pesticides results in substantial health and environmental threat. Despite the instructions provided on pesticide containers, there are several hurdles that restrict their implementation. Lack of education among farmers and farmworker is the foremost barrier.

A recent review by Dhananjayan and Ravichandran [30] highlighted the detrimental health effects of farmworkers associated with pesticide exposures. Farmers involved in the handling of pesticides are at a high risk of exposure to pesticides through contact with pesticide residues on treated crops, unsafe handling, storage and disposal practices, poor maintenance of spraying equipment, and the lack of protective equipment or failure to use it properly [31]. Studies in developing countries reported low to moderate levels of knowledge about pesticides among farmworkers [32, 33].

Sodavy et al. [34] reported that the pesticide distributors used to distribute products with labels written in foreign language, and in developing countries, farmers usually get information from pesticide vendors and from other farmers. The adverse risk was linked with lack of information about the products handled [35]. The most common practices of farmers in agricultural sectors are non-usage of personal protective equipment (PPE) and usage of worn-out PPE [36], exposure to pesticide due to unsafe pesticide storage [37] and improper way of disposal of empty pesticide containers [32]. The lack of education and poor knowledge on understanding the safe practices in pesticide application including storage, handling and disposal makes farmer at risk [36]. Higher level of education gives pesticide users better access to information and more knowledge of the risks associated with pesticides, and how to avoid exposure. In general, poorly educated farmers hampered in their ability to understand the hazard warnings on pesticide labels and guidelines recommended for safety application [38, 39]. Earlier studies have found that 68% of agrarian population reported the symptoms of sickness after routine pesticide applications and pesticide-related illness including skin problems and neurological symptoms [27, 40].

The Food and Agriculture Organisation (FAO) of the United Nations recommends that governments in developing countries promote pesticides that require little personal protective equipment [41, 42]. The improper use of pesticides can cause human poisonings, build up as residues in food and the environment, and lead to the development of resistance in pests [43]. Our recent studies have reported the presence of residues in various matrices [44] and reduced cholinesterase activities and DNA damage in farm workers [45–47]. The most frequently used pesticides in agricultural community belonged to WHO class II. Since farmers had poor knowledge about pesticide toxicity, the majority did not use appropriate PPE or good hygiene when handling pesticides. Several occupational uses of WHO class II pesticides and those of lower toxicity are seen in combination with inadequate knowledge and practice among the farmers. This poses a danger of acute intoxications, chronic health effects and environmental pollution. Training of farmers in integrated pest management (IPM) methods, use of proper hygiene and PPE when handling pesticides should be promoted [48].

1.7 Fate of Pesticides in the Environment

Unsafe use of pesticides can contaminate soil, water, air, vegetation and the entire environment. Despite killing insects or weeds, pesticides can also be toxic to other non-target organisms including birds, fish, beneficial insects and non-target plants. The fate of pesticides in the environment and its toxic effects are reviewed and described in the following section.

1.7.1 Contamination of Air

Pesticide sprays can directly hit non-target vegetation, or can drift or volatilize from the treated area and contaminate air, soil and non-target plants. The potential for exposure due to atmospheric transport and deposition of pesticides may create risks to living organisms including humans [49]. Airborne chemical contaminants are responsible for millions of annual premature deaths globally. Hazards associated with airborne exposure are several times greater than the hazards expected from other contamination [50]. Pesticide drift account for a loss of 2–25% of pesticide applied on field during every application, which can spread and travel several miles far away from the source of pesticide drift [51]. A large volume of applied pesticide can be volatilized within a few days of application. Despite the fact that only a few research studies have been carried out on these issues, studies continuously find pesticide residues in air [49, 52–54]. According to the USGS, the occurrence of pesticides in the atmosphere was detected in air samples collected in several areas of the USA [55]. Nearly every pesticide used in agricultural and other purposes has been detected in rain, air, fog or snow across the nation at different times of the year. Photochemical oxidation of pesticides in air will be very rapid, and these chemicals' interaction in the atmosphere influence the lifetimes [56]. Experimental designs and computer-based modelling related to emissions and dispersion of pesticides in air and ambient monitoring of air particularly in relation to opportunities to improve the data quality and their use in risk assessment will be most essential [57, 58].

1.7.2 Water Contamination

The atmospheric trace pesticide can interact with the moisture, and deposit from air during wet precipitation leads to contamination of surface water system [59]. Contamination of water by pesticides is widespread. A recent study conducted in Hooghly River basin in West Bengal, India indicates the presence of organochlorine pesticides and chlorpyrifos and its ecologic risk on aquatic animals. Traces of pesticides have been identified in water. About 44% of surface water was detected above the limit of quantification, and among these, 13% of samples exceeded the level of 100 ppb. The research investigations on river water across the word indicated that more than 90% of the samples tested for pesticides had one, or more often, several pesticides [60].

Several studies have reported the importance of pesticide monitoring in surface and groundwater system in order to regulate and assess the status of pesticides [61, 62]. Different types of pesticides were detected in waterways across the world, including 17 herbicides. The herbicides, namely 2,4-D, diuron and prometon, and the insecticides, chlorpyrifos and diazinon, are commonly used in urban home and school, as a result most often found in water. The USGS also found that concentrations of insecticides in urban streams commonly exceeded guidelines for protection of aquatic life. The herbicide 2,4-D was the most commonly found pesticide. Organochlorine pesticides such as endrin aldehyde, total BHCs and heptachlor exceeded the regulatory limits prescribed for groundwater in Pampanga River, Philippines. Surface water samples from Bhandara and Yavatmal region exceeded the EU (European Union) limit of 1.0 μg/L for sum of pesticide levels in surface water. Surface water often was found to be more contaminated with organophosphate and organochlorine pesticide than groundwater [63]. According to USEPA [64], pesticides exceeded standards for human health in ~10% of agricultural streams, ~7% of urban streams and ~1% of groundwater tested, whereas aquatic health exceeded in 57% of agricultural and 83% of urban streams. Similarly, 12% of groundwater in northern Spain exceeded the European Union (EU) directives regulatory standards [65].

1.7.3 Effect on Soil Fertility

Soil contaminated with pesticides can cause decline in populations of beneficial soil microorganisms [66, 67]. In general, loss of both bacteria and fungi leads to degradation of soil. Indiscriminate uses of chemical fertilizers and pesticides have effects on the soil organisms that are similar to human overuse of antibiotics. Overuse of those chemicals might work for a few years, but after awhile, there are not enough beneficial soil organisms to hold onto the nutrients [55]. Plants depend on a variety of soil microorganisms to transform atmospheric nitrogen into nitrates, which plants can use. Pesticide and herbicide applied in agricultural sector disrupt the soil condition and the normal process of plant growth. Many pesticides including 2,4-D reduce the efficiency of nitrogen fixation by the bacteria that live on the roots of bean plants [68, 69], reduce the growth and activity of nitrogen-fixing blue-green algae [44, 70] and also inhibit the transformation of ammonia into nitrates by soil bacteria [71]. The adverse effect of organochlorine and carbamate pesticides on beneficial microorganism including fungi is also reported [70].

1.7.4 Non-target Organisms (Fish and Wildlife)

A raising public awareness has come into force regarding the impact of pesticides after the publication of Book, Silent Spring by Rachel Carson [5]. This book highlighted the impact of pesticides on non-target organisms, birds, and contributed to the reversal of national pesticide policy. The nationwide ban on DDT and other similar pesticides stirred environmental movement that led to the creation of the US Environmental Protection Agency (US EPA) in 1970. Pesticides are found as most common contaminants in the environment including soil, air and water, and also on non-target organisms. As a consequence, they can harm non-target plants, fish, birds and other wildlife. Several recent studies highlighted the persistent nature of pesticides and its accumulation in various environmental matrixes such as fish [44, 71], birds [72, 73], bird eggs [74] and humans [75].

Toxicity of agrochemical such as atrazine affects the lipid peroxidation and activities of antioxidant enzymes in the freshwater fish [76] and increased immunoglobulin production in silver catfish [77]. Herbicide exposure to laboratory animal shows mitochondrial dysfunction and insulin resistance [78], and it also gives multiple stress effects on wetland zooplankton [79], fish and amphibian [80]. It was also reported that herbicide atrazine exposure affects the longevity, development time and body size in Drosophila [81]. Several cases studies on pesticide accumulation on dolphins have been reported [82–84]. Dolphins inhabiting riverine ecosystems are particularly vulnerable because of their habitats closer to point sources of pollution usualy caused by humans. Our recent study on fresh water fishes of bird santuaries in Southern India highlighted the presence of organochloirne pesticide residues [85].

1.8 Health Effects of Pesticide on Humans

Overwhelming evidences are available across the world on pesticide usage in agricultural field and its potential risk to humans and other life forms. Humans are exposed to pesticides through contact with the skin, ingestion or inhalation, and numerous negative health effects have been associated with these pesticides [30]. The acute toxicity of pesticides on humans is mainly due to interference with neural conduction by targeting voltage-gated ion channels or Na^+/K^+ ATPase, interference with neural transmission by inhibiting acetylcholine esterase, stimulating respiratory sensory neurons or initiating pro-inflammatory signals. Pesticides are broadly classified into three categories based on their neurotoxic effects in mammals: 1. neural conduction interferer (organochlorine pesticides: DDE, DDE, cyclodienes, toxaphene, hexachlorocyclohexane, chlordecone and pyrethroid pesticide: pyrethrin and tetramethrin), 2. acetylcholine esterase inhibitor (organophosphate pesticides: parathion, malathion, methyl parathion, chlorpyrifos, diazinon and carbamate pesticides: aldicarb, carbofuran, carbaryl, ethienocarb and fenobucarb) and 3. pro-inflammatory stimulator (chlorophenoxy herbicide: 2,4-dichlorophenoxyacetic acid) [86]. No part of the population is completely away from pesticide exposure. The central and peripheral neural systems of the humans are affected who are exposed to high dose of OCs, OPs and pyrethroids [87]. The numerous negative health effects that have been associated with chemical pesticides (Fig. 1.1.) include dermatological, gastrointestinal, neurological, carcinogenic, respiratory, reproductive and endocrine effects. Additionally, occupational, accidental or intentional exposure to pesticides also leads to hospitalization and death [8, 88].

Occupational exposure to pesticide is considered as one of the high-risk groups. This group of workers are exposed to pesticides during production, formulation, spraying, mixing, loading and agricultural farm work [11, 89]. Farmers are routinely exposed to various types of agrochemicals, but there is no evidence about adverse effects posed by these chemicals on human health. Studies across the globe had shown adverse health effects of commonly applied pesticides in the agricultural sector. A significant relationship between occupational exposure to agricultural pesticides and the development of several diseases has been identified. It is quite common in agricultural community that they exposed to pesticide even when they are not directly performing tasks related to pesticide use [90]. Although studies on occupational exposure do not include all variables which can contribute to risk assessment, the real effects associated with agrochemicals are well established. It is more obvious that the agricultural community needs the implementation of a new

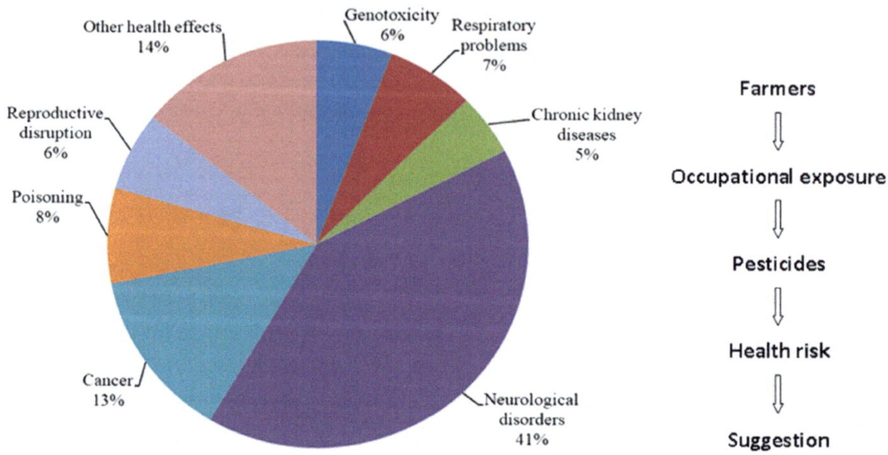

Fig. 1.1 Health associated with pesticide exposure among farmers. Adapted from Dhananjayan and Ravichandran [30]

agricultural concept regarding food production, which is safer for farmers, farm workers and the environment [30].

This section of review summarizes the recent findings interconnecting the association between occupational exposure to pesticides and related health effects on farmers in agricultural activities. Ye et al. [91] reported the evidence for an association between occupational pesticide exposure and asthma, and a perfect link between occupational pesticide exposure and chronic bronchitis or COPD among agricultural occupation. One of the studies at rural farming villages in Tanzania found a high potential for pesticide exposure with high frequency of self-reported acute pesticide poisoning and poor recording in hospital records [92]. Similar study at south-west Ethiopia reported health complications of pesticide exposure such as headache, nausea and vomiting, skin rash and irritation and abdominal pain [93].

Greenhouse workers exposed to pesticides showed increased counts of erythrocytes, leucocytes, platelets, haemoglobin, and genotoxicity and decreased cholinesterase, glucose, creatinine, total cholesterol, triglyceride and alkaline phosphatase [94–96]. A case-control study in the USA and Canada found an association between pesticide use and blood cancer risk [97]. An increased risk of death was reported due to non-Hodgkin lymphoma (NHL) among young farm-workers in southern Brazil [98]. Farmers in agricultural field found to have the highest prevalence of allergies, nasal congestion, wheezing and acute symptoms after pesticide use [99, 100]. Agricultural sector workers have reported numerous respiratory diseases such as chronic obstructive pulmonary disease, asthma [101] and interstitial lung diseases [102]. Children in close proximity to farms showed adverse respiratory system and the urinary metabolites of pesticides were associated with a decrease in lung function [103, 104].

Several findings provided important evidences on pesticide exposure and changes in neurobehavioral functioning in Chinese farmworkers [105] and neurodegenerative and neurodevelopment disorders [106]. Studies have also highlighted the alteration in thyroid function after long-term exposure to pesticides among male pesticide applicator [106], and poorer sperm morphology and lower luteinizing hormone (LH) among men [107]. Waheed et al. [108] suggested that dust contaminated with pesticides engenders significant health risk to the nervous and endocrine system, not only for occupational workers exposed to direct ingestion but also for nearby residential community.

Renal disease due to pesticide exposure provided reassuring findings of significantly increased risks among pesticide applicators [109, 110]. Migrant farmworkers in Thailand involved in spraying of pesticides have the increased risk of developing acute or chronic illness [111]. Over the past two decades, there has been an increase in chronic interstitial nephritis in agricultural communities. Existing studies provide scarce evidence for an association between pesticides and regional chronic kidney disease epidemics but, given the poor pesticide exposure assessment in the majority, a role of nephrotoxic agrochemicals cannot be conclusively discarded. Future research should be made with an assessment of lifetime exposures to specific pesticides with other major risk factors [112]. Several studies have reported that the low-level use of insecticides does not pose a risk to DNA in general [113]; however, recent studies have shown oxidative stress, DNA damage and lipoprotein peroxidation [47, 114].

Numerous studies have reported the associations between specific pesticides and bladder cancer risk, lung cancer, laryngeal cancer [101, 106, 115] and skin melanoma and multiple myeloma [116]. Further studies regarding this occupational exposure of farmers are needed, to determine the causes for the increased risk of this cancer, particularly in regions where there is an intense agricultural activity and where extensive pesticides are used. However, further studies warranted given the high burden and the ubiquity of these chemicals.

1.9 Controlled Release of Pesticides

Pesticide molecules are used worldwide for improving the agricultural productivity. However, these compounds and their degradation products show varying degrees of persistence and mobility in the environment and can have toxic, carcinogenic, mutagenic and teratogenic potentials, as well as effects on the endocrine systems of non-target organisms, including humans [117]. Variety of release systems applicable to the bioactive compounds in agriculture is given in Table 1.2.

Therefore, it is important for researchers to improve the usage rate of pesticides and to extend their duration of activity in the environment [118, 119]. Despite the fact that a sufficient amount of pesticides are typically applied during spraying, pesticide microemulsions, water dispersible granules and other traditional pesticide formulations usually below the effective concentration for controlling pests within a

Table 1.2 Variety of release systems applicable to the bioactive compounds in agriculture

Bioactive compounds with chemicals	Purpose of addition and observation	References
Silica and clay such as bentonites and sepiolite in agrochemicals	Activate the degradation	Grillo et al. [123] Barik et al. [124] Maqueda et al. [125]
Silica nanoparticles in drug delivery	Non-toxicity, low cost, high surface area and high reactivity as carriers	Dizaj et al. [126] Yang et al. [127] Wu et al. [128]
Nanotechnology in phytotechnology	For safe use and social acceptance of phytonanotechnology, the adverse effect and transfer of NPs in food	Wang et al. [129]
Lignin-based controlled release of pesticides	Lignin fragmentation and re-polymerization allow more control with flexibilities and improved properties of the modified lignin materials, and help achieve the desired control release outcomes	Chowdhury [130]
Emamectin benzoate (EMB)	EMB slow-release microspheres are an attractive candidate for improving pesticide efficacy and prolonging the control effects	Wang et al. [131]
Porous silica nanoparticles with abamectin	Improvement in photostability, water solubility, bioavailability and reduction in the residues of pesticides	Wang et al. [132]
Polymers: alginate and lignin, and synthetic polymer: polyhydroxyalkanoates (PHAs) with herbicides	Complete biodegradable, inexpensive	Reis et al. [133]
Polyhydroxybutyrate (PHB) and its hydroxyvalerate (PHBV) with herbicides	Complete biodegradable, isostatic and highly crystalline, and decrease in the speed of degradation	Amass and Tighe [134] Sudesh et al. [135]
Polyhydroxybutyrate (PHB) and its hydroxyvalerate (PHBV) with ametryn	The amount of herbicide released in the same period of time was significantly reduced (75–87%)	Grillo et al. [136]
Avermectin/polysuccinimide with glycine methyl ester nanoparticles (AVM-PGA) with avermectin (AVM)	To improve bioactivity and transportation	Wang et al. [137]

short period of time [118, 120, 121]. The loss of large amounts of pesticides in non-target areas is a serious threat to food security and the environment [122]. To mitigate the toxicity of these compounds in the environment, new and improved controlled release systems are emerging to increase the effectiveness of pesticide usage while minimizing their environmental impacts and aiding sustainable agricultural development.

The advanced controlled release systems can offer: (a) reduction in the amount of chemical substance required to protect crops; (b) diminished risk of environmental contamination; (c) reduction in energy consumption, since fewer applications are needed compared to conventional formulations; and (d) increased safety of the individuals who apply the product in the field.

Pesticide slow-release formulations provide a way to increase the efficiency of active components by reducing the amount of pesticide that needs to be applied. Slow-release formulations also increase the stability and prolong the control effect of photosensitive pesticides [131]. Currently, the silica nanoparticles as pesticide carriers did not show controllable porous surface properties, and release of pesticides was generally controlled by adjusting pesticides concentration and the thickness of coating layer, resulting in the relatively limited tuning range of release rate. Moreover, the size uniformity of silica nanocarriers needs to be further improved because monodisperse carriers are favourable to promote adhesion and permeability of the pesticide on target crops [131]. Nanomaterials (NM) serve equally as additives (mostly for controlled release) and active constituents. Product efficiencies possibly increased by NM should be balanced against enhanced environmental NM input fluxes. The dynamic development in research and its considerable public perception are in contrast with the currently still very small number of NM-containing products on the market. Nano-risk assessment and legislation are largely in their infancies [138].

1.10 Conclusion

Information on the occurrence of pesticide-related illnesses among agricultural populations in developing countries is insufficient. Therefore, descriptive epidemiological studies on the development of intervention strategies to lower the incidence of acute poisoning and periodic surveillance studies on high-risk groups are needed. Valuable information can be collected by monitoring the end product of human exposure in the form of residue levels in body fluids and tissues of the general population. The periodical education and training of workers ensure the safe use of pesticides. The occurrence of pesticide residues in various environmental matrixes contributes to the problems associated with pesticides. Pesticide contamination poses significant risks to the environment and non-target organisms ranging from beneficial soil microorganisms to insects, plants, fish and birds. The total cost–benefit picture from pesticide use differs appreciably between developed and developing countries. There is a need to convey the message that prevention of adverse health effects and promotion of health are profitable investments for employers and employees as a support to sustainable development of economics. Therefore, modern technology-assisted development in the application of pesticides is warranted in order to get the sustainable development in the agricultural sector and human well-being.

References

1. Abate T, van Huis A, Ampofo JK (2000) Pest management strategies in traditional agriculture: an African perspective. Annu Rev Entomol 45:631–659
2. Mohanty MK, Behera BK, Jena SK, Srikanth S, Mogane C, Samal S, Behera AA (2013) Knowledge attitude and practice of pesticide use among agricultural workers in Puducherry, South India. J Forensic Leg Med 20:1028–1031
3. Peshin R, Zhang W (2014) Integrated pest management and pesticide use. In: Pimentel D, Peshin R (eds) Integrated pest management: pesticide problems, vol 3. Springer, The Netherlands, pp 1–46
4. Sharma R, Peshin R (2016) Impact of integrated pest management of vegetables on pesticide use in subtropical Jammu, India. Crop Prot 84:105–112
5. Carson R (1962) Silent spring. Boston: Houghton Mifflin Company; Cambridge, MA
6. Sharma R, Peshin R, Shankar U, Kaul V, Sharma S (2015) Impact evaluation indicators of an integrated pest management program in vegetable crops in the subtropical region of Jammu and Kashmir, India. Crop Prot 191–199
7. Kesavan PC, Swaminathan MS (2007) Strategies and models for agricultural sustainability in developing Asian countries. Philos Trans R Soc Lond B Biol Sci 363:877–891
8. World Health Organization (1990) Public health impact of pesticides used in agriculture. World Health Organization, England
9. Groot MJ, Van't Hooft KE (2016) The hidden effects of dairy farming on public and environmental health in the Netherlands, India, Ethiopia, and Uganda, considering the use of antibiotics and other agro-chemicals. Front Public Health 4:12
10. Gangemi S, Miozzi E, Teodoro M, Briguglio G, De Luca A, Alibrando C, Polito I, Libra M (2016) Occupational exposure to pesticides as a possible risk factor for the development of chronic diseases in humans (review). Mol Med Rep 14(5):4475–4488
11. Maroni M, Fanetti AC, Metruccio F (2006) Risk assessment and management of occupational exposure to pesticides in agriculture. Med Lav 97:430–437
12. Khan M, Mahmood HZ, Damalas CA (2015) Pesticide use and risk perceptions among farmers in the cotton belt. Crop Prot 67:184–190
13. Sarkar A, Aronson KJ, Patil S, Hugar LB, Van Loon GW (2012) Emerging health risks associated with modern agriculture practices: a comprehensive study in India. Environ Res 115:37–50
14. Kumar S, Verma AK, Bhattacharya S, Rathore S (2013) Trends in rates and methods of suicide in India. Egypt J Forensic Sci 3:75–80
15. Zadjali S, Morse S, Chenoweth J, Deadman M (2014) Factors determining pesticide use practices by farmers in the Sultanate of Oman. Sci Total Environ 476–477:505–512
16. Abhilash PC, Singh N (2009) Pesticide use and application: an Indian scenario. J Hazard Mater 65:1–12
17. Bateman R, Matthews GA, Miller P, (2014) Pesticide application methods, 4th edn. Wiley-Blackwell Publication
18. Fan L, Niu H, Yang X, Qin W, Bento CPM, Ritsema CJ, Geissen V (2015) Factors affecting farmer's behavior in pesticide use: insights from a field study in northern China. Sci Total Environ 537:360–368
19. Houbraken M, Bauweraerts I, Fevery D, Labeke MCV, Spanoghe P (2016) Pesticide knowledge and practice among horticultural workers in the Lam Dong region, Vietnam: a case study of chrysanthemum and strawberries. Sci Total Environ 550:1001–1009
20. Hill R (2008) Detergents in agrochemical and pesticide applications. Surf Sci 301–330. https://doi.org/10.1201/9781420018165.ch12
21. Gil E (2015) Calibration, maintenance and operation, BTSF Training activities on inspection and calibration of pesticide application equipment in professional use. S3, Braunschweig

22. Gil E, Kole J (2015) Workshop inspections of sprayers following EN ISO 16122, BTSF Training activities on inspection and calibration of pesticide application equipment in professional use. S3, Braunschweig
23. Pimentel D (2005) Environmental and economic cost of the application of pesticides primarily in the United States. Environ Dev Sustain 7:229–252
24. Jallow MF, Awadh DG, Albaho MS, Devi VY, Thomas BM (2017) Pesticide knowledge and safety practices among farm workers in Kuwait: results of a survey. Int J Environ Res Public Health 14(4):340
25. Gross K, Rosenheim JA (2011) Quantifying secondary pest outbreaks in cotton and their monetary cost with causal inference statistics. Ecol Appl 21:2770–2780
26. Arias-Estévez M, López-Periago E, Martínez-Carballo E, Simal-Gándara J, Mejuto JC, García-Río L (2008) The mobility and degradation of pesticides in soils and the pollution of groundwater resources. Agric Ecosyst Environ 123:247–260
27. Sekiyama M, Tanaka M, Gunawan B, Abdoellah O, Watanabe C (2007) Pesticide usage and its association with health symptoms among farmers in rural villages in West Java, Indonesia. Environ Sci 14:23–33
28. Konradsen F, Vander Hoek W, Cole DC, Daisley G, Singh H, Eddleston M (2003) Reducing acute poisoning in developing countries—options for restricting the availability of pesticides. Toxicology 192:249–261
29. Issa Y, Sham'a FA, Nijem K, Bjertness E, Kristensen P (2010) Pesticide use and opportunities of exposure among farmers and their families: cross-sectional studies 1998–2006 from Hebron governorate, occupied Palestinian territory. Environ Health 9:63
30. Dhananjayan V, Ravichandran B (2018) Occupational health risk of farmers exposed to pesticides in agricultural activities. Curr Opin Environ Sci Health 4:31–37
31. Matthews GA (1998) Application techniques for agrochemicals. Chem Technol Agrochem Formulations 302–336. https://doi.org/10.1007/978-94-011-4956-3_10
32. Ibitayo OO (2006) Egyptian farmers' attitudes and behaviors regarding agricultural pesticides: implications for pesticide risk communication. Risk Anal 26(4):989–995. https://doi.org/10.1111/j.1539-6924.2006.00794.x
33. Nalwanga E, Ssempebwa JC (2011) Knowledge and practices of in-home pesticide use: a community survey in Uganda. J Environ Publ Health. https://www.hindawi.com/journals/jeph/2011/230894. Accessed on 20 Nov 2011
34. Sodavy P, Sitha M, Nugent R, Murphy H (2000) Situation analysis on farmers' awareness and perceptions of the effect of pesticides on their health. Field document. FAO Community IPM Program, Cambodia
35. Lekei EE, Uronu AB, Mununa FT (2004) Pesticides labels and risk reduction in developing countries. Afr Newsl Occup Health Saf 14(3):57–60
36. Mekonnen Y, Agonafir T (2002) Pesticide sprayers' knowledge, attitude and practice of pesticide use on agricultural farms of Ethiopia. Occup Med 52(6):311–315
37. Ajayi OC, Akinnifesi FK (2007) Farmers' understanding of pesticide safety labels and field spraying practices: a case study of cotton farmers in northern Côte d'Ivoire. Sci Res Essays 2 (6):204–210
38. Damalas CA, Hashemi SM (2010) Pesticide risk perception and use of personal protective equipment among young and old cotton growers in Northern Greece. Agrociencia 44: 363–371
39. Blanco-Muñoz J, Lacasaña M (2011) Practices in pesticide handling and the use of personal protective equipment in Mexican agricultural workers. J Agromed 16:117–126
40. Kesavachandran CN, Fareed M, Pathak MK, Bihari V, Mathur N, Srivastava AK (2009) Adverse health effects of pesticides in agrarian populations of developing countries. Rev Environ Contam Toxicol 200:33–52
41. Dsfs Dinham B (2003) Growing vegetables in developing countries for local urban populations and export markets: problems confronting small-scale producers. Pest Manag Sci 59:575–582

42. Food and Agriculture Organization (2002) International code of conduct on the distribution and use of pesticides. Rome
43. Ntow WJ, Gijzen HJ, Kelderman P, Drechsel P (2006) Farmer perceptions and pesticide use practices in vegetable production in Ghana. Pest Manag Sci 62(4):356–365
44. Dhananjayan V, Ravichandran B (2010) Organochlorine pesticide residues in foodstuffs, fish, wildlife, and human tissues from India: historical trend and contamination status. In: Masood F, Grohmann E, Akhtar R, Malik A (eds) Environmental deterioration and human health. Springer, Netherlands
45. Dhananjayan V, Ravichandran B, Anitha N, Rajmohan HR (2012) Assessment of acetylcholinesterase and butyrylcholinesterase activities in blood plasma of agriculture workers. Indian J Occup Environ Med 16(3):127–130
46. Dhananjayan V, Ravichandran B, Rajmohan HR (2012) Organochlorine pesticide residues in blood samples of agriculture and sheep wool workers in Bangalore (rural), India. Bull Environ Contam Toxicol 88(4):497–500
47. Dhananjayan V, Ravichandran B, Panjakumar K, Kalaiselvi K, Kausic Rajasekar, Mala A, Avinash G, Shridhar K, Manju A, Rajesh Wilson (2019) Assessment of genotoxicity and cholinesterase activity among women workers occupationally exposed to pesticides in tea garden. Mutat Res Genet Toxicol Environ Mutagen 841:1–7
48. Oesterlund AH, Thomsen JF, Sekimpi DK, Maziina J, Racheal A, Jørs E (2014) Pesticide knowledge, practice and attitude and how it affects the health of small-scale farmers in Uganda: a cross-sectional study. Afr Health Sci 14(2):420–433
49. Woodrow JE, Gibson KA, Seiber JN (2018) Pesticides and related toxicants in the atmosphere. Rev Environ Contam Toxicol. https://doi.org/10.1007/398_2018_19
50. Landrigan PJ, Froines JR, Mahaffey KR (1985) Body lead burden: epidemiology data and its relation to environmental sources and toxic effects. In: Dietary and environmental lead: human health effects. Elsevier, Amsterdam
51. Huang H, Ding Y, Chen W, Zhang Y, Chen W, Chen Y, Mao Y, Qi S (2019) Two-way long-range atmospheric transport of organochlorine pesticides (OCPs) between the Yellow River source and the Sichuan Basin, Western China. Sci Total Environ 651(Pt 2):3230–3240
52. Raherison C, Baldi I, Pouquet M, Berteaud E, Moesch C, Bouvier G, Canal-Raffin M (2018) Pesticides exposure by air in vineyard rural area and respiratory health in children: a pilot study. Environ Res 169:189–195. https://doi.org/10.1016/j.envres.2018.11.002
53. Ruge BZ, Muir D, Helm P, Lohmann R (2018) Concentrations, trends, and air-water exchange of PCBs and organochlorine pesticides derived from passive samplers in Lake Superior in 2011. Environ Sci Technol 52(24):14061–14069. https://doi.org/10.1021/acs.est.8b04036
54. Mai C, Theobald N, Hühnerfuss H, Lammel G (2016) Persistent organochlorine pesticides and polychlorinated biphenyls in air of the North Sea region and air-sea exchange. Environ Sci Pollut Res Int 23:23648–23661
55. Savonen C (1997) Soil microorganisms object of new OSU service. Good Fruit Grower. http://www.goodfruit.com/archive/1995/6other.html
56. Socorro J, Lakey PSJ, Han L, Berkemeier T, Lammel G, Zetzsch C, Poschl U, Shiraiwa M (2017) Heterogeneous OH oxidation, shielding effects, and implications for the atmospheric fate of terbuthylazine and other pesticides. Environ Sci Technol 51(23):13749–13754
57. Brooks L (2012) Department of pesticide regulation air monitoring shows pesticides well below health screening levels. California Department of Pesticide Regulation, Sacramento. http://www.cdpr.ca.gov/docs/pressrls/2012/120719.htm. Accessed 3 Aug 2015
58. Crosby DG (1998) Environmental toxicology and chemistry. Wiley, New York
59. NPIC, National Pesticide Information Center (2016) Pesticides and water resources. http://npic.orst.edu/envir/waterenv.html. Accessed 22 Feb 2018
60. András S, Mária M, Béla D (2015) Monitoring pesticide residues in surface and ground water in Hungary: surveys in 1990–2015. J Chem. Article ID 717948, 15
61. Tsaboula A, Menexes G, Papadakis EN, Vryzas Z, Kotopoulou A, Kintzikoglou K, Papadopoulou-Mourkidou E (2018) Assessment and management of pesticide pollution at a

river basin level part II: optimization of pesticide monitoring networks on surface aquatic ecosystems by data analysis methods. Sci Total Environ
62. Chronopoulou EG, Vlachakis D, Papageorgiou AC, Ataya FS, Labrou NE (2018) Structure-based design and application of an engineered glutathione transferase for the development of an optical biosensor for pesticides determination. Biochim Biophys Acta Gen Subj 1863(3):565–576
63. Lari SZ, Khan NA, Gandhi KN, Meshram TS, Thacker NP (2014) Comparison of pesticide residues in surface water and ground water of agriculture intensive areas. J Environ Health Sci Eng 12(1):11. https://doi.org/10.1186/2052-336X-12-11
64. US Environmental Protection Agency (2012) 2006–2007 pesticide market estimates: usage
65. Hildebrandt A, Guillamón M, Lacorte S, Tauler R, Barceló D (2008) Impact of pesticides used in agriculture and vineyards to surface and groundwater quality (North Spain). Water Res 42(13):3315–3326
66. Shao H, Zhang Y (2017) Non-target effects on soil microbial parameters of the synthetic pesticide carbendazim with the biopesticides cantharidin and norcantharidin. Sci Rep 7:5521. https://doi.org/10.1038/s41598-017-05923-8
67. Doolotkeldieva T, Maxabat K, Saykal B (2018) Microbial communities in pesticide-contaminated soils in Kyrgyzstan and bioremediation possibilities. Environ Sci Pollut Res Int 25(32):31848–31862
68. Aigner EJ, Leone AD, Falconer RL (1998) Environ Sci Technol 32:1162–1168
69. Fox JE, Gulledge J, Engelhaupt E, Burow ME, McLachlan JA (2007) Pesticides reduce symbiotic efficiency of nitrogen-fixing rhizobia and host plants. Proc Natl Acad Sci U S A 104(24):10282–10287
70. Staley ZR, Harwood VJ, Rohr JR (2015) A synthesis of the effects of pesticides on microbial persistence in aquatic ecosystems. Crit Rev Toxicol 45(10):813–836
71. Feld L, Hjelmsø MH, Nielsen MS et al (2015) Pesticide side effects in an agricultural soil ecosystem as measured by amoA expression quantification and bacterial diversity changes. PLoS One. https://doi.org/10.1371/journal.pone.0126080
72. Dhananjayan V, Muralidharan S (2012) Organochlorine pesticides and polychlorinated biphenyls in various tissues of waterbirds in Nalabana bird sanctuary, Chilika Lake, Orissa, India. Bull Environ Contam Toxicol 89:197–201
73. Dhananjayan V (2013) Accumulation pattern of persistent organochlorine pesticides in liver tissues of various species of birds from India. Environ Sci Pollut Res 20:3149–3156
74. Dhananjayan V, Muralidharan S, Ranapratab S (2011) Organochlorine pesticide residues in eggs and tissues of house sparrow, Passer domesticus, from Tamil Nadu, India. Bull Environ Contam Toxicol 87:684–688
75. Dhananjayan V, Ravichandran B, Rajmohan HR (2012) Organochlorine pesticide residues in blood samples of agriculture and sheep wool workers in Bangalore (rural), India. Bull Environ Contam Toxicol 88:497–500
76. Nwani CD, Wazir Singh L, Naresh SN, Kumar R, Basdeo K, Satish Kumar S (2010) Toxicity of the herbicide atrazine: effects on lipid peroxidation and activities of antioxidant enzymes in the freshwater fish *Channa punctatus* (Bloch). Int J Environ Res Public Health 7 (8):3298–3312
77. Kreutz LC, Pavan TR, Alves AG, Correia AG, Barriquel B, dos Santos ED, Barcellos LJG (2014) Increased immunoglobulin production in silver catfish (*Rhamdia quelen*) exposed to agrichemicals. Braz J Med Biol Res 47(6):499–504
78. Lim S, Sun YA, In Chan S, Myung HC, Hak CJ, Kyong SP, Lee K-U, Youngmi KP, Hong KL (2009) Chronic exposure to the herbicide, atrazine, causes mitochondrial dysfunction and insulin resistance. PLoS One 4(4):e5186
79. Chen CY, Hathaway KM, Thompson DG, Carol LF (2008) Multiple stressor effects of herbicide, PH, and food on wetland zooplankton and a larval amphibian. Ecotoxicol Environ Saf 71(1):209–218

80. Jayawardena UA, Rohr JR, Navaratne AN, Amerasinghe PH, Rajakaruna RS (2016) Combined effects of pesticides and trematode infections on hourglass tree frog Polypedates cruciger. Ecohealth 13(1):111–122
81. Marcus SR, Fiumera AC (2016) Atrazine exposure affects longevity, development time and body size in Drosophila melanogaster. J Insect Physiol 91–92:18–25
82. Senthilkumar K, Kannan K, Sinha RK, Tanabe S, Giesy JP (1999) Bioaccumulation profiles of polychlorinated biphenyl congeners and organochlorine pesticides in Ganges River dolphins. Environ Toxicol Chem 18:1511–1520
83. Kannan K, Ramu K, Kajiwara N, Sinha RK, Tanabe S (2005) Organochlorine pesticides, polychlorinated biphenyls and polybrominated diphenyl ethers in Irrawaddy dolphins from India. Arch Environ Contam Toxicol 49:415–420
84. Sinha RK, Kannan K (2014) Ganges River dolphin: an overview of biology, ecology, and conservation status in India. Ambio 43(8):1029–1046
85. Jayakumar S, Muralidharan S, Dhananjayan V (2018) Levels of organochlorine pesticide residues in fresh water fishes of three bird sanctuaries in Tamil Nadu, India. Environ Sci Pollut Res Int 26(2):1983–1993
86. Louis J, Casarett JD, Curtis D, Klaassen C, Doull S (2008) Toxicology: the basic science of poisons, 7th edn. McGraw-Hill Professional, New York, NY, USA
87. Keifer MC, Firestone J (2007) Neurotoxicity of pesticides. J Agromed 12:17–25. https://doi.org/10.1300/J096v12n01_03
88. Gunnell D, Eddleston M, Phillips MR, Konradsen F (2007) The global distribution of fatal pesticide self-poisoning, systematic review. BMC Public Health 7:357. https://doi.org/10.1186/1471-2458-7-357
89. Damalas CA, Eleftherohorinos IG (2011) Pesticide exposure, safety issues, and risk assessment indicators. Int J Environ Res Public Health 8:1402–1419
90. Bradman A, Salvatore AL, Boeniger M, Castorina R, Snyder J, Barr DB, Jewell NP, Kavanagh-Baird G, Striley C, Eskenazi B (2009) Community-based intervention to reduce pesticide exposure to farmworkers and potential take-home exposure to their families. J Expo Sci Environ Epidemiol 19:79–89
91. Ye M, Beach J, Martin JW, Senthilselvan A (2013) Occupational pesticide exposures and respiratory health. Int J Environ Res Public Health 10(12):6442–6471. Published 28 Nov 2013. https://doi.org/10.3390/ijerph10126442
92. Lekei EE, Ngowi AV, London L (2014) Farmers' knowledge, practices and injuries associated with pesticide exposure in rural farming villages in Tanzania. BMC Public Health 14:389. https://doi.org/10.1186/1471-2458-14-389
93. Gesesew HA, Woldemichael K, Massa D, Mwanri L (2016) Farmers knowledge, attitudes, practices and health problems associated with pesticide use in rural irrigation villages, Southwest Ethiopia. PLoS One 11(9):e0162527. https://doi.org/10.1371/journal.pone.0162527
94. García-García CR, Parrón T, Requena M, Alarcón R, Tsatsakis AM, Hernández AF (2016) Occupational pesticide exposure and adverse health effects at the clinical, hematological and biochemical level. Life Sci 145:274–283
95. Riaz S, Manzoor F, Mahmood N, Shahid S (2017) Molecular detection of *M. tuberculosis* and *M. bovis* and hematological and biochemical analyses in agricultural sprayers exposed to pesticides: a cross-sectional study in Punjab, Pakistan during 2014–2016. J Expo Sci Environ Epidemiol 27(4):434–443
96. Ahmadi A, Shadboorestan A (2016) Oxidative stress and cancer; the role of hesperidin, a citrus natural bioflavonoid, as a cancer chemoprotective agent. Nutr Cancer 68:29–39
97. Presutti R, Harris SA, Kachuri L, Spinelli JJ, Pahwa M, Blair A, Zahm SH, Cantor KP, Weisenburger DD, Pahwa P, McLaughlin JR, Dosman JA, Freeman LB (2016) Pesticide exposures and the risk of multiple myeloma in men: an analysis of the North American Pooled Project. Int J Cancer 139(8):1703–1714

98. Boccolini PM, Boccolini CS, Chrisman JR, Koifman RJ, Meyer A (2017) Non-Hodgkin lymphoma among Brazilian agricultural workers: a death certificate case-control study. Arch Environ Occup Health 72(3):139–144
99. Mazurek JM, Henneberger PK (2017) Lifetime allergic rhinitis prevalence among US primary farm operators: findings from the 2011 Farm and Ranch Safety survey. Int Arch Occup Environ Health 90(6):507–515
100. Kongtip P, Techasaensiri B, Nankongnab N, Adams J, Phamonphon A, Surach A, Sangprasert S, Thongsuksai A, Srikumpol P, Woskie S (2017) The impact of prenatal organophosphate pesticide exposures on Thai infant neurodevelopment. Int J Environ Res Public Health. https://doi.org/10.3390/ijerph14060570
101. Bonner MR, Freeman LE, Hoppin JA, Koutros S, Sandler DP, Lynch CF, Hines CJ, Thomas K, Blair A, Alavanja MC (2017) Occupational exposure to pesticides and the incidence of lung cancer in the agricultural health study. Environ Health Perspect 125 (4):544–551
102. Nordgren TM, Bailey KL (2016) Pulmonary health effects of agriculture. Curr Opin Pulm Med 22(2):144–149
103. Raanan R, Balmes JR, Harley KG, Gunier RB, Magzamen S, Bradman A, Eskenazi B (2016) Decreased lung function in 7-year-old children with early-life organophosphate exposure. Thorax 71(2):148–153
104. Raanan R, Gunier RB, Balmes JR, Beltran AJ, Harley KG, Bradman A, Eskenazi B (2017) Elemental sulfur use and associations with pediatric lung function and respiratory symptoms in an agricultural community (California, USA). Environ Health Perspect 125(8):087007. https://doi.org/10.1289/EHP528
105. Zhang X, Wu M, Yao H, Yang Y, Cui M, Tu Z, Stallones L, Xiang H (2016) Pesticide poisoning and neurobehavioral function among farm workers in Jiangsu, People's Republic of China. Cortex 74:396–404
106. Bolognesi C, Holland N (2016) The use of the lymphocyte cytokinesis-block micronucleus assay for monitoring pesticide-exposed populations. Mutat Res 770(Pt A):183–203
107. Cremonese C, Piccoli C, Pasqualotto F, Clapauch R, Koifman RJ, Koifman S, Freire C (2017) Occupational exposure to pesticides, reproductive hormone levels and sperm quality in young Brazilian men. Reprod Toxicol 67:174–185
108. Waheed S, Halsall C, Sweetman AJ, Jones KC, Malik RN (2017) Pesticides contaminated dust exposure, risk diagnosis and exposure markers in occupational and residential settings of Lahore, Pakistan. Environ Toxicol Pharmacol 56:375–382
109. Calvert GM (2016) Agricultural pesticide exposure and chronic kidney disease: new findings and more questions. Occup Environ Med 73(1):1–2
110. Lebov JF, Engel LS, Richardson D, Hogan SL, Hoppin JA, Sandler DP (2016) Pesticide use and risk of end-stage renal disease among licensed pesticide applicators in the Agricultural Health Study. Occup Environ Med 73(1):3–12
111. Thetkathuek A, Yenjai P, Jaidee W, Jaidee P, Sriprapat P (2017) Pesticide exposure and cholinesterase levels in migrant farm workers in Thailand. J Agromedicine 22(2):118–130
112. Valcke M, Levasseur ME, Soares da Silva A, Wesseling C (2017) Pesticide exposures and chronic kidney disease of unknown etiology: an epidemiologic review. Environ Health 16 (1):49. https://doi.org/10.1186/s12940-017-0254-0
113. Zeljezic D, Vinkovic B, Kasuba V, Kopjar N, Milic M, Mladinic M (2017) The effect of insecticides chlorpyrifos, α-cypermethrin and imidacloprid on primary DNA damage, TP 53 and c-Myc structural integrity by comet-FISH assay. Chemosphere 182:332–338
114. Lerro CC, Beane Freeman LE, Portengen L, Kang D, Lee K, Blair A, Lynch CF, Bakke B, De Roos AJ, Vermeulen R (2017) A longitudinal study of atrazine and 2,4-D exposure and oxidative stress markers among iowa corn farmers. Environ Mol Mutagen 58(1):30–38
115. Boulanger M, Tual S, Lemarchand C, Guizard AV, Velten M, Marcotullio E, Baldi I, Clin B, Lebailly P (2017) Agricultural exposure and risk of bladder cancer in the AGRIculture and CANcer cohort. Int Arch Occup Environ Health 90(2):169–178

116. Lemarchand C, Tual S, Levêque-Morlais N, Perrier S, Belot A, Velten M, Guizard AV, Marcotullio E, Monnereau A, Clin B, Baldi I, Lebailly P (2017) Cancer incidence in the AGRICAN cohort study (2005–2011). Cancer Epidemiol 49:175–185
117. Armas ED, Monteiro RTR, Antunes PMMAP, Santos F, Camargo PB (2007) Uso de agrotoxicos em cana-de ac¸ ucar na bacia do rio Corumbatai e o risco de poluic¸ ao hidrica. Quim Nova 30:1119–1127
118. Liu B, Wang Y, Yang F, Wang X, Shen H, Cui H, Wu D (2016) Construction of a controlled-release delivery system for pesticides using biodegradable PLA-based microcapsules. Colloids Surf B Biointerfaces 144:38–45
119. Sekhon BS (2014) Nanotechnology in agri-food production: an overview. Nanotechnol Sci Appl 7:31–53. https://doi.org/10.2147/NSA.S39406
120. Anamika R, Sunil KS, Jaya B, Anil KB (2014) Controlled pesticide release from biodegradable polymers. Cent Eur J Chem 12:453–469
121. Cui B et al (2015) Evaluation of stability and biological activity of solid nanodispersion of lambda-cyhalothrin. PLoS One 10:e0135953
122. Alan K (2008) Recent developments of safer formulations of agrochemicals. Environmentalist 28:35–44
123. Grillo R, De Melo NFS, De Araujo DR, De Paula E, Rosa AH, Fraceto LF (2010) Polymeric alginate nanoparticles containing the local anesthetic bupivacaine. J Drug Target 18: 688–699
124. Barik TK, Sahu B, Swain V (2008) Nanosilica—from medicine to pest control. Parasitol Res 103:253–258
125. Maqueda C, Villaverde J, Sopena F, Undabeytia T, Morillo E (2008) Novel system for reducing leaching of the herbicide metribuzin using clay-gel-based formulations. J Agric Food Chem 56:11941–11946
126. Dizaj S, Jafari S, Khosroushahi A (2014) A sight on the current nanoparticle-based gene delivery vectors. Nanoscale Res Lett 9:252
127. Yang G, Gong H, Qian X, Tan P, Li Z, Liu T, Liu J, Li Y, Liu Z (2014) Mesoporous silica nanorods intrinsically doped with photosensitizers as a multifunctional drug carrier for combination therapy of cancer. Nano Res. https://doi.org/10.1007/s12274-014-0558-0
128. Wu X, Wang Z, Zhu D, Zong S, Yang L, Zhong Y, Cui Y (2013) A pH- and thermo dual stimuli responsive drug carrier based on mesoporous silica nanoparticles encapsulated in copolymer-lipid bilayers. ACS Appl Mater Interfaces 5(21):10895–10903
129. Wang P, Lombi E, Zhao FJ, Kopittke PM (2016) Nanotechnology: a new opportunity in plant sciences. Trends Plant Sci 21(8):699–712
130. Chowdhury MA (2014) The controlled release of bioactive compounds from lignin and lignin-based biopolymer matrices. Int J Biol Macromol 65:136–147
131. Wang Y, Wang A, Wang C, Cui B, Sun C, Zhao X, Zeng Z, Shen Y, Gao F, Liu G, Cui H (2017) Synthesis and characterization of emamectin-benzoate slow-release microspheres with different surfactants. Sci Rep 7(1):12761
132. Wang Y, Cui H, Sun C, Zhao X, Cui B (2014) Construction and evaluation of controlled-release delivery system of Abamectin using porous silica nanoparticles as carriers. Nanoscale Res Lett 9(1):2490. https://doi.org/10.1186/1556-276X-9-655
133. Reis KC, Pereira J, Smith AC, Carvalho CWP, Wellner N, Yakimets I (2008) Characterization of polyhydroxybutyrate-hydroxyvalerate (PHB-HV)/maize starch blend films. J Food Eng 89:361–369
134. Amass W, Tighe B (1998) A review of biodegradable polymers: use, current developments in the synthesis and characterization of biodegradable polyesters, blends of biodegradable polymers and recent advances in biodegradable studies. Polym Int 47:89–144
135. Sudesh K, Fukui T, Iwata T, Doi Y (2000) Factors affecting the freeze-fracture morphology of in vivo polyhydroxyalkanoate granules. Can J Microbiol 46:304–311
136. Grillo R, Pereira Ado E, de Melo NF, Porto RM, Feitosa LO, Tonello PS, Dias Filho NL, Rosa AH, Lima R, Fraceto LF (2011) Controlled release system for ametryn using polymer

microspheres: preparation, characterization and release kinetics in water. J Hazard Mater 186 (2–3):1645–1651

137. Wang G, Xiao Y, Xu H, Hu P, Liang W, Xie L, Jia J (2018) Development of multifunctional avermectin poly (succinimide) nanoparticles to improve bioactivity and transportation in rice. J Agric Food Chem 66(43):11244–11253
138. Gogos A, Knauer K, Bucheli TD (2012) Nanomaterials in plant protection and fertilization: current state, foreseen applications, and research priorities. J Agric Food Chem 60 (39):9781–9792

Chapter 2
Methods for Determination of Pesticides and Fate of Pesticides in the Fields

Angel G. Polanco Rodríguez and Jesús Alfredo Araujo León

Abstract Actually, the globalization with the high industrialization and development has a high impact on diverse sector, such as agricultural, livestock, urban development, and public health. It is of high concern, because of the uncontrolled use and application of carcinogenic pesticides. Several modern extraction techniques have been developed and may be divided into two groups: fluid-phase partitioning methods and sorptive and membrane-based extraction methods. The selection of the extraction method is done based on the type of pesticide (physicochemical properties) and the matrix sample. The most popular in the last year were microwave-assisted extraction (MAE), liquid–liquid partitioning, solid-phase extraction (SPE), solid-phase micro-extraction (SPME), and membrane extraction techniques. The last step in the method to select the pesticides analysis, when you know the chemical and physicochemical properties, the matrix sample and the extraction method, you can select the method of determination. In this chapter, we will focus on the analysis by chromatographic methods. The chromatographic methods used in pesticide analysis were divided into two groups: gas chromatography and liquid chromatography. The first of them has been proved to be a largely used, most versatile, and sensitive, coupled with high-selectivity and sensitivity detectors such as electron capture detector (ECD), mass spectrometry (MS), MS/MS, and nitrogen–phosphorus detector (NPD).

Keywords Pesticides · Method · Fate · Human health · Cancer

A. G. P. Rodríguez (✉)
Laboratorio de Medicina Social y Salud Pública, Centro de Investigaciones Regionales, Universidad Autónoma de Yucatán, Mérida, CP 97000, Mexico
e-mail: polanco07@gmail.com

J. A. A. León
Laboratorio de Cromatografía, Facultad de Química, Universidad Autónoma de Yucatán, Mérida, CP 97000, Mexico
e-mail: jesus.araujo@correo.uady.mx

Rakhimol K. R. et al. (eds.), *Controlled Release of Pesticides for Sustainable Agriculture*, https://doi.org/10.1007/978-3-030-23396-9_2

2.1 Introduction

Historically, we are experiencing the greatest transition in terms of population health, since the profile of indicators of the causes of illness and death is being transformed. Health and humanity prospects depending on an uncertain process of globalization and the emergence of global environmental changes are produced in response from the global economic development. Therefore, researchers and public health professionals must address both traditional and new challenges for the health of the population as a vision of environmental impacts, taking into account the social and natural aspects, which have represented greatly assessment and measured the dynamics of the incidence of diseases emergent over time, because the ecological interactions with the human health are interdependent. The processes of globalization and economic development have been unprecedented and currently have a high impact on the environment, social development, and human health [1–3].

The globalization and spreading of pollution process for food production with high use of pesticides in agricultural and livestock activities, it is a global concern because are cause of a large number of environmental negative effects, has increased the chemical burden on natural ecosystems with a high impact on the public health [4].

Pesticide contamination of groundwater is a subject of international importance because groundwater is used as drinking water for population; however, this especially concerns people living in the agricultural areas where pesticides are most often used [5] and even more so in developing countries where environmental norm and legislation is not respected. The scenery of the pesticide pollution is clear, and there is a legal framework consolidated that coexists with disrespect related to the national and international companies with the official sector in the use and distribution of carcinogenic pesticides [6], reducing by one half of population in the world without the access to safe drinking water that is one of the key millennium development goals [7].

Urban populations in developing countries are exposed not only to traditional diseases such as diarrheal diseases and respiratory infections, but also to modern danger emergent diseases such as physical and chemical risks, which are the products of poorly regulated industrialization and development with high impact on the water. This has increased the chemical burden of contamination on water, soil, and wind in the world [8]; however, in developed countries, the impact on the public health is smaller [9] due to greater regulation for development.

In recent years, the effects of environmental pollutants and chronic degenerative diseases have been given high importance at the global level. There is ample evidence on the relationship between exposure to pollutants such as pesticides and the high rate of chronic diseases, different types of cancers, diabetes, neurodegenerative disorders, birth defects, and reproductive disorders. The evidence in the association of exposure to pesticides with the incidence of chronic diseases shows genetic damage, epigenetic modifications, endocrine disruption, mitochondrial dysfunction, oxidative stress, and endoplasmic reticulum stress [10].

Now, we are sure that human health depends on the processes of globalization and regulation, which impact on the social and environmental conditions. These changes include air pollution, degradation of land and water, high deforestation, impact on fisheries, and loss of biodiversity. This degradation of the systems that support the life and the planet poses high impacts on the human health. A very important indicator for development is that governments provide water infrastructure for sanitation and water purification, and it is not permissible to say only in a globalized world "improved water" instead of drinking water of good biological and chemical quality. Water quality monitoring in developing countries should put more effort for auditing this process and whether the health benefits are being achieved [11].

2.2 Agrochemicals of High Impact on the Environment

Due to the population growth worldwide, during the last decades, the need to produce more food and the use of toxic agrochemicals have increased. The increase in the intensive use of agrochemicals in this process has increased the levels of agrochemicals in soil, air, and water, with high impacts on human health and terrestrial and aquatic ecosystems. Although little by little they have been replaced by less toxic agrochemicals, the residues of persistent organic pollutant still continue to impact the human being, the production of food for human consumption, water, and air [12, 13].

Several studies have been documented on the pollution of the water of cenotes (sinkholes) with high concentrations of organochlorine pesticides (OCP) [14], as well as their bioaccumulation in the blood of women with cancer and in breast milk [15, 16]. Coupled with the use of carcinogenic pesticides such as DDT, lindane, heptachlor, endosulfan, aldrin, endrin, and dieldrin (organochlorines), with the high impacts on human health and the environment, in Yucatan, glyphosate herbicide (organophosphate herbicide) is used in the production of soybeans in the east and south of the state, and Mexican producers and authorities are currently going through various legal procedures due to the high impact of glyphosate on ecosystems and on public health [17], with significant human rights violations regarding informed procedures to indigenous peoples. In this regard, the "Precautionary Principle," adopted by European governments, should be adopted at international level, mainly in developing countries. The Rotterdam Convention [18] on the Prior Informed Consent Procedure (PIC), where governments, have the necessary information on hazardous chemicals, to assess the risks and take informed decisions on their imports, and the Stockholm Convention [19], for the prohibition of toxic carcinogenic agrochemicals.

The use of glyphosate globally for the production of transgenic crops has increased in the last decade. Transgenic or genetically modified (GM) crops are the result of the application of DNA technology in agriculture with the aim of improving seeds or creating resistance to pests at the genetic level. Transgenic soybeans with the use of glyphosate herbicide are currently a subject of broad discussion, considering the high impacts on public health. The World Health Organization classified glyphosate as a probable carcinogen for humans.

In 2010, 148 million hectares were planted, cultivated by 15.4 million farmers in 29 countries. The Working Group of the International Agency for Research on Cancer (IARC) [20] also reported the herbicide glyphosate as a probable human carcinogen. The company that produces the glyphosate at an international level is Monsanto, and in recent years, it has received lawsuits in several countries for the high toxicity of its product.

In the state of Yucatan, farmers maintain the traditional existence of the milpa for the cultivation of corn as a central component and crops such as squash, tomatoes, habanero peppers; however, soy is being cultivated in the agricultural and livestock area. From the recent studies on farmers in the main agricultural area of southern of Yucatan, 55% of farmers reported using the glyphosate herbicide for the production of transgenic soybean [21].

2.2.1 Environmental Pollution Routes

What happens with the organochlorine pesticides (OCP) that are sprayed on the soils in agricultural activities? The pesticides remain in the environment for a shorter or longer period; this depends on the type of soils, the degree of permeability to the groundwater, the climate factors such as dry (concentration of pollutants) or rainy (flow of pollutants) seasons and in some cases due to remote pollution and transport of pollutants.

The pesticides sprayed on fields can be transported to the surrounding, affecting the soil and water. Surface runoff can carry pesticides out into rivers and lakes, to the groundwater and finally into the coast and sea.

The filtration of pesticides is aggravated when there are karstic soils of easy filtration of contaminants to the underground aquifer; this increases the vulnerability to the aquifer when there are other natural conditions such as deforestation, lack of regulation, and application of the regulations for environmental conservation such as the case of the region of karstic aquifer in Yucatan, Mexico.

In the karstic scenario of Yucatan, Mexico, water contamination studies have been conducted detecting high levels of OCP and their metabolites, during rainy and dry seasons. Concentrations of 6.5 ppm of α-lindane and 10.86 of δ-lindane were founded during the dry season, in the main water recharge zone called the Ring of Cenotes (sinkholes); this is a Ramsar site [22] and decreed as the main water conservation zone in Yucatan [23]. However, this environmental legislation in Yucatan, Mexico, for the water conservation is not applied, damaging the aquifer with very high impacts for the public health. In this zone, at the northwest zone of this Ring of Cenotes, in the Celestun coast, were detected levels of 4.42 ppm of heptachlor and 2.56 ppm of γ-lindane; highest concentrations of 13.61 and 12.54 ppm of heptachlor were founded in the municipality of Dzilam, in the dry period [24]. High concentrations have been found in similar studies, such as one carried out in the estuaries of southern Honduras, were detected 23 ppm of lindane and 45.8 ppm of aldrin, due to the high use of toxic pesticides in agriculture [25].

Pesticide properties affect the risk of unwanted spread in the environment. The volatility of substances (vapor pressure) determines how much of a pesticide will evaporate to the atmosphere during and after application. This in turn affects the risk of a compound spreading over a wide area. Water solubility and adsorption capacity (capacity to bind to soil particles, particularly humus and clay) greatly affect the risk of compounds ending up in surface water and groundwater. High water solubility increases the risk of the compounds accompanying the water on its path down to the groundwater and to subsurface drains. High adsorption capacity, on the other hand, means that the compounds are not carried away easily by water, although in conditions causing erosion and surface runoff they can easily end up in watercourses. The persistence of a compound (its endurance) determines how long a pesticide will remain in the environment without being broken down. Substances with a low adsorption capacity (poor binding to the soil) and those with slow rate of breakdown pose the greatest risk of spread [26].

2.2.2 *Fate of Pesticides*

Fate of pesticides when they enter to the atmosphere, the pesticides and their photodegradation products may be transported long distances before the removal processes of atmospheric wet and dry deposition return them to the earth's surface [27, 28]. Initial distribution describes the proportion of pesticide that is on or in the air, soil, water, plants, after application, and affects fate of a pesticide in the environment. This amount is determined by the formulation, method of application, topography, type of landscape, vegetation and ground cover, and weather conditions. With time, the pesticide may be redistributed within the application site or may move off site of the target area. Pesticides that move off site represent an economic loss and may pollute groundwater or surface water [29].

The high use of pesticides during the past decades has highlighted the potential risk to the environment and the public health. The studies of the presence of pesticides in soil, sediment, and water have generated great concern. Pesticides in the environment can be degraded by the influence of physical, chemical, and biological factors and can be volatilized, adsorbed by soil colloids and transported offsite by surface runoff and leaching, and then transported to the coast. The quantity of each residue of pesticide that can be found in soil, sediment, water bodies and sea depends on these processes. There are several routes through which pesticides can contaminate the environment. Pointsource or diffuse contamination and pesticide dispersion to different environmental compartments is affected by several processes. Pesticides are deliberately or unintentionally released to nearly all biospheres. Actually, there is evidence that soils, sediments, surface and groundwater bodies, and wind, are contaminated with diverse levels of pesticides [30].

Pesticide transport is in the atmosphere that includes volatilization, spray drift, and terrestrial as wash and runoff, leaching, and lateral drainage; pesticide degradation encompasses photolysis and biotic and abiotic process such as environmental contamination for pesticides may also occur via dust [31]. On the other hand, climate change is an important factor that affects the fate of pesticides in the soil, sediment, water, and wind. The large dry periods followed by intense rainfall events are expected to increase the mobility of pesticides through runoff and preferential routes of flow to surface and groundwater bodies, respectively [32].

Due to their physicochemical properties, the pesticides can be widely spread for atmospheric transport of pollutants all over the globe. They represent a serious threat to both human being and wildlife/natural resources. It has been only speculated that sorption to aerosol particles may increase significantly the half-lives of pesticides in the atmosphere. The pesticides particles in the lower boundary layer of the atmosphere are an important impact on the air quality and human health [33].

The air pollutants remain in the atmosphere long enough to be transported over the planet across intercontinental distances. Research has shown that there is important transport of natural and anthropogenic material from Africa to North and South America, from Asia to North America, from North America to Europe, and from Europe and Asia to the Arctic. For example, satellite images and aircraft measurements have shown that large quantities of Saharan dust are transported across the North Atlantic. African dust is detected routinely at a monitoring site in Miami. This dust affects large areas of the Eastern and Southern USA and has implications for particulate matter (PM). Further, biomass burning in Africa contributes large quantities of ozone to the atmosphere of the Southern Hemisphere. Ozone, ozone precursors, and particulate matter from pollution sources along the East Coast of North America have been observed in the remote Atlantic and at European ground stations. Investigators at the Mauna Loa Observatory, on the island of Hawaii, routinely see evidence of long-range transport of natural and anthropogenic aerosols and trace gases originating over Asia. Asian pollution has been observed off the West Coast of the USA. Researchers have shown that haze in the Arctic is caused by anthropogenic emissions traveling 8000–10,000 km from their source regions in Eastern Europe. Investigators have shown that biomass burning, caused by both human activities and natural processes, results in significant regional pollution and may have impacts on atmospheric chemistry and global climate. The development of a global economy has effects beyond raising standards of living. We are in an era of increasing environmental as well as economic interdependence. Industrialization of the developing world means higher energy consumption and more pollution. The implications of this growth for human health, global climate, regional environmental quality, and sustainable economic development are significant. The review of "The Intercontinental of Air Pollution, The Air Quality Research Subcommittee of the Committee on Environment and Natural

Resources (CENR)" studies the source and fate relationships of pollutants related to the study of Intercontinental Air Pollution about what is the impact of pollution (ozone, particle Matter (PM), mercury, and POPs) from Africa to North America, Africa to Southern Hemisphere, North America to Atlantic and Europe, Asia to remote Pacific and Western USA, and Eastern Europe and Asia to Alaska, Canada, and the Arctic [34].

In addition to the remote transport of pollutants, the fate and impact of pesticides produce local (or punctual) pollution in soils, surface and underground water, coasts and seas, and air, and finally their bioaccumulation in the fat tissue of the human being, aggravating the pollution on the planet. Dispersion of pesticide residues in the environment and species, such as bees, birds, amphibians, fish, and mammals, been reported [35].

This problem of widespread contamination is accentuated in developing countries, which does not possess and apply public policies for the conservation of the environment, especially water, lack adequate infrastructure for water sanitation and technologies for the treatment and obtaining of drinking water. Under this scenario, fundamental human rights of access to a healthy environment, access to safe drinking water, and access to health are violated.

The bioaccumulation of pesticides in seas and marine species is currently a subject of high concern, due to its potential damage to the ecosystem and risks to human health due to the consumption of marine species. Studies in Mumbai have been evaluated, the levels of pollutants of total HCHs in seawater varied from 0.16 to 15.92 ng/L, and concentrations of total DDT varied from 3.01 to 33.21 ng/L. The total HCH in the sediment samples was from 3.8 to 16.2 ng/g range. The concentration of total HCHs in some marine species varied from 0.87 to 33.73 ng/g, and concentrations of total DDT varied from 0.38 to 34.1 ng/g [36].

2.2.3 *Impact to the Environment and Public Health*

The theme is of high concern, because of the uncontrolled use and application of carcinogenic pesticides with effects as endocrine disruptors are chemicals that cause a large number of diseases and diverse cancer in the human being. Although there exist international agreements such as the Stockholm and Rotterdam Convention for the prohibition and restriction of these pesticides due to their high toxicity, the main problem is that all the countries that have signed these international agreements do not respect the agreements and continue the use of these agrochemicals, affecting the environment and the public health. There are countries with high cancer rates such as India, Argentina, and Mexico, with an excessive use of carcinogenic agrochemicals. Many scientific publications have reported the relationship of levels of pesticides in humans and diverse cancers [37–41].

The global burden of diseases (GBD) in humans can be related to the exposure of environmental pollutants and continues to be an important source of high risk to human health globally. These risks are higher in developing countries, where poverty, lack of investment in technologies for sanitation, and access to safe drinking water, as well as weak policies on environmental legislation that are not meet combine to cause high levels of pollution that affect human population health. Developed countries have adequate high-tech sanitation infrastructure programs and systematized monitoring programs to monitor that contaminants do not exceed permitted limits and affect human health. However, the associations between environmental pollution, sanitary level, and health are complex and generally poorly characterized. The above is due to the lack of results for the measurement of exposure and contamination levels; this is often unknown because of the lack of systematic monitoring programs. Environmental contaminants can bioaccumulate and relate to various diseases. In recent years, several attempts have been made to assess the global burden of diseases because of environmental contamination, either in terms of mortality or in terms of disability-adjusted life years (DALYs). About 8–9% of the total burden of the disease can be attributed to pollution, but considerably higher in developing countries. Contaminated water and lack of sanitation are the main sources of exposure, along with indoor air pollution [42].

Cancer is a complex multifactorial disease where there are interactions between the environment and the genes [43]. The human being is exposed to many environmental carcinogens, and the increase in the prevalence of cancer can be related to occupational risk at work [44].

Gene–environment interaction (GEI) is the interaction of genetic and environmental factors that intervene in the human health disease process [45]. Exposure to pollutants under specific conditions has effects at the level of genetic polymorphisms that promote the onset of the disease. GHG has different effects on the exposure of environmental contaminants in the genotypes of different people with different exposure histories [46], and these interactions can be important determinants for the development of cancer [47].

Nowadays, the bioaccumulation of environmental contaminants can play an important role in the etiology of various diseases in humans, having high public health impacts, such as mutagenicity and carcinogenicity [48]. Environmental pollutants are endocrine disruptors (EDCs) that interfere with the synthesis, transport, storage, binding, and natural hormonal activity, associated with the risk of various cancers [49–52], and affect the female reproductive system [53].

Persistent organic pollutants (POPs) accumulate in fatty tissue because they are lipophilic, and have high resistance to biodegradation. Although the use of POPs was restricted several decades ago due to its harmful effects on human health and the environment, POPs still affect human health because of their exposure and persistence in the environment [54].

POPs include polychlorinated biphenyls, dioxins, and organochlorine pesticides, such as dichlorodiphenyltrichloroethane (DDT). These were originally called chemicals that alter the endocrine system; they are endocrine disruptors, because they interfere with the natural functions of hormones, such as sex hormones, glucocorticoids, and thyroid.

In a prospective 19-year follow-up study, the incidence of type 2 diabetes increased with plasma levels of hexachlorobenzene (lindane). In a meta-analysis, the incidence of type 2 diabetes in subjects with a high concentration of polychlorinated biphenyl in plasma was 1.7 times higher than in subjects with a low concentration. Although epidemiological studies have shown positive associations between POPs and type 2 diabetes, in vitro studies in animals should clarify the pathogenesis and mechanisms by which POPs affect the development of type 2 diabetes and metabolic disorders. A recent study showed that rats fed a diet of crude fish oil containing a low dose of POPs developed visceral obesity, hepatosteatosis, and insulin resistance. The risk factors traditionally associated with diabetes are family history, lifestyle, obesity, lack of exercise, and age. However, these traditional factors alone cannot explain the rapid increase in their prevalence worldwide. Recently, new risk factors for obesity and diabetes have been reported, such as environmental chemical contaminants. Among them, heavy metals such as cadmium, mercury, and arsenic are reported, as well as persistent organic pollutants, such as organochlorine pesticides, these chemical contaminants are considered new risk factors for diabetes [55].

In the state of Yucatan, Mexico, studies were conducted to determine levels of organochlorine pesticides, finding high blood levels of 72 Mayan women with uterine cervical cancer in three different zones of Yucatan, agricultural area, livestock area, and metropolitan area [56]. These high impacts on public health and water are due to the continued use of carcinogenic OCPs prohibited by international conventions such as Stockholm and Rotterdam. The problem is complex and worrisome, since there is legislation of Official Mexican Standards on the Maximum Residue Limits (MRLs) allowed for pesticides in water, but the official norms are not met and there is no infrastructure for sanitation and obtaining potable water for human consumption. In effect, Yucatan is in the last place at the national level in terms of infrastructure for water sanitation [57] (CONAGUA-Semarnat 2015).

2.3 Analysis of Pesticides

The pesticides classification is one of the most important parameters to evaluate to decide the method analysis. Focused on the analysis pesticides method, the most popular classification is based on the mode of action, target pest species, chemical structure, and physicochemical properties such as solubility, polarity, boiling point, thermosensibility, molecular weight, vapor pressure, and finally the Henry's constant. Another parameter to evaluate to decide the pesticides method analysis is the matrix in which, for example, all types of water (drinking, lake, sea, groundwater, river, costal, etc.), sediments, plants, agricultural products such as vegetables and fruits mainly, and more complex matrix as biological human fluids (urine, total blood, serum, plasma, women milk, etc.), human and animal tissues; commercial foods (milk, butter, yogurt, eggs, oils, etc.) and finally pesticides in the air are available. If you know the matrix in which pesticides are available, this will help to select the extraction method. Actually, several modern extraction techniques have been developed, and they may be divided into two groups: fluid-phase partitioning methods and sorptive and membrane-based extraction methods. The selection of the extraction method is done based on the type of pesticide (physicochemical properties) and the matrix sample. The most popular in the last year were microwave-assisted extraction (MAE), liquid–liquid partitioning, solid-phase extraction (SPE), solid-phase micro-extraction (SPME), and membrane extraction techniques.

The last step in the method to select the pesticides analysis, when you know the chemical and physicochemical properties, the matrix sample and the extraction method, you can select the method of determination. In this chapter, we will focus on the analysis by chromatographic methods.

The chromatographic methods used in pesticide analyses were divided into two groups: gas chromatography and liquid chromatography. The first one, has been largely used, most versatile, and sensitive, coupled with high selectivity and sensitivity detectors as electron capture detector (ECD), mass spectrometry (MS), MS/MS, and nitrogen–phosphorus detector (NPD), the main rule to use gas chromatography in pesticides analysis is that the analyte be thermostable and have a good vapor pressure with boiling point less than 400 °C.

The liquid chromatography, specifically the high-performance liquid chromatography (HPLC), may be classified as a modern method on pesticide analysis, mainly depending on the partition coefficient of pesticide on the mobile and stationary phase. The heart of the HPLC depends of the column used, because the separation occurs here, is chemically bonded to the support with different surface adsorption activity and porosity of the material. A quick classification may be divided into normal phase and reverse phase. The HPLC coupled with detectors as UV, diode array detector (DAD), fluorescence, ion-trap MS, simple quad MS, and triple quad MS have been used to determine pesticides in different matrices.

Another parameter to evaluate the pesticides analysis method is the matrix in which, for example, all types of water (drinking, lake, sea, groundwater, river, costal, etc.); sediments, plants, agricultural products as vegetables and fruits mainly, and more complex matrix such as biological human fluids (urine, total blood, serum, plasma, breast milk, etc.); human and animal tissues; commercial foods (milk, butter, yogurt, eggs, oils, etc.), and finally pesticides in the air are available. If you know the matrix in which pesticides are available, this will help to select the extraction method.

In this chapter, we will study all aspects of pesticide analysis, since sample collection, storage, sample pre-treatment and treatment in the laboratory, then, how to choose the best extraction method according with the matrix and pesticides properties, a clean-up step, to finally choice which chromatographic technique is better to have results with high trueness.

2.3.1 Sample Preparation for Pesticide Analysis

Currently, more than 1100 chemical compounds are used as pesticides. For this reason, several organizations have set maximum residue limits for pesticide residues in different matrices. The development of pesticide methods detection based mainly on chromatographic techniques, due to allows the execution in a single analysis, is a worldwide strategy in pesticide analysis. However, the analysis requires pretreatment to extract the target compounds before the identification and quantification of the selected pesticides. The sensitivity, selectivity, and specificity of the test are the most critical factors in pesticide analysis. Thus, several modern extraction techniques have been developed, and they may be divided into the following two groups: partition methods on liquid phase and sorptive extraction methods. In the two cases, a cleanup step is desirable for the elimination of impurity compounds to improve the chromatographic analysis.

2.3.2 Methods of Pesticide Determination

Chromatographic method is the most common to identify and quantify pesticides. This analytical technique was divided in two: gas chromatography and liquid chromatography; each one depends on the physicochemical properties of the analyte. If the analyte has a boiling point less than 350 °C, the best option is gas chromatography; however, if the analyte has thermostability problems or the boiling point is more than 400 °C, the best option is liquid chromatography.

2.3.3 Gas Chromatography

Gas chromatography (GC) has been proved to be a primarily used, most versatile, and sensitive method for pesticide residue analysis. This is because of the high separation efficiency of GC and the full range of analyzers that it can be coupled to simplify interpretation of pollutants in complex environmental matrices.

The GC process comprises several steps. First, the sample is injected, to do it is necessary a syringe to broken the septum in the inlet. The solvent is transferred to inlet zone in GC injector with a liner (glass vial), it is in this zone when the sample is volatized, the GC injector is recommended to have a temperature 50 °C higher than the boiling point of the analyte less volatile around 250–320 °C. The injection sample can be of two main modes: split and splitless; this depends on septum purge gas. Split mode is recommended when the sample has a high concentration, and splitless mode is recommended when there is a need of more sensibility into the analysis.

The second step is the temperature rate, which is the most important because the temperature gradient defines the analyte separation. In GC, according to the van Deemter curves, the hydrogen is the better carrier gas; however, this is explosive, and thus, helium is preferred because it has a similar theoretical plates with same velocity and viscosity. This step is related with type of capillary column, for no polar compounds, the main column is methyl siloxane o methyl-phenyl siloxane until polar columns as polyethylene glycol or cyanopropyl siloxane. The codes of these columns depend on the provider. The final step is the detection of the pesticides; in this chapter, we talk about three detectors: electron capture detector (ECD), nitrogen–phosphorus detector (NPD), and mass spectrometer (MS).

2.3.4 Partition Methods on Liquid Phase

The partition methods on liquid phase include a single drop, liquid micro-extraction, and microwave-assisted extraction (MAE), mainly. Szarka et al. [58] developed an economical and rapid method using dispersive liquid–liquid micro-extraction (DLLME) to extract and determine forty pesticides with recoveries between 70 and 120% with relative standard deviations (RSD) less than 20% and detection limits around from 0.001 to 0.910 μg/L. A switchable solvent liquid-phase micro-extraction was developed by Chormey et al. [59], for the determination of selected endocrine disruptors, pesticides, and hormones with a limit of detection around 0.20–13 ng/mL, and percent recoveries recorded were between 91 and 110%, and RSD values were all below 10%.

Currently, Farajzadeh et al. [60] in 2018 developed a new temperature-controlled liquid phase micro-extraction using a deep eutectic solvent for extraction and pre-concentration of diazinon, metalaxyl, bromopropylate, oxadiazon, and fenazaquin pesticides. Dispersion of the extraction solvent into the aqueous phase was performed by temperature changes. The enrichment factors and extraction recoveries were obtained in the ranges of 280–465 and 56–93%, respectively. Low limits of detection and quantification were obtained around of 0.13–0.31 and 0.45–1.1 ng/mL, respectively. Relative standard deviations were less than 9%.

MAE has several similarities with single-drop and liquid micro-extraction; Wang et al. [61] developed a novel method for simultaneous determination of pyrethroids residues using microwave-assisted extraction (MAE) coupled with ultrasonic-assisted dispersive liquid–liquid micro-extraction (UADLLME). The results demonstrated that the proposed method had good performance with linearity of 0.0050–4.98 mg/L, recovery of 83.3–91.5%, RSDs below 5.6%, and detection limit of 1.15–2.46 μg/L, concluding that MAE-UADLLME provided a sensitive and efficient alternative to determination of trace amounts of pesticide residues.

2.3.5 Sorptive Extraction Methods

The sorptive extraction methods include solid-phase extraction, solid-phase micro-extraction, and a novelty method named as QUECHERS (Quick, Easy, Cheap, Effective, Rugged, and Safe).

The solid-phase extraction (SPE) dated to approximately five decades ago with a hundred applications on environmental analysis. SPE can be used as an extraction and preconcentration tool in environmental matrices as well as for cleaning up of the samples. This technique is divided into three stationary phases: a normal phase such as silica, alumina, and Florisil, an ion-exchange adsorbent, and finally, a reverse-phase adsorbent from the conventional alkyl-modified silica materials (C8 and C18 nonpolar phase).

Nowadays, a novel magnetic copper-based metal–organic framework composed of nanocomposite for solid-phase extraction was developed to isolate five triazole pesticides from water samples. The limits of detection of the triazole pesticides are 0.05–0.1 with correlation coefficients larger than 0.992. The recoveries in spiked water samples are 72.3–91.53% with RSD ranging from 1.5 to 9.1% [62]. Also recently, they have synthesized novel magnetic nanoparticles, coated with octadecyl group-modified silica (nano-SiO2C18) containing carbon nanotubes under super-critical CO2 medium. These nanoparticles were used for the determination of five neonicotinoids and four sulfonylureas pesticides; the application as an adsorbent has a higher recovery with detection limit ranging from 0.07 to 0.60 μg/mL, recoveries from 78 to 95%, and RSD <5% [63].

SPE is currently exploring the synthesis of nanomaterials as fullerenes, carbon nanotubes, graphene, graphene oxide as excellent candidates for proper sorption media because of their unique structural and surface properties with noticeably enhanced sorption capability compared with traditional SPE [64]. The studies mentioned above about SPE show that the technique remains highly applicable, especially for concentration, isolation, and cleanup, for mainly chromatography analysis.

The solid-phase micro-extraction (SPME) is a solvent-free extraction technique. The SPME involves the use of a fiber coated with a polymeric phase that extracts the analytes. The fiber is exposed to the sample until equilibrium is reached or for a rigorously controlled period, and the analytes are then thermally desorbed by transferring the fiber into the GC injection port. The main strategies consist of wetting the matrix with water and then submitting the slurry to an elevated temperature to partition the analytes to the headspace, where the SPME fiber is placed. All these procedures afford a cleanup effect simultaneously, due to the selectivity of the SPME coating and a further concentration factor.

Currently, new solid-phase micro-extraction fibers were synthesized from hyper-cross-linked polymer nanoparticles to identify organochlorine pesticides. These coated fiber demonstrated excellent adsorption performance, with a good accuracy and precision to established a method that was finally applied to the analysis of the organochlorine pesticides from cucumber, Chinese cabbage and tomato samples [65]. Other fibers were fabricated of graphene-oxide-incorporated monolithic polymer in the fiber for determination of organophosphate esters in soil samples. These have a large surface area (536 m^2 /g), pore capacity (0.694 cm^3/g), good thermal stability (up to 320 °C), long service life, and good reproducibility and was applied for the analysis of organophosphate esters in environmental soil samples, and the relative recoveries were found to be in the range from 80.1 to 105.6% [66]. The most recent on solid-phase micro-extraction was the development of magnetic fibers for determination of the endocrine disrupting chemicals using Fe_3O_4 nanoparticles and graphite powder to extract the target analytes and provided excellent cleanup for simultaneous extraction of both nonpolar and polar analytes [67].

The last technique is a QUECHERS, the name of this provision of Quick, Easy, Cheap, Effective, Rugged, and Safe. The QUECHERS method offers the advantages of high recoveries, precise results, fast treatment, and little use of solvent and glass material and also requires small laboratory space and few reagents. Moreover, the process is robust and reliable. The QUECHERS method consists in two steps: extraction and cleanup sample; the extraction step uses $MgSO_4$ to reduce the water content in the sample and NaCl for most applications or anhydrous sodium acetate for compounds sensitive to an alkaline medium. The cleanup stage uses a primary/secondary amine (PSA) to the removal of organic acids and polar pigments. Some products besides PSA have C18 to eliminate most of the lipids and sterols or graphitized carbon to clean sterols and pigments such as chlorophyll.

When the sample is a complex matrix, in the QUECHERS method, it is possible to combine with another sorptive extraction method; this is the case of a study of Zhang et al. [68], combined with dispersive solid-phase extraction were both investigated and applied for the determination of 60 pesticide residues in cinnamon. In Huang et al. [69], a modified QUECHERS method was used for green tea sample preparation in pesticide residue analysis to determine 102 pesticides. And finally, QUECHERS method used in a very complex matrix as fatty biological tissue [70], the adapted extraction procedure was tested on biological samples of varying fat content, including fish muscle tissue and two previously untested matrices of coral and adipose tissue. These reports showed that QUECHERS methods are a trending analytical technique, mainly for soil samples, and offer excellent traceability, precision, and accuracy.

Acknowledgements The author and coauthor of this chapter agree to recognize the international contamination of the world and their high impact on the environment and public health. We believe that this book will contribute to the decision making of international organizations, academic groups, and society.

We sincerely appreciate the kind consideration for inviting us for the edition of this international book, especially to Dr. Sabu Thomas, Pro-Vice Chancellor, Mahatma Gandhi University, Kottayam, Kerala, India; Prof. Tatiana Volova, Institute of Biophysics of Russian Academy of Sciences, Siberian Federal University, Akademgorodok, Krasnoyarsk, Russia; Prof. Jayachandran K, School of Biosciences, Mahatma Gandhi University, Kottayam, Kerala, India; Rakhimol KR, Research Scholar, IIUCNN, Mahatma Gandhi University, Kottayam, Kerala, India.

References

1. McMichael AJ, Beaglehole R (2000) The changing global context of public health. Lancet 356(9228):495–499
2. McLaren L, Hawe P (2005) Ecological perspectives in health research. J Epidemiol Community Health 59:6–14. https://doi.org/10.1136/jech.2003.018044
3. McMichael AJ (2000) The urban environment and health in a world of increasing globalization: issues for developing countries. Bull World Health Organ 78(9)
4. Nicolopoulou-Stamati P, Maipas S, Kotampasi C, Stamatis P, Hens L (2016) Chemical pesticides and human health: the urgent need for a new concept in agriculture. Front Public Health 4:148. Published online 2016 https://doi.org/10.3389/fpubh.2016.00148
5. USGS (2016) U.S. geological survey. Pesticides in groundwater. https://water.usgs.gov/edu/pesticidesgw.html
6. Chapagain K (2011) International regulations of the use of pesticides: the costarican experience. Rev Costarr Salud Pública 20:124–129
7. Sachs JD (2005) Investing in development: a practical plan to achieve the millennium development goals. UN Development Programme, New York
8. McMichael AJ (2000) The urban environment and health in a world of increasing globalization: issues for developing countries. Bull World Health Organ 78(9)
9. Martens P, Akin S-M, Maud H, Mohsin R (2010) Is globalization healthy: a statistical indicator analysis of the impacts of globalization on health. Glob Health 6:16. https://doi.org/10.1186/1744-8603-6-16
10. Mostafalou S, Abdollahi M (2013) Pesticides and human chronic diseases: evidences, mechanisms, and perspectives. Toxicol Appl Pharm 268(2):157e177. http://dx.doi.org/10.1016/j.taap.2013.01.025
11. Hunter PR, Zmirou-Navier D, Hartemann P (2009) Estimating the impact on health of poor reliability of drinking water interventions in developing countries. Sci Total Environ 407(8):2621–2624
12. EEA (2013) Late lessons from early warnings: science, precaution, innovation. European Environment Agency. Report no 1/2013. EEA, Copenhagen
13. Carvalho FP (2017) Pesticides, environment, and food safety. Food Energy Secur 6(2):48–60. https://doi.org/10.1002/fes3.108
14. Polanco AG, Alberto JAN, Sánchez JS, Rejón GJM, Gómez JMM, Del Valls Casillas TA (2015) Contamination by organochlorine pesticides in the aquifer of the ring of Cenotes in Yucatán, México. Water Environ J 29:140–150. http://dx.doi.org/10.1111/wej.12080
15. Rodríguez ÁGP, López MIR, Casillas TAD, León JAA, Mahjoubd O, Prusty AK (2017a) Monitoring of organochlorine pesticides in blood of women with uterine cervix cancer. Environ Pollut 220:853–862
16. Rodríguez ÁG, López MI, Casillas TA, León JA, Prusty BA, Cervera FJ (2017b) Levels of persistent organic pollutants in breast milk of Maya women in Yucatán, México. Environ Monit Assess 189:59. https://doi.org/10.1007/s10661-017-5768-y
17. Rodriguez AG, Alfredo J, Leon A, Manrique JM, Gil AM (2018). The glyphosate herbicide in Yucatan, Mexico. MOJ Bioequivalence Bioavailab 5(6):284–286
18. Rotterdam Convention (2010) Convenio de Rotterdam sobre el Procedimiento de Consentimiento FundamentadoPrevio Aplicable a Ciertos Plaguicidas y Productos Químicos Peligrosos Objeto de Comercio Internacional. http://www.pic.int/ElConvenio/Generalidades/TextodelConvenio/tabid/1980/language/es-CO/Default.aspx
19. Stockholm (2009) The new POPs under the stockholm convention. http://chm.pops.int/Convention/ThePOPs/tabid/2509/Default.aspx
20. IARC (2015) International agency of cancer research. Glyphosate Monography. https://www.iarc.fr/featured-news/media-centre-iarc-news-glyphosate/
21. Rodríguez AGP, Magaña Castro V, CetzUitz J, Quintal López R (2019) Use of carcinogenic agrochemicals in the agricultural region of Yucatan, Mexico

22. Ramsar Convention (2012) The annotated Ramsar list: Mexico. http://www.ramsar.org/cda/en/ramsar-pubs-annolistannotated-ramsar-6517/main/ramsar/1-30-168%5E16517_4000_0__
23. Decreto 117, Diario Oficial (2013) Diario Oficial del Gobierno del Estado de Yucatán. Decreto Número 117. Decreto que establece el área natural protegida denominada Reserva Estatal Geohidrológica del Anillo de Cenotes. http://www.yucatan.gob.mx/docs/diario_oficial/diarios/2013/2013-10-28_1.pdf
24. Polanco AG, Alberto JAN, Sánchez JS, Rejón GJM, Gómez JMM, Del Valls Casillas TA (2015) Contamination by organochlorine pesticides in the aquifer of the ring of Cenotes in Yucatán, México. Water Environ J 29(2015):140–150. https://doi.org/10.1111/wej.12080
25. Pratt L, Quijandría G (1997) La industria del camarón enHonduras: análisis de sostenibilidad. http://www.incae.ac.cr/ES/clacds/investigacion/pdf/cen742.pdf
26. Shen L, Wania F (2005) Compilation, evaluation, and selection of physical–chemical property data for organochlorine pesticides. J Chem Eng Data 50(3):742–768
27. Cessna AJ, Wolf TM, Stephenson GR, Brown RB (2005) Pesticide movement to field margins: routes, impacts and mitigation. In: Field boundary habitats: implications for weed, insect and disease management, vol 1, pp 69–112
28. Cessna AJ, Larney FJ, Kerr LA, Bullock MS (2006) Transport of trifluralin on wind eroded sediment. Can J Soil Sci 86:545–554
29. Kerle EA, Jenkins JJ, Vogue PA (2007) Understanding pesticide persistence and mobility for groundwater and surface water protection. Oregon State Univ Extension Service, EM8561-E
30. Vryzas Z (2018) Pesticide fate in soil-sediment-water environment in relation to contamination preventing actions. Environ Sci Health 4:5–9
31. Krupke CH, Hunt GJ, Eitzer BD, Andino G, Given K (2012) Multiple routes of pesticide exposure for honey bees living near agricultural fields. PLoS One 7:29268
32. Masiá A, Campo J, Vázquez-Roig P, Blasco C, Picó Y, (2013) Screening of currently used pesticides in water, sediments and biota of the Guadalquivir river basin (Spain). J Hazard Mater 263:95–104
33. Socorro J, Durand A, Temime-Roussel B, Gligorovski S, Wortham H, Quivet E (2016) The persistence of pesticides in atmospheric particulate phase: an emerging air quality issue. Sci Rep 6. Article no 33456
34. CENR (2001) Intercontinental transport of air pollution: a review of federal research and future needs. Air quality research subcommittee of the committee on environment and natural resources CENR. https://www.esrl.noaa.gov/csd/aqrsd/reports/itap.pdf
35. Chandra YI, Linthoingambi DN, Hussain SJ, Cheng Z, Li J, Zhang G, Jones KC (2015) Current status of persistent organic pesticides residues in air, water, and soil, and their possible effect on neighboring countries: a comprehensive review of India. Sci Total Environ 511(2015):123–137
36. Pandit GG, Sahu SK, Sharma S, Puranik VD (2006) Distribution and fate of persistent organochlorine pesticidesin coastal marine environment of Mumbai. Environ Int 32 (2006):240–243
37. Alavanja M, Hoppin J, Kamel F (2004) Health effects of chronic pesticide exposure: cancer and neurotoxicity. Annu Rev Publ Health 25(1):155–197
38. Mathur V, John PJ, Soni I, Bhatnagar P (2008) Blood levels of organochlorine pesticide residues and risk of reproductive tract cancer among women from Jaipur. India Adv Exp Med Biol 617:387–394
39. Valavanidis A, Vlachogianni T (2010) Agricultural pesticides: ecotoxicological studies and environmental risks assessment. Sci Adv Environ Toxicol Ecotoxicol Issues. http://chem-toxecotox.org/wp/wp-content/uploads/2010/11/07-Pesticides-06_11_2010.pdf
40. Mnif W, Hassine AI, Bouaziz A, Bartegi A, Thomas O, Roig B (2011). Effect of endocrine disruptor pesticides: a review. Int J Environ Res Pub Health 8(6):2265–2303. http://dx.doi.org/10.3390/ijerph8062265

41. Brophy JT, Keith MM, Watterson A, Park R, Gilbertson M, Maticka-Tyndale E, Beck M, Abu-Zahra H (2012) Breast cancer risk in relation to occupations with exposure to carcinogens and endocrine disruptors: a Canadian case–control study. Environ Health 11:87
42. Briggs D (2003) Environmental pollution and the global burden of disease. Br Med Bull 68:1–24
43. Tabrez S, Ahmad M (2009) Toxicity, biomarker, genotoxicity and carcinogenicity of trichloroethylene (TCE) and its metabolites: a review. J Environ Sci Health C 27(3):178–196
44. Soffritti M, Belpoggi F, Esposti DD, Falcioni L, Bua L (2008) Consequences of exposure to carcinogens beginning during developmental life. Basic Clin Pharmacol Toxicol 102:118–124
45. Tabrez S, Priyadarshini M, Priyamvada S, Khan MS, Arivarasu NA (2014) Gene-environment interactions in heavy metal and pesticide carcinogenesis. Mutat Res 760:1–9
46. Berwick M (2000) Gene–environment Interaction in Melanoma. In: Forum (Genova), vol 10, pp 191–200
47. Brennan P (2002) Gene–environment interaction and etiology of cancer: what does it mean and how can we measure it? Carcinogenesis 23:381–387
48. Waliszewski S, Caba M, Arroyo SG, Pietrini RV, Martínez A, Quintana RV, Flores ME, Torres MA (2013) Niveles de plaguicidas organoclorados en habitantes de México. Rev Int Cont Amb 29:121–131. http://www.encuentra.gob.mx/resultsAPF.html?q=Yucat'n&client=senasica
49. Falconer IR, Chapman HF, Moore MR, Ranmuthugala G (2006) Endocrine-disrupting compounds: a review of their challenge to sustainable and safe water supply and water reuse. Environ Toxicol 21(2):181–191. http://dx.doi.org/10.1002/tox.20172
50. Mustafa AM (2010) Endocrine disrupting chemicals (EDCs): it's impact on health. Health Environ J 1(1)
51. Mnif W, Hassine H, Bouaziz A, Bartegi A, Thomas O, Roig B (2011) Effect of endochrine disruptors: a review. Int J Environ Res Publ Health 8:2265–2303
52. European Environment Agency (EEA) (2010) The impacts of endocrine disrupters on wildlife, people and their environments. European Environment Agency. Technical report no 2, 2012, 978-92-9213-307-8, 1725-2237. http://dx.doi.org/10.2800/41462
53. Costa EM, Spritzer PM, Hohl A, Bachega TA (2014) Effects of endocrine disruptors in the development of the female reproductive tract. Arq Bras Endocrinol Metab 58(2)
54. Wang M, Wang A, Leung J, Chan M, (2005) An overview on the accumulation, distribution, transformations, toxicity and analytical methods for the monitoring of persistent organic pollutants. Chemosphere 60:740
55. Jeon JY et al (2015) New risk factors for obesity and diabetes: environmental chemicals. J Diabetes Invest 6(2)
56. Rodríguez ÁG, López MI, Casillas TÁ, León JA, Mahjoub O, Prusty AK (2017) Monitoring of organochlorine pesticides in blood of women with uterine cervix cancer. Environ Pollut 220:853–862
57. CONAGUA-Semarnat (2015) Situación del Subsector Agua Potable, Drenaje y Saneamiento. http://www.cmic.org/comisiones/Sectoriales/infraestructurahidraulica/estadisticas/Estadisticas_2014-2018/Publicaciones_2015/DSAPAS%202015.pdf
58. Szarka A, Turková D, Hrouzková S (2018) Dispersive liquid-liquid microextraction followed by gas chromatography–mass spectrometry for the determination of pesticide residues in nutraceutical drops. J Chromatogr A 1570:126–134
59. Chormey DS, Bodur S, Baskın D, Fıra, M, Bakırdere S (2018) Accurate and sensitive determination of selected hormones, endocrine disruptors, and pesticides by gas chromatography–mass spectrometry after the multivariate optimization of switchable solvent liquid-phase microextraction. J Sep Sci 41(14):2895–2902
60. Farajzadeh MA, Shahedi Hojghan A, Afshar Mogaddam MR (2018) Development of a new temperature-controlled liquid phase microextraction using deep eutectic solvent for extraction and preconcentration of diazinon, metalaxyl, bromopropylate, oxadiazon, and fenazaquin

pesticides from fruit juice and vegetable samples followed by gas chromatography-flame ionization detection. J Food Compos Anal 66:90–97
61. Wang W, Wang W, Zhang S, Li Z, Wang C, Wang Z (2018) Hyper-crosslinked polymer nanoparticles as the solid-phase microextraction fiber coating for the extraction of organochlorines. J Chromatogr A 1556:47–54
62. Liu G, Li L, Huang X, Zheng S, Xu D, Xu X, Zhang Y, Lin H (2018) Determination of triazole pesticides in aqueous solution based on magnetic graphene oxide functionalized MOF-199 as solid phase extraction sorbents. Microporous Mesoporous Mater 270:258–264
63. Moreno V, Llorent-Martínez EJ, Zougagh M, Ríos A (2018) Synthesis of hybrid magnetic carbon nanotubes—C18-modified nano SiO2 under supercritical carbon dioxide media and their analytical potential for solid-phase extraction of pesticides. J Supercrit Fluids 137:66–73
64. Azzouz A, Kailasa SK, Lee SS, Rascón AJ, Ballesteros E, Zhang M, Kim KH (2018) Review of nanomaterials as sorbents in solid-phase extraction for environmental samples. TrAC Trends Anal Chem 108:347–369
65. Wang K, Xie X, Zhang Y, Huang Y, Zhou S, Zhang W, Lin Y, Fan H (2018) Combination of microwave-assisted extraction and ultrasonic-assisted dispersive liquid-liquid microextraction for separation and enrichment of pyrethroids residues in Litchi fruit prior to HPLC determination. Food Chem 240:1233–1242
66. Jian Y, Deng J, Zhou H, Cheng J (2018) Fabrication of graphene oxide incorporated polymer monolithic fiber as solid phase microextraction device for determination of organophosphate esters in soil samples. J Chromatogr A
67. Gorji S, Biparva P, Bahram M, Nematzadeh G (2019) Development of magnetic solid phase microextraction method for determination of the endocrine disrupting chemicals leached from reused plastic bottles. Talanta 194:859–869
68. Zhang Z, Dong M, Hao X, Ha L, Song S, Yao W (2019) Evaluation of cleanup procedures in pesticide multi-residue analysis with QuEChERS in cinnamon bark. Food Chem 276:140–146
69. Huang Y, Shi T, Luo X, Xiong H, Min F, Chen Y, Nie S, Xie M (2019) Determination of multi-pesticide residues in green tea with a modified QuEChERS protocol coupled to HPLC-MS/MS. Food Chem 275:255–264
70. Buah-Kwofie A, Humphries MS (2019) Validation of a modified QuEChERS method for the analysis of organochlorine pesticides in fatty biological tissues using two-dimensional gas chromatography. J Chromatogr B 1105:85–92

Chapter 3
Chemistry and Toxicology Behind Insecticides and Herbicides

S. Sousa, M. L. Maia, L. Correira-Sá, V. C. Fernandes, C. Delerue-Matos, C. Calhau and V. F. Domingues

Abstract Pesticides are chemicals used around the globe to kill, reduce, or repel whatever organism capable of threatening public health or economy. Although essential to maintain agricultural production, these compounds target systems or enzymes on the intended organism that are often similar or even the same present in other living beings. Furthermore, a considerable range of pesticides is environmentally persistent and has bioaccumulation characteristics. Even after some of the pesticides have been banned and prohibited from use, these can still be found accumulated in the environment, threatening the health and even life. Hence, the understanding of how the insecticides and herbicides mode of action work and how

S. Sousa · M. L. Maia · L. Correira-Sá · V. C. Fernandes · C. Delerue-Matos · V. F. Domingues (✉)
REQUIMTE/LAQV, Instituto Superior de Engenharia,
Instituto Politécnico do Porto, Porto, Portugal
e-mail: vfd@isep.ipp.pt

S. Sousa
e-mail: sara_sousa_23@hotmail.com

M. L. Maia
e-mail: mariadaluz.maia@gmail.com

L. Correira-Sá
e-mail: mariacs31@gmail.com

V. C. Fernandes
e-mail: vicru@gmail.com

C. Delerue-Matos
e-mail: cmm@isep.ipp.pt

C. Calhau
Center for Research in Health Technologies and Information Systems
(Centro de Investigação em Tecnologias e Serviços de Saúde—CINTESIS),
Porto, Portugal
e-mail: ccalhau@nms.unl.pt

C. Calhau
Nutrition and Metabolism, NOVA Medical School, Universidade Nova de Lisboa, Campo dos Mártires da Pátria, 130, 1169-056 Lisbon, Portugal

Rakhimol K. R. et al. (eds.), *Controlled Release of Pesticides for Sustainable Agriculture*, https://doi.org/10.1007/978-3-030-23396-9_3

these chemicals can negatively affect their surrounding is essential. In this chapter, an overview of chemical aspects, uses, environmental persistence, and human toxicity concerning herbicides and insecticides will be presented.

Keywords Insecticides · Herbicides · Chemical aspects · Human toxicity · Environmental persistence

3.1 Introduction

Pesticides are used since ancient times (1000 BC) to kill, reduce, or repel whatever organism capable of threatening public health or economy. The World War II trigged pesticides development and application, essentially to control the spread of diseases such as typhus, river blindness, and malaria. Furthermore, over 4 million tons of pesticides were recorded to be used in 2016 [1]. Rachel Carson was the first person to alert the public on the pesticides environmental impact through her book "Silent Spring" published in 1960 [2].

Pesticides can be classified mainly in two ways: by target pest (e.g., insecticides, fungicides, rodenticides, and pediculicides) or by chemical identity (e.g., pyrethroids, organophosphates, organochlorines, carbamates, glyphosate, and triazoles) [3].

Intensive pesticide application allied with the biodegradation-resistance shown by some pesticides often results in environmental accumulation and bioaccumulation [4, 5]. Pesticides target systems or enzymes in pests which may be the same or similar in other non-target organisms, as human beings [3].

Exposure to pesticides is constant since these transfer to air, soil, and water. Through this way, pesticides enter in living organisms where biological magnification then occurs. The awareness of pesticides impact on biota led to great concern about their continuous and sometimes excessive application. Even after some of the pesticides have been banned and prohibited from use, these can still be found and accumulate on the environment, threatening the health and even life [6].

Hence, the understanding of how pesticides mode of action works and how these chemicals can negatively affect their surrounding is essential. This chapter will focus on the chemical aspects, uses, environmental persistence, and human toxicity of chemical classes of herbicides and insecticides.

3.2 Herbicides

As the name suggests, herbicides are phytotoxic chemicals used on weed to growth prevention or control [7, 8]. Herbicides represent closely 48% of pesticides consumption worldwide [9]. In this group of pesticides, the followed chemical family included phenoxy acid derivatives, bipyridyl derivatives, ureas and thioureas, organic phosphorus/phosphonomethyl amino acids or inhibitors of aromatic acid biosynthesis, protoporphyrinogen oxidase inhibitors, triazines, and triazoles [10]. These chemicals are able to inhibit biological processes (i.e., photosynthesis, mitosis, cell division, enzyme function, root growth, and leaf formation), interfere with the synthesis of biomolecules (e.g., pigments, proteins, DNA), destroy cell membranes, and promote uncontrolled growth [11].

Herbicides can be selective or non-selective [7]. Selective herbicides are specific to a certain plant presenting low toxicity for mammals. Non-selective herbicides (e.g., arsenicals, chlorates, and dinitrophenols) show a toxic effect on animals and humans [12]. Improper use and disposal are the primary sources of health problems from exposure to animals and humans. Moreover, runoff from agricultural fields is an additional concern to human health as a possible entry point of herbicides into drinking water supply [9].

3.2.1 Amides and Acetamides

These groups of herbicides are commonly used and include compounds such as propanil, bensulide, and dimethenamid-P [10]. Propanil (N-(3,4-dichlorophenyl) propenamide) was discovered in 1957 by Rohm and Haas and is classified as post-emergence herbicide with highly selectivity [13]. On the other hand, bensulide [O, O-diisopropyl S-2-phenylsulfonylaminoethyl phosphorodithioate] is a pre-emergent organophosphate herbicide [14]. Dimethenamid [2-Chloro-N-(2,4-dimethyl-3-thienyl)-N-(2-methoxy-l-methylethyl) acetamide] is a racemic mixture of the M and P stereoisomers, the P isomer is the one with valuable herbicidal activity [10].

3.2.1.1 Chemical Aspects

As an example of the group of the amides, is the propanil (N-(3,4-Dichlorophenyl) propenamide), a synthetic anilide herbicide (Fig. 3.1) that when pure is an odorless, brown crystalline solid. This herbicide is characteristically non-volatile, denser than water with a reasonable affinity for soils. Propanil is quite soluble in organic solvents; its solubility in water is reduced [13].

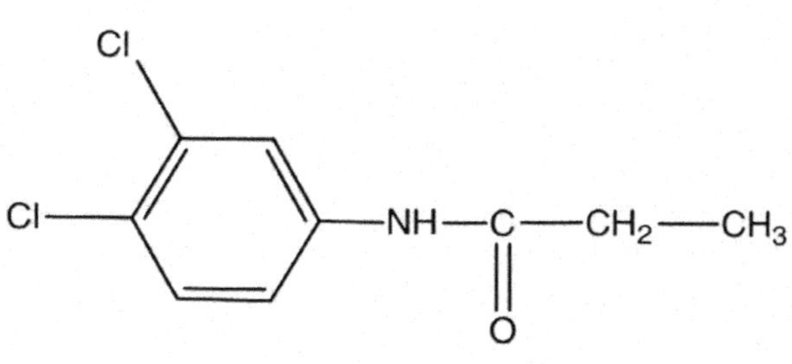

Fig. 3.1 Structure of propanil [13]

One degradation pathway of propanil that has been described is photolysis. Under this direct condition occurs the breakage of amine bonds, as a consequence of the elimination of chlorine atoms [13].

3.2.1.2 Uses and Environmental Persistence

Propanil and bensulide are used as regulators of plant growth [12]. Propanil is a contact post-emergence herbicide used worldwide to control grasses and broad-leaf weeds primarily in rice production. Its mobility in soils can be amplified when other pesticides are used in combination. Although this herbicide has the capacity to be adsorbed to soil, it has been discussed that it has the potential to leach into groundwater [13]. Propanil acts through the inhibition of photosynthesis, and it inhibits the carbon fixation that leads to the stop of plant growth. Within plant leaves, this compound inhibits the photosystem II disrupting the electron transport chain. It has been shown that propanil has a short period of activity, being degraded rapidly into its major degradation compound, 3,4-dichloroaniline (DCA), that has been highly detected in groundwater and surface water in the world [13, 15]. This degradation is directly dependent on the amount of herbicide applied and environmental conditions. Being the major degradation pathway known is through enzymatic hydrolysis [15].

Bensulide is a pre-emergent herbicide used globally in agricultural crops, such as cole crops, cucurbits, leafy, legumes, vegetables, onion, and garlic. Typically, this herbicide is applied to the soil before planting the crops. Bensulide acts inhibiting cell division in meristematic root tissues, and it also has activity as a cholinesterase inhibitor [14].

Another example of chloroacetamide herbicides is the dimethenamid that can be used in several field crops, such as grass, large crabgrass, and some broadleaf weeds. It primarily used in corn, soybean, and sorghum. Dimethenamid-P (2-chloro-N-(2,4-dimethyl-3-thienyl)-N-(2-methoxy-1-methylethyl) acetamide) has shown greater activity on weeds than the R-isomer, and this allows the use of dimethenamid-P at lower rates. These components can have some toxicity, can be persistent and reactive, so through agricultural circle and washing, they can reach and endanger the aquatic environment [16, 17].

3.2.1.3 Toxicity on Human

Generally, these groups are slightly or moderately hazardous, and they are of relatively low toxicity [10].

In the case of propanil, studies have shown that acute exposure may lead to methemoglobinemia, central nervous system depression, and respiratory failure. The ingestion of this herbicide causes a burning sensation through the digestive system, resulting in fever, nausea and vomiting, and dizziness and sleepiness. Their inhalation can cause throat and nose irritation. Long-term effects of propanil are not fully established. However, repeated or prolonged exposure may result in cyanosis. Exposure via dermal contact may cause slight skin irritation [18].

In the case of dimethenamid, the studies have shown some evidence of different toxicological profiles between racemic dimethenamid and dimethenamid-P. Dimethenamid can reduce fetal birth weight but has no evidence of been teratogenic [12].

3.2.2 Benzoic Acids

Benzoic acids herbicides have low toxicity and include chloramben, dicamba, and naptalam. These are often combined with other herbicides [10]. Dicamba, the best known benzoic acid herbicide, has been used since 1967 in the USA [19].

3.2.2.1 Chemical Aspects

Benzoic acids herbicides are highly water soluble [20]. Dicamba (3,6-Dichloro-2-methoxybenzoic acid) (Fig. 3.2), is soluble in water also and shows resistant to hydrolysis and oxidation under normal environmental conditions [19].

3.2.2.2 Uses and Environmental Persistence

Dicamba, as other benzoic acids, has a wide range of applications in weeds control, being used as a pre-emergent, post-emergent, or pre-plant herbicide in cereal crops, pastures, forest lands, fence rows, lawns, and golf courses. It mimics auxins effects leading to uncontrolled growth rates, alterations in plant function homeostasis and death, and is more persist in the environment than phenoxy acids [8, 19, 21]. Furthermore, this benzoic acid can persist in soils for 7–10 months [19].

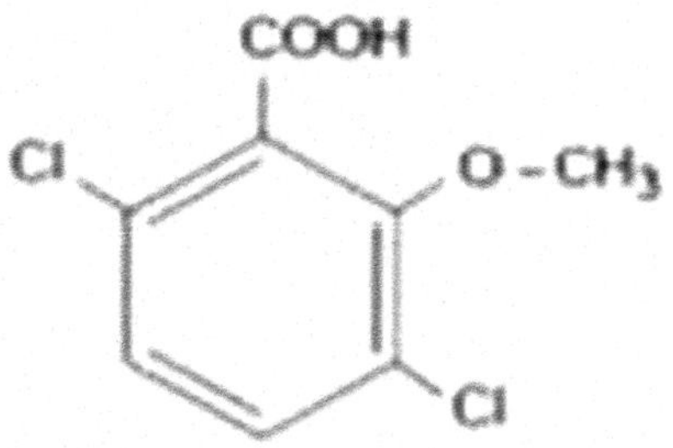

Fig. 3.2 Structure of dicamba [8]

3.2.2.3 Toxicity on Humans

Benzoic acid is not classified as very toxic, although the development of immune and endocrine defects may occur in the presence of these chemicals [20].

For instance, dicamba is rapidly absorbed after ingestion but is also quickly excreted from the organism without being metabolized. Dermal exposure shows low absorption. In mammals, skin and eye irritation can happen but presents low acute toxicity and does not cause dermal sensitization. Nonetheless, severe or permanent ocular damage is a possibility primarily in high concentrations. Breathing difficulty, vomiting, cyanosis, loss of appetite, depression, weight loss, incontinence, and muscular weakness/exhaustion subsequent to muscle spasms (myotonia) are known intoxication symptoms of this benzoic acid. However, only limited data are available on dicamba exposure since it is often used combined with other herbicides [19].

3.2.3 *Bipyridyl Derivatives*

Paraquat and diquat are the herbicides included in this group. First labeled in 1882 as a chemical redox indicator by Weidel and Russo, paraquat is by far the most toxic of the frequently used herbicides [9, 10].

Despite being banned in 32 countries, paraquat is one of the most widely used herbicides in the USA. On the other hand, other great producer, Brazil decided to ban paraquat in 2017, that will be forced only in 2020 [22].

3.2.3.1 Chemical Aspects

The toxicity of Paraquat (1,1′-dimethyl-4,4′-bipyridinium ion) depends upon the formulation, which is typically formulated as dichloride salt. Whereas, Diquat (1,1′-ethylene-2,2′-bipyridinium ion) is formulated as dibromide salt (Fig. 3.3) [10, 12].

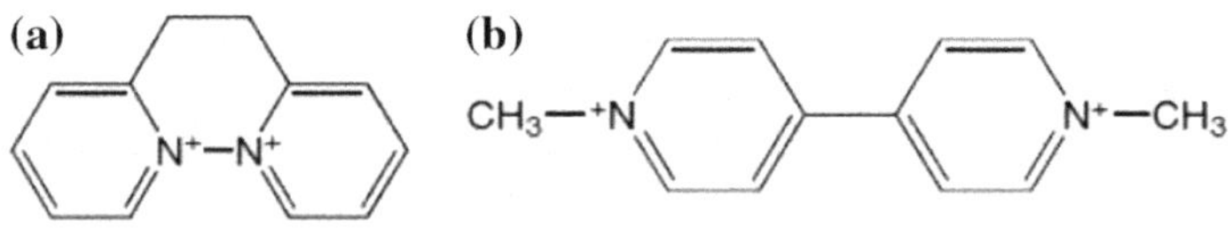

Fig. 3.3 Structure of diquat (**a**) and paraquat (**b**) [23, 24]

3.2.3.2 Uses and Environmental Persistence

Bipyridyl derivatives are non-selective, fast-acting herbicides capable to kill or affect nearly any plant foliage they contact. Paraquat is often used prior to plating to control existing vegetation, on weed management and in some crops as pre-harvest desiccants to speed drying. However, the options of other herbicides less toxic and persistent lessened the use of paraquat. Whereas, diquat is effective on broadleaves and used in ponds, lakes, and drainage ditches to control cattails and submersed aquatic species [8]. Since diquat is quickly absorbed by plant leaves and does not translocate, the toxicity effects on plants are swiftly shown [23].

Bipyridyl derivatives are rapidly adsorbed and persistent in soil. Paraquat is resistant to hydrolysis, microbial degradation, and photodegradation in water. Contrary to diquat that in water soluble and susceptible to microbial degradation and photodegradation [23, 24].

Paraquat and diquat mode of action present similarities as interfering with cyclic reduction-oxidation reactions. With paraquat, nicotinamide adenine dinucleotide phosphate (NADPH) depletion and the generation of oxygen free radicals occurs, whereas with diquat the formation and aerobic autooxidation of free radicals lead to superoxide production and oxidative damage. The affected or target organs are also different, though both affect the kidneys paraquat also target the lungs and brain and diquat affect also the liver, gastrointestinal tract, plasma and adrenal gland [9, 12, 23, 24].

3.2.3.3 Toxicity on Humans

Bipyridyl derivatives are toxic to humans either by skin contact, inhalation, or ingestion [8]. However, these do not present a tendency to accumulate in the organism. After ingestion, a high amount of paraquat is excreted either by urine or feces without being nearly metabolized [9].

The presence of paraquat in the lungs leads to pulmonary lesions as consequence of type I and type II pneumocytosis. Furthermore, paraquat can make the cation to form a free radical, stable in the absence of oxygen by enduring a single electron reduction. Present in the brain via the neural amino acid transporter is thought paraquat to be a factor for Parkinson's disease occurrence [9]. Within 24 h after ingestion, inhalation, or dermal exposure, all organ systems may be severely affected. Acute toxic signs include burning in the mouth and throat with vomiting and diarrhea, subsequent fluid and electrolyte loss, esophageal perforation, renal failure, cardiac arrhythmias, convulsions, and even coma. Death is this stage often occurs because of hepatic and renal toxicities, the estimated lethal dose (LD) in humans is 40 mg per kg. After a few days of exposure, progressive pulmonary fibrosis often develops, several weeks later pulmonary edema occurs and death happens by asphyxia [12, 24]. Although no teratogenic effect was demonstrated by this herbicide [9].

On the other hand, diquat can be found in liver, kidneys, gastrointestinal tract, plasma, and adrenal gland. No signs of fertility effects or teratogenic were shown by diquat, however, fetotoxicity can happen at doses that are maternally toxic. Acute symptoms include nausea, emesis, diarrhea, nosebleed, skin irritation, cough, and chest pain. Hence, gastrointestinal tract ulceration, acute renal failure, hepatotoxicity, and breathing difficulties are a link to diquat overdose [9, 12, 23].

3.2.4 Dinitro Compounds

Initially introduced in 1933 in the USA, this group of dinitrophenols includes pesticides as 2,4-dinitrophenol (DNP), dinitro-orthocresol (DNOC), dinoseb, and dinoterb [10, 25].

3.2.4.1 Chemical Aspects

Dinitro compounds can be used as herbicides either alone or as salts, and these show some solubility in water [10, 12]. Dinoseb or 2,4-Dinitro-6-(1-methyl propyl) phenol (Fig. 3.4) is produced in emulsifiable concentrates or as water-soluble ammonium or amine salts [26].

3.2.4.2 Uses and Environmental Persistence

Dinitro compounds were first used as an obesity treatment; later, these were banned from the US market owing to mortality rate and cataracts occurrence [10, 25]. Dinoseb was applied in soybeans, a sort of vegetables, fruits, nuts, citrus trees, and with other field crops for a weeds control. These herbicides not suffer hydrolysis in the environment but are suggestible to abiotic degradation. Normally, dinitro compounds do not persist in the soil, however, when spills occur, persistence in soil and groundwater contaminant was found. Even though low bioaccumulation potential is shown, these chemicals are highly toxic to birds, fish, and invertebrates [25, 27].

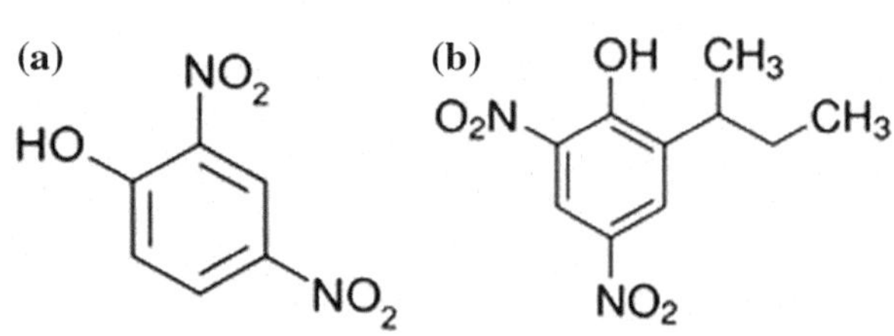

Fig. 3.4 Structure of DNP (**a**) and dinoseb (**b**) [25, 27]

The mode of action of these chemicals is by discharging phenolic protons in the mitochondrial matrix compromising oxidative phosphorylation. After exposure, oxidative metabolism increases leading to pyrexia, tachycardia, dehydration, ultimately depleting carbohydrate, and fat stores [25–27].

3.2.4.3 Toxicity on Humans

Dinitro compounds are quickly absorbed and very toxic to humans via inhalation, ingestion, and dermal exposure. These chemicals can be slowly excreted by urine in the original form and as metabolites [10, 12]. Even so, fatal cases after DNP exposure were reported. First signs of DNP toxicity include fever, the rise in blood pressure, heart rate and body temperature; a headache, mental confusion, nausea, vomiting, abdominal cramps and following coma in severe cases. After 12–72 h, hepatic and renal damage have been reported. Chronic exposure can cause peripheral nerve damage and cataracts. No carcinogenic effects have been documented [25]. Regarding, dinoseb early signs of toxicity include hyperthermia, profuse sweating, headache, confusion followed by restlessness, hyperactivity, seizures, fatigue, thirst, insomnia, weight loss, flushing of the face, nausea, vomiting, abdominal pain, occasional diarrhea, methemoglobinemia, hemolytic anemia, and coma in severe cases. Ocular exposure can cause cataracts, secondary glaucoma, paresis of accommodation and nystagmus. Dermal exposure causes irritation, yellow stains, burns, and dermatitis. Following 12–72 h of exposure, liver and kidney damage was reported. Furthermore, sinus tachycardia, ventricular tachycardia, ventricular fibrillation, and immune system damage may happen [26, 27].

3.2.5 Dintroaniline

This group of herbicides are aromatic amines that include compounds such as trifluralin and tridiphane [10].

3.2.5.1 Chemical Aspects

Dinitroaniline herbicides are characteristically a yellow liquid formulation; they have low water solubility and little leaching [8]. They comprise fluoride, propylamine, or sulfonamide prosthetic groups [28].

These herbicides act preventing cell division because they bind to alpha-tubulin (a protein present in the microtubules). This binding inhibits tubulin polymerization preventing that way the alignment and separation of chromosomes during mitosis.

The trifluralin herbicide, an example of this group presented in Fig. 3.5, is related chemically, but not functionally, to the explosive trinitrotoluene (TNT) [8].

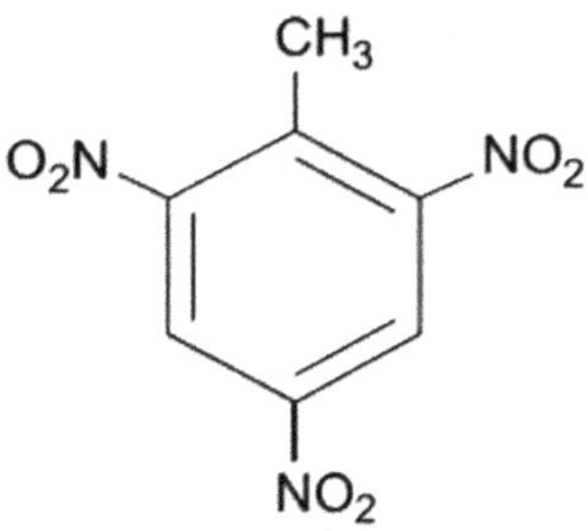

Fig. 3.5 Structure of trifluralin [8]

3.2.5.2 Uses and Environmental Persistence

Dinitroaniline group are mainly used as pre-emergence herbicides. These herbicides are generally applied prior to planting, the ones used post-planting need to be applied prior to seed germination to have effective activity. The application with good incorporation of these herbicides, that are root growth inhibitors, is placing them in the weed seed germination, improving this way their activity, and in this case usually, plants do not emerge from the soil. The primary use of these herbicides is in cotton, soybean, dry bean, potato, canola, and many horticultural crops [8].

3.2.5.3 Toxicity on Human

The acute exposure of dinitroanilines through inhalation may result in throat pain and coughing. The most predominant effect of exposure to this group of herbicides is methemoglobinemia. Data describing the effects of chronic exposure in human are lacking [28]. For example, dinitroaniline trifluralin is classified as practically non-toxic for dermal irritation and acute oral toxicity and slightly toxic for eye irritation, acute dermal toxicity, and acute inhalation toxicity. Chronic exposure to trifluralin may result in eye irritation, mild skin irritation, and respiratory depression [29].

3.2.6 Imidazolinones

Imidazolinones herbicides were discovered in the 1970s, being the first US patent obtained in 1980 [9]. This group embrace imazapyr, imazethapyr, imazamethabenz-methyl, imazapic, imazamox and imazaquin [10].

3.2.6.1 Chemical Aspects

Each imidazolinone has a carboxylic acid group or a carboxylic ester group (in imazamethabenz-methyl case), linked to a backbone group (Fig. 3.6). For instance, imazapic is linked to a methyl group, imazethapyr to an ethyl group, imazamox to a methoxymethyl group, and imazaquin to a benzene ring fused to the pyridine ring [30].

3.2.6.2 Uses and Environmental Persistence

Imidazolinones are often used to control grasses and broadleaf weeds in wheat, barley, sunflowers, alfalfa, edible legumes (e.g., dry beans), corn, soybean, and peanuts crops. Some being more selective (as imazamethabenz-methyl) and persistent than others (as imazethapyr). The persistence shown by some imidazolines applied repeatedly in the same site can cause weed resistance [8].

Imidazolinones act by inhibiting acetohydroxyacid synthase (AHAS), the first enzyme in the biosynthesis of valine, leucine and isoleucine, the branched-chain amino acid. These chemicals bind to the enzyme outside of the active site of the enzyme making these extraneous target site inhibitors [31].

Following application, the imidazolinone travels through the foliage and/or roots of the susceptible weed, becoming translocated throughout the plant. The target weed stop growing and competing with the specific crop soon after translocation and within several weeks after application, the weed dies [30].

3.2.6.3 Toxicity on Humans

Imidazolinones are considered relatively non-toxic and not represent a threat to human chronic exposure. After ingestion, these herbicides are quickly absorbed and excreted unchanged and in metabolite form, showing small bioaccumulation potential in mammals. So far, no case of either accidental or deliberate poisoning in humans by imidazolines was reported [9, 30]. Furthermore, imidazolines do not have reproductive and developmental toxicity [10].

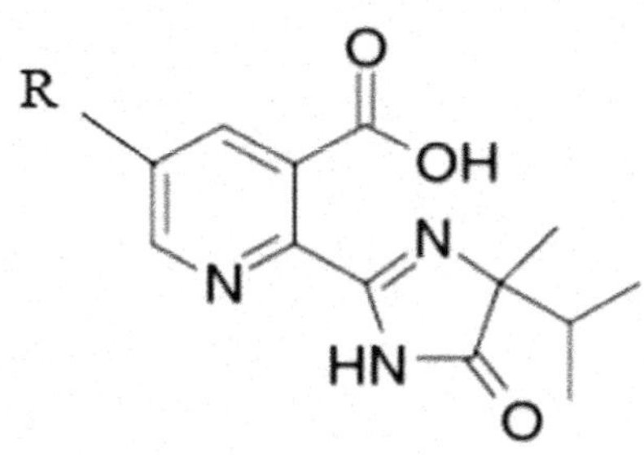

Fig. 3.6 Structure of imidazolinones herbicides [31]

3.2.7 *Methyl Uracil*

These group of herbicides include compounds such as bromacil (5-Bromo-3-sec-butyl-6-methyluracil) and terbacil (3-Cyclohexyl-5,6-trimethyleneuracil) [10].

3.2.7.1 Chemical Aspects

The elementary structure of uracil is identical to the core structure of the building blocks of nucleic acids (uracil, thymidine, cytosine and guanine). Bromacil and terbacil are composed by the basic uracil structure, an asymmetrical ring with two nitrogens, with the presence of a methyl group located at the sixth position on the ring as shown in Fig. 3.7. Bromacil is not a volatile herbicidal, although is moderately leachable [8, 32].

3.2.7.2 Uses and Environmental Persistence

The uses for these herbicides include pre-emergence control of perennial weeds. Bromacil and terbacil have their primarily absorption point in roots; they have both been extensively and repeatedly used for several years [8, 34]. The primary uses of bromacil are in citrus, pineapple, brush on non-cropland alfalfa, mint, pecan, and sugarcane. In the case of terbacil, the primary uses are in alfalfa, cotton, sugarcane, pineapple, grapes, tree fruits, peppermint, spearmint, sugarcane, small fruits, and deciduous tree fruits [8].

The herbicides from the uracil group are known to be strong inhibitors of photosynthesis; they inhibit photosystem II (PSII)5 activity. The phytotoxic residues are capable to persist in the soil for one year, so they can be used as soil sterilant [8, 34].

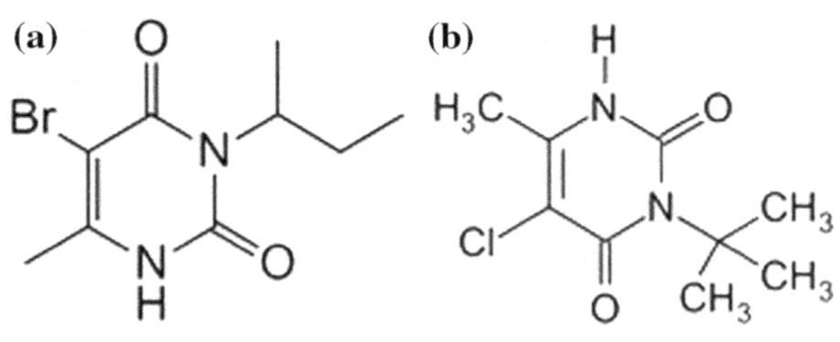

Fig. 3.7 Structure of bromacil (**a**) and terbacil (**b**) [33]

3.2.7.3 Toxicity on Human

Bromacil can be a cause of concern due to its broad use, low degradation rate, and persistence in the environment. Although, the studies reported that bromacil has very low toxicity in mammalian species [32].

3.2.8 *Nitriles*

The group of nitrile herbicides is composed of: hydroxybenzonitriles, such as ioxynil (3,5-Diiodo-4-hydroxybenzonitrile) and Bromoxynil (3,5-Dibromo-4-hydroxybenzonitrile); and benzonitriles such as dichlobenil (2,6-Dichlorobenzonitrile) [8, 10]. The hydroxybenzonitriles group were first used in the 1960s in the UK and the USA [8].

3.2.8.1 Chemical Aspects

Ioxynil and bromoxynil are structurally similar to dichlobenil. Figure 3.8 is presented the chemical structure of bromoxynil. Dichlobenil is volatile, highly absorbed, little leaching, and primarily soil active [8].

3.2.8.2 Uses and Environmental Persistence

Ioxynil and bromoxynil are inhibitors of photosynthesis contrarily to dichlobenil. The first two act in the photosynthetic electron transport system, specifically in D-1 quinone protein. On the other hand, dichlobenil act inhibiting the cellulose synthesis, seed germination, and the actively dividing meristems in roots and shoots.

As primary use, ioxynil is effective in sugarcane, rice, flax, and in pastures. Bromoxynil primary use is weed control in field and sorghum, popcorn, peppermint, and spearmint. These two hydroxybenzonitriles are contact herbicides, so they do not have soil activity. In order to extend the weed-management spectrum and

Fig. 3.8 Structure of bromoxynil [35]

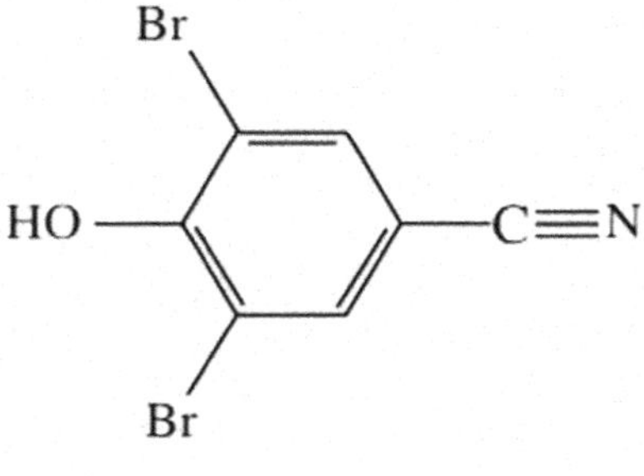

reduce the costs, both herbicides are frequently sold in combination with a phenoxyacid herbicide. In the case of dichlobenil herbicide, the primary uses are in ornamentals, turf, cranberries, and in aquatic environment [8].

3.2.8.3 Toxicity on Human

Studies have shown that bromoxynil and the product buctril fed to rats or applied to their skin during pregnancy can lead to developmental effects in offspring at the higher doses tested. According to studies EPA considered that both ioxynil and bromoxynil are moderately toxic to mammals [8].

3.2.9 Phenoxy Acid Derivatives

Phenoxy herbicides have been commercially available for over 60 years and are the most widely used family of herbicides worldwide. Phenoxy herbicides are a family of chemicals related to the growth hormone indoleacetic acid. When sprayed on broad-leaf plants, they induce rapid, uncontrolled growth. 2,4-dichlorophenoxyacetic acid (2,4-D), the most common of the phenoxy herbicides, is one of the best-studied agricultural chemicals [36].

3.2.9.1 Chemical Aspects

The wide variety of phenoxies in use today can be grouped into the phenoxyacetic, phenoxybutyric, and phenoxypropionic subtypes, the latter containing the aryloxyphenoxypropionic subtype with the greatest number of commercial variants. Chemically, they are carboxylic acids, typically applied in an ester or salt form. The commonly used herbicide 2,4-D, for example, is stable to its melting point at 140.5 °C. It is an acid herbicide, characterized by low vapor pressure, moderate water solubility. Figure 3.9 showed a phenoxy acid example, 2,4-D [37].

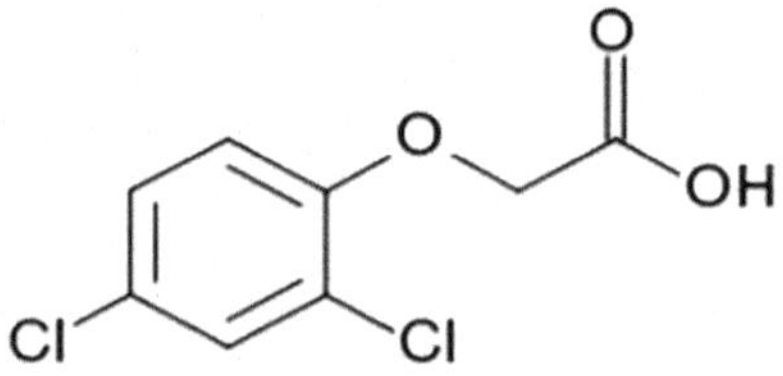

Fig. 3.9 Structure of 2,4-D, a phenoxy herbicide [37]

3.2.9.2 Uses and Environmental Persistence

Due to their relatively high polarity and stability, phenoxy acid herbicides easily enter the surface or groundwater through natural drainage or infiltration, leading to slow accumulation to a toxic level for human health. As a result, the maximum residue limits for phenoxy acid herbicides in drinking water have been set in European Union, the USA, Japan, China, and other nations to protect the health of human beings [37]. The fate of 2,4-D may be affected by several processes including runoff, adsorption, chemical and microbial degradation, photodecomposition, and leaching. The potential mobility of 2,4-D in soil is indicated by the water solubility and the soil adsorption coefficient; while the persistence in soil is indicated by the aerobic and anaerobic soil metabolism, hydrolysis half-lives, and field dissipation rate [38].

3.2.9.3 Toxicity on Humans

Exposure to phenoxy herbicides can occur via inhalation, ingestion, and dermal contact. The humans can ingest phenoxy acid herbicide residues residing on food. The widespread long-term use of phenoxy acid herbicides and the high potential for contamination have highlighted concerns about the long-term adverse effects on human health [39]. Based on epidemiological studies, the highest exposures were obtained in occupational settings for forestry workers, commercial applicators, and farmers [40].

A study showed that the phenoxy acid herbicides are well absorbed from the gastrointestinal tract [41]. Phenoxy acid herbicides have been shown to act as potent peroxisome prolifereators and peroxisome prolifereator activated receptor (PPAR) gene activators, which may have important implications for their developmental and metabolic toxicity [42]. It has been reported that this kind of herbicide (2,4-D) may be potentially teratogenic and carcinogenic, and some of them are endocrine disruptors [43].

3.2.10 Polycyclic Alkanoic

Polycyclic alkanoic acids (PCAs) herbicides can be separated in two groups: oxyphenoxy alkanoic acids, usually called fops (e.g., diclofop); and benzoyl-N-phenyl phenoxy propanoic acids, usually called props (e.g., flamprop). Some examples of compounds comprised in this group are diclofop, fenoxaprop ethyl, fenthiaprop, fuazifop, and haloxyfop [10]. The first of these group developed for the use as herbicides were props, and they were created due to a particular weed problem with wild oats [44].

3.2.10.1 Chemical Aspects

The structure of PCAs is diverse that their physical and chemical properties are difficult to generalize. This group is constituted usually for an alkanoic acid with more than one ring structure, which one is generally a phenyl ring, adjacent to asymmetric noncarbonyl carbon, as demonstrated in Fig. 3.10. However, PCAs with grass-killing activity have an asymmetric carbon attached to the carboxyl groups of the alkanoic acid [44].

3.2.10.2 Uses and Environmental Persistence

The fops group are generally used as foliar graminicides that selectively remove annual grasses from broadleaved crops and grass crops (such as wheat and barley). For example, fluazifop-P is a post-emergence herbicide selective in broadleaved crops [8].

3.2.10.3 Toxicity on Human

In general, PCAs have moderately low toxicity. An exception is haloxyfop-methyl that has high toxicity. They tend to be more toxic if exposure is dermal [12].

3.2.11 Protoporphyrinogen Oxidase Inhibitors (PROTOX) DPE or Non-DPE

The protoporphyrinogen oxidase inhibitors or PROTOX could be either diphenyl ether (DPE) or non-diphenyl ether (non-DPE) herbicides. Diphenylethers, phenylpyrazoles, oxadiazoles, triazolinones, thiadiazoles, pyrimidindiones, oxazolidinedione, and N-phenylphthalimides are included in this group [10, 45]. Nitrofen

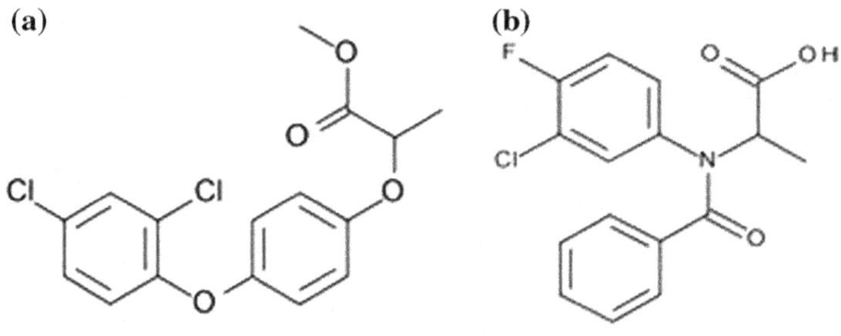

Fig. 3.10 Structure of diclofop (**a**) and flamprop (**b**) [44]

was the first PROTOX commercialized in the 1960s, and these herbicides are now a great and increasing part of the herbicide market [9, 10, 46].

3.2.11.1 Chemical Aspects

The majority of the PROTOX, with the exception of DPE, consists of heterobicyclic structures with one phenyl ring linked to a heterocyclic ring (Fig. 3.11) [31]. The attachment can be a carbon–carbon bridge or a carbon–nitrogen bridge. Whereas DPE herbicides have a p-nitrophenyl substitution or p-trifluoromethyl phenyl substitution, these last are heterocyclic phenyl ethers, with one of the phenyl rings of ether replaced by an aromatic heterocycle [46].

3.2.11.2 Uses and Environmental Persistence

PROTOX are applied on corn, soybean, rice, and sunflower crops as pre-emergent or post-emergent. PROTOX inhibit the enzyme protoporphyrinogen oxidase, a step in the porphyrin pathway that produces half of the chlorophyll molecule. This metabolic process works independently of photosynthesis but needs light. The inhibition of protoporphyrinogen oxidase causes accumulation of the phytotoxic molecule protoporphyrin, causing peroxidation of the unsaturated bonds of fatty acids found in cell membranes. Subsequently, loss of membrane integrity and leakage, pigment breakdown and necrosis of the leaf occurs, resulting in the plant death [8, 31].

These herbicides are degradable by microbial activity and DPE are also photodegradable. Additionally, PROTOX are either low soluble in water or dissipate quickly, being of low risk to groundwater or surface water runoff. Hence, these demonstrated little threat to the environment [46].

3.2.11.3 Toxicity on Humans

PROTOX have low acute toxicity and when used properly are doubtful to pose an acute hazard. So far, no health problems were associated with human consumption of crops treated with these chemicals [10, 46]. Furthermore, this class of herbicides does not bioaccumulate, if due to poor absorption or fast degradation by

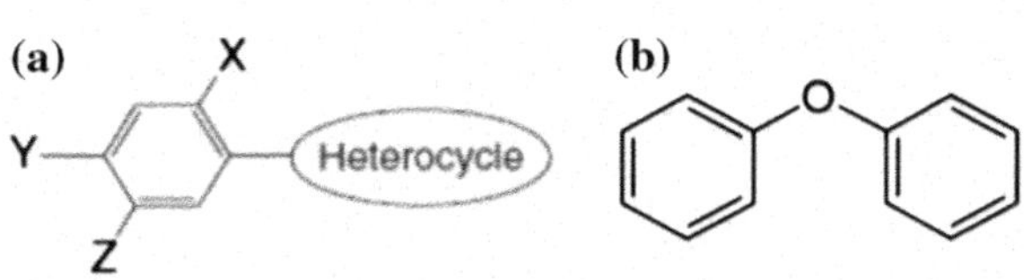

Fig. 3.11 Structure of PROTOX non-DPE (**a**) and DPE (**b**) herbicides [31, 46]

metabolism or excretion. Hence, when used appropriately these are not a threat. Furthermore, most of the PROTOX are not mutagenic nor carcinogenic in nature and have low acute toxicity [9, 12].

3.2.12 Substituted Anilines: Anilides

Anilides are a class of amides that have been extensively studied because of their various biological activities, such as antimycobacterial, antifungal, larvicidal, turbeculostatic, insecticidal and mainly in herbicidal activities. Anilide herbicides are used to control weeds on hard surfaces, such as roads, railway tracks, paths, forestry, and in crops [47].

3.2.12.1 Chemical Aspects

Currently, many anilides are used as herbicides in agriculture, as they are easily obtained by reactions of aromatic amines with acyl chlorides or with carboxylic acid anhydrides, known as "one-pot" reactions [48]. Figure 3.12 showed a general chemical structure of an anilide.

3.2.12.2 Uses and Environmental Persistence

A subgroup of anilide herbicides, the chloroacetanilide, has been extensively reported in the literature. The chloroacetanilides inhibit long-chain fatty acid biosynthesis, i.e., cell division and protein synthesis. Chloroacetanilides cause the death of plants that are susceptible to this mechanism of action [49].

The repeated and indiscriminate use of this herbicides, careless handling, accidental spillage, or discharge of untreated effluents into natural waterways has harmful effects on the fish and other aquatic organisms [50]. The toxicity of these herbicides to fish and other aquatic organisms have been reported by several authors [51, 52].

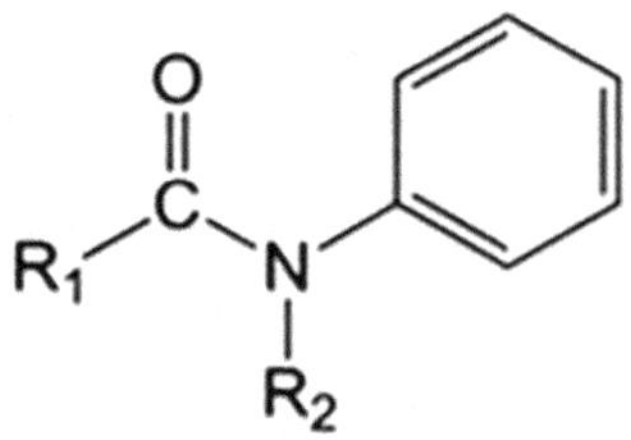

Fig. 3.12 Structure of an anilide [48]

3.2.12.3 Toxicity on Humans

The toxicological potential and mechanisms of the anilide herbicides remain poorly understood. Acetanilide herbicides including metolachlor, alachlor, and acetochlor are among the most widely used pesticides in the USA, with collective usage of >60 million pounds per year, and are primarily used for agricultural weed control [53]. Exposure to these agents has been associated with various adverse health effects in rodent models including cancer, developmental and reproductive abnormalities, and dysregulation of thyroid and liver function, but the mechanisms underlying these or other toxicological effects are poorly understood [54].

3.2.13 Symmetrical and Asymmetrical Triazines

Triazines can be asymmetrical or symmetrical: asymmetrical triazines include thiomethyl-s-triazines (e.g., ametryn, prometryn, metribuzin, and terbutryn) and the methoxy-s-triazines (e.g., prometon); symmetrical triazines are chloro-s-triazines (e.g., simazine, atrazine, propazine, and cyanazine). Atrazine and metribuzin are the most studied and more applied triazine herbicides [10, 55]. Regardless being used since the 1970s, some triazine herbicides continue to be agronomical and commercially important, essentially as a pre-emergent control of broadleaf weeds [56].

3.2.13.1 Chemical Aspects

Triazines have a three-nitrogen aromatic ring (Fig. 3.13). For symmetrical triazines could be chlorine, thiomethyl or methoxy group at the 2-position of the ring and frequently the 4- and 6-positions have an alkylamino group. The known symmetrical triazines include atrazine (6-Chloro-N-ethyl-N0-isopropyl-1,3,5-triazine-2,4-diamine), simazine (2-Chloro-4,6-bis (ethylamino)-s-triazine), propazine (2-Chloro-4,6-bis (isopropylamino)-s-triazine) and terbuthylazine (2-(Tertbutylamino)-4-chloro-6-(ethylamino)-s-triazine). Whereas, for asymmetrical triazines once the nitrogen atoms are unequally spaced, aromaticity is maintained by the presence of the

Fig. 3.13 Structure of triazine herbicides [58]

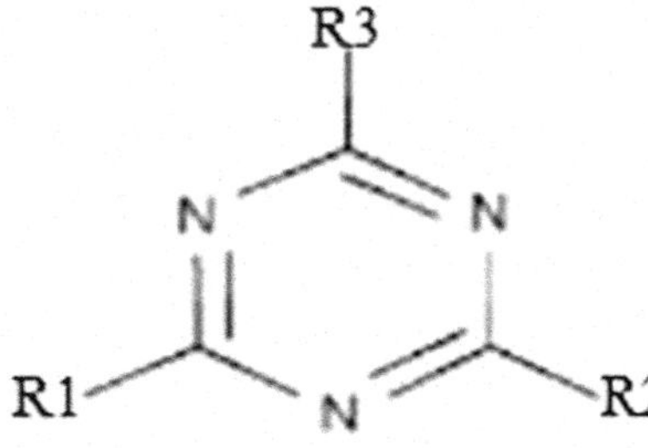

carbonyl group, in metribuzin (4-Amino-6-(1,1-dimethylethyl)-3-(methylthio)-1,2,4-triazin-5(4H)-one) case. Metribuzin also has a thiomethyl-substituent on the ring [57–59].

3.2.13.2 Uses and Environmental Persistence

Used on broad-leafed weeds and grasses control as pre- and post-emergent herbicide, triazines act by inhibiting the photosynthetic electron transport in certain plants [55]. Triazines and its metabolites have high contamination potential for groundwater since their water solubility, soil adsorption coefficient, vapor pressure, field half-life, and the amount of use [58].

The broadly worldwide used in agriculture for more than 50 years, specifically for atrazine lead to surface, ground, and drinking water contamination henceforward this herbicide was restricted or banned in Europe 20 years ago [9]. Resistance to abiotic hydrolysis and to direct aqueous photolysis, couple with limited volatilization potential and moderately susceptible to aerobic biodegradation are the factors that grant atrazine a high contamination potential [60]. Metribuzin is used since 1973 on several crops, turf grasses and as selective control of broadleaf weeds and grassy weed species. The triazine is photodegradable in soil and water, however, is poorly sorbed by soil hence susceptible to leaches [59].

3.2.13.3 Toxicity on Humans

Excepting for metribuzin, atrazine, ametryn and cyanazine, triazines herbicides are considered low toxic, without acute hazard in normal use nor development toxicity [61].

Atrazine appear to be well absorbed through gastrointestinal tract, skin or by inhalation and rapidly excreted from the body [58, 60]. This is metabolized primarily by dealkylation, followed by glutathione conjugation and conversion to mercapturic acids. Available data on the effects of atrazine in humans is limited. Nonetheless, increased cases of small-for-gestational-age were associated with the consumption of water contaminated with atrazine during pregnancy and alteration to semen qualities of adult males after atrazine exposure [58]. There is a record of one death after extensive dermal exposure, which causes skin rash, erythema, blisters, and edema. When ingested, inhaled, or ocular contact may occur different symptoms such as ocular irritation, chest pains, chest tightness, nausea, and dizziness. Is not clear if atrazine has carcinogenicity potential, this triazine chronic impact on humans seems to be stronger on reproduction and development via effects on endocrine signaling [60]. Regarding metribuzin, no data was found on its exposure to humans [59].

3.2.14 *Triazolopyrimidines*

The triazolopyrimidines are pre-emergent, and post-emergent herbicides used to control broadleaf weeds in a variety of crops. Discovered in the early 1980s, this chemical class includes cloransulam-methyl, diclosulam, florasulammethyl, flumetsulam, metosulam, penoxsulam, and pyroxsulam [31, 62].

3.2.14.1 Chemical Aspects

The generic chemical structure of triazolopyrimidines is a substituted triazolopyrimidine connected to a substituted phenyl or pyridine ring through a sulfonamide bridge (Fig. 3.14) [62].

3.2.14.2 Uses and Environmental Persistence

Triazolopyrimidines are used in several crops as weeds control, these are selective and have a diversified activity: cloransulam-methyl in soybean, diclosulam in peanuts, florasulam in cereal crops and turf and flumetsulam in corn and soybeans [8].

Triazolopyrimidines act by inhibiting the AHAS, a fundamental enzyme to the synthesis of branched-chain aliphatic amino acids leucine, isoleucine, and valine in plants. The inhibition of this enzyme leads to cell growth and division cessation, and consequently the death of the plant [62].

3.2.14.3 Toxicity on Humans

Once the AHAS is inexistent in humans and other mammals, triazolopyrimidines are regarded as low toxic to humans. Acute, short-term, long-term (chronic), genotoxicity, reproduction, developmental, carcinogenicity, and neurotoxicity tests support this claim. These chemicals are rapid adsorbed and excreted, with no evidence of accumulation [10, 62]. Liver and kidneys have been identified as target organs with effects that were often adaptive in nature [9, 12].

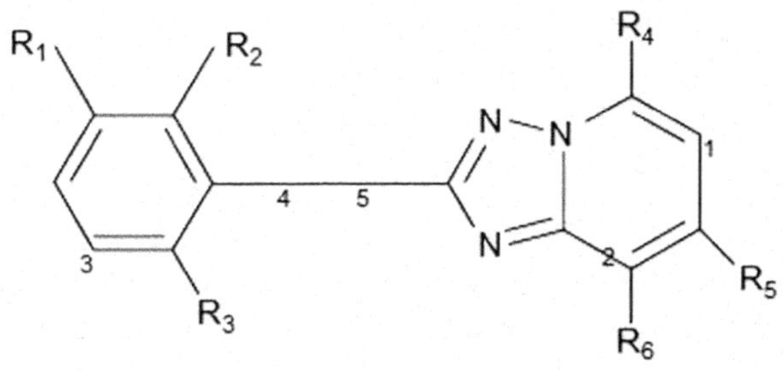

Fig. 3.14 Structure of triazolopyrimidine herbicides [62]

3.3 Insecticides

Insecticides are broadly used in agriculture, horticulture, and forestry to control or kill insects. However, these are not only toxic for insects, but their numerous mechanisms of action can also affect other organisms and by doing so negatively affect non-target species (including humans). The main target is the nervous system, however, other systems and organs can be affected (skeletal muscles, digestive, cardiovascular, respiratory, ophthalmic, reproductive, endocrine, dermal, immune systems) [63, 64].

Insecticides can be classified according to their chemical family, namely organochlorines, organophosphates, carbamates, pyrethroids, formamidines, avermectins, neonicotinoids, spinosyns, and insect growth regulators (IGR) [64].

3.3.1 Avermectins

Avermectins, discovered in 1967, are produced by *Streptomyces avermitilis* fungus. Although, neither categorized as antibacterial nor as antifungal, avermectins have nematocidal, acaricidal, and insecticidal activity [65, 66]. By being antiparasitic agents and highly effective, these are broadly applied in agriculture and medicine, particularly in veterinary medicine [66, 67]. This family of insecticides includes moxidectin, milbemycin oxime, doramectin, selamectin, eprinomectin, abamectin, and ivermectin, being the last two the more often used [66].

3.3.1.1 Chemical Aspects

Chemically, avermectins are a class of 16-membered macrocyclic lactones possessing a bisoleandrosyloxy substituent located at the carbon 13 (Fig. 3.15). Eight structural combinations are possible with A (methoxy group at position 5) and B (hydroxy group at position 5), then combined with 1 (double bond between the position 22 and 23); 2 (single bond with a hydroxy group at position 23); a (secondary butyl side chain at position 25) and b (isopropyl substituent at position 25). Group a and b components belong to sec-butyl and isopropyl homologs, respectively [66, 68].

Regarding avermectin derivatives, these are often classified as cis or trans isomers that can something be as active as the parental molecule [66].

3.3.1.2 Uses and Environmental Persistence

Avermectins are used as a veterinary drug in livestock, to protect against a broad variety of parasites [69]. Eprinomectin is used mostly in cattle to collect parasites.

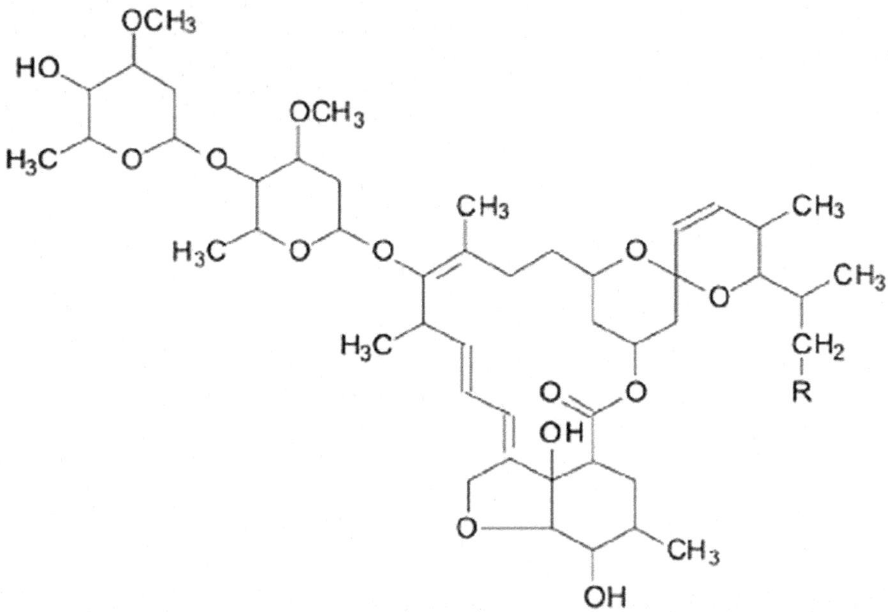

Fig. 3.15 Structure of avermectin [66]

Moxidectin and ivermectin are antiparasitic drugs that are also used in chemotherapeutic. Abamectin is very lipophilic and causes paralysis by blocking the nervous system of the target since it stimulates the release of γ-aminobutyric acid. Ivermectin also affects the nervous system and muscle function [67]. Abamectin is intensively used worldwide in ornamental, horticultural, and agricultural crops [67, 69].

Avermectins can accumulate in the environment even though their short half-life. Persistence in water, air, sediment, soil, and food products due to spray drift or field drainage, or even their presence in livestock feces are strong possibilities [66, 69, 70]. For instance, abamectin is reported to persist between 14 and 70 days in the soil at concentrations with toxic effects [69]. Studies show that abamectin present toxic effects at low concentrations. Runoff of these insecticides from the application site leads to contamination of non-target organisms [69, 71]. The US Environmental Protection Agency (EPA) reports avermectins to have a strong toxic effect in fish and bees and to be even more toxic for mammals and aquatic invertebrates [70, 72]. In addition, with the increase in avermectins resistance, there is an excessive application of these chemicals [72].

Concerning the mode of action, avermectins are considered neurotoxins. Acting on the glutamate and/or g-amino butyric acid (GABA)-gated chloride channel, a neurotransmitter that functions between nerve and muscle cells [66]. However, the action mode of avermectins is not specific to parasitic arthropods and nematodes and thus could affect other organisms [69].

Contrary to other insecticides, avermectins are able to degradation, mostly by photodegradation. This phenomenon takes place at specific wavelengths below the UV-B range, degradation by reactive oxygen species (ROS) pathways are also possible. Abamectin and ivermectin degraded rapidly from the soil, water, and plants in the presence of light, but the degradation products vary according to environmental conditions [66].

3.3.1.3 Toxicity on Humans

The toxicity of avermectins in humans is thought to be low since the neurons affected by GABA are kept in the mammalian central nervous system and avermectins do not readily cross the blood-brain barrier (BBB) [72, 73]. In fact, these are used, in humans, to treat onchocerciasis or river blindness [69]. Although considered inactive against bacteria, the testing of ivermectin, selamectin, and moxidectin against several mycobacterium tuberculosis strains showed a powerful antibacterial activity. These were thereafter approved to treat tuberculosis [70].

Even though avermectins are degradable, nearly no metabolism happens in the organism of the target and the dose offered to the target is discharged as the parent compound, posing a serious potential risk to human health [72].

Little is known about the magnitude of their toxicity for humans and further research are needed. Nevertheless, for instance, abamectin is very toxic if inhaled or swallowed. Moreover, decrease in sperm quality and motility was reported in humans follow by abamectin exposure. It was observed a decreased sperm quality in farmers who applied abamectin, without the personal protective equipment [66].

Additionally, high doses of avermectins can pass through the BBB and have neurotoxic effects on animals and possibly humans. These effects can manifest as mydriasis, emesis, diarrhea, drooling, depression, ataxia, stupor, coma, tremors, or death [73].

3.3.2 Formamidines

Formamidines, discovered in the early 1960s, are a class of insecticides widely used to treat *Lepidoptera*, *Hemiptera*, *phytophagous* mites, and cattle ticks. Chemicals such as amitraz, chlordimeform, formetanate, formparanate, medimeform, and semiamitraz are included in this group [74, 75].

3.3.2.1 Chemical Aspects

Amitraz, chlordimeform, and formetanate are the most commonly known formamidines (Fig. 3.16) [76]. Chlordimeform presents high volatility and low water solubility [77] and structurally presents a tautomeric form [78].

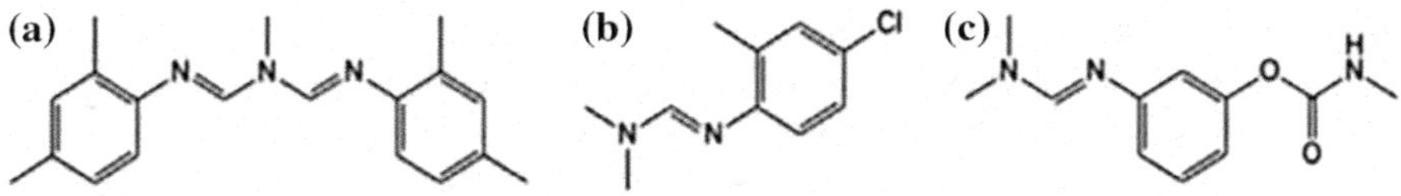

Fig. 3.16 Structure of amitraz (**a**), chlordimeform (**b**) and formetanate (**c**) [76]

3.3.2.2 Uses and Environmental Persistence

The formamidines acaricide characteristics make them be extensively used in agriculture and livestock production. Amitraz, for instance, is used on fruit trees and livestock (against scabies, ticks, and lice) due to its efficiency and a wide range of action. Formetanate is used on fruit crops and uncultivated soils [76, 79]. Chlordimeform, once used against mites, ticks, Lepidoptera insects and on fruits, rice and cotton crops, was taken out of the US market in 1992 [77]. Formamidines act by target the adrenergic nervous system, in insects, activating the octopamine receptors and in mammals the alpha2-adrenergic receptors [77], presenting toxic effects toward a wide range of plants, animals and bacteria species [78]. Notwithstanding amitraz is use as acaricide in cattle, sheep, goat, and pigs, the use of this formamidine is inadvisable in horses, cats, and small breed dogs. Low blood pressure, anorexia, hypothermia, high blood glucose levels, lethargy, vomiting, and digestive tract disorders may occur in exposed animals [79]. Chlordimeform is rapidly degraded in plants and soil, and so far no bioaccumulation has been found [77].

3.3.2.3 Toxicity on Humans

As mentioned before, formamidines can have toxic effects on mammals. Cases of formamidines poisoning caused either by incidental exposure or suicide attempt have been reported. Metabolites of amitraz, chlordimeform, and formetanate have been identified in human being these rapid degraded and metabolized in the human body. Formetanate inhibits the enzyme acetylcholinesterase which grants it with a risk for neurotoxicity [76].

Amitraz is reported to cause vomiting, tachypnea/dyspnea, mydriasis, intestinal distension, lethargy, sedation, ataxia, bradycardia, central nervous system depression, hypersalivation, dehydration, hypoperfusion, seizures, gastrointestinal stasis, hypertension, and hypotension in human beings through oral or dermal contact [76, 79].

Chlordimeform common effects on humans include severe hypotension, cardiac toxicity, nausea, vomiting, central nervous system depression, and blurred vision. Impair kidney and liver functions were reported in some cases, and hematuria was reported in chlordimeform packaging plant workers. Chlordimeform is classified as a probable human carcinogen by either the International Agency For Research On Cancer (IARC) and EPA [77], consequently their agricultural used has been limited in many countries [76]. Studies in individuals exposed to chlordimeform show an increased risk of bladder cancer, mainly when exposure to 4-Chloro-o-toluidine (a chlordimeform metabolite), proven to be highly carcinogenic [76, 77].

3.3.3 Insect Growth Regulators (IGR)

IGR was first used by Schneiderman in 1972 to refer hormones that control insect growth and development "such as juvenile hormones and ecdysones". From these became the third generation of insecticides, synthetic or natural chemicals that affect specific insects' growth, development, or reproductivity [80, 81]. More specific, these chemicals can inhibit or accelerate the regulatory process fundamental to the normal development or reproductivity of the target [82]. IGR include chitin synthesis inhibitors (buprofezine), juvenile hormone analogs/agonists, ecdysone agonists (tebufenozide, methoxyfenozide, halofenozide), and anti-juvenile hormone analogs [81, 83, 84].

3.3.3.1 Chemical Aspects

Mostly synthetic versions of natural hormones, IGR can include different chemical classes [81, 85] and hence these do not have a generic chemical structure (Fig. 3.17). Chitin synthesis inhibitors include pyrimidinenucleoside peptides, benzoylurea, oxazolines, thiazolidines, tetrazines, thiadiazines, thiophthalimides, certain chromo- and fluorophores [86]; and juvenile hormone agonists include methoprene, hydroprene, pyriproxyfen and fenoxycarb [80].

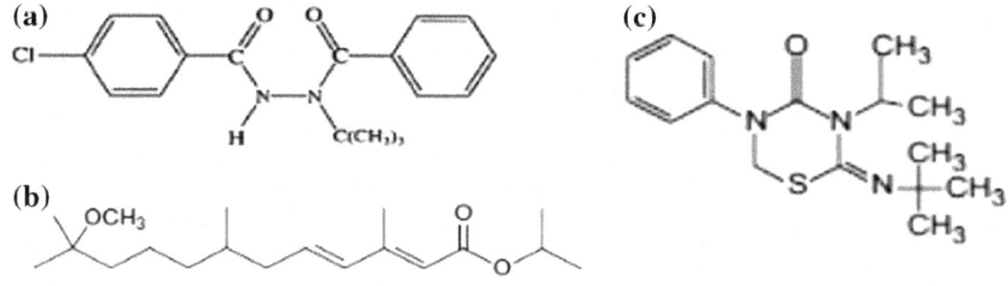

Fig. 3.17 Structure of some IGR: halofenozide ecdysone agonists (**a**), methoprene juvenile hormone agonists (**b**) and buprofezin chitin synthesis inhibitors (**c**) [84]

3.3.3.2 Uses and Environmental Persistence

IGR are used as insect control and act by disrupting key regulatory process as metamorphosis regulation or reproduction impairing the insect survival without necessarily having a toxic effect on the target [81, 82]. Moreover, metamorphosis disruptors stop the formation of reproductive adults and reproduction disruptors can lessen the reproductive potential by promoting morphogenetic abnormalities in developing adults. IGR are highly selective and effective [81].

Juvenile hormone agonists disrupt insect development and increase larval mortality by mimicking and boost the juvenile hormone mode of action. Although juvenile hormone agonists show advantages as low toxicity to non-targets and to the environment, they have a low range of application [83].

Chitin synthesis inhibitors affect chitin synthesis frequently inducing malformations of the cuticle and decreasing chitin amounts. Impair septum and cell wall formation or abortive molting and defects in egg hatching are some consequences of the use of these chemicals [86]. Chitin synthesis inhibitors are used as green insecticides in agricultural crop protection and sanitary insect pest control [87].

The mode of action of these chemicals is not the same, some chitin synthesis inhibitors affect the catalytic site by interfering with the enzyme's nucleotide binding sites, others block a post-catalytic step of chitin synthesis and some may react with sulfhydryl groups of cysteines close to the catalytic site [86]. Regardless of the action mode, these insecticides are considered unharmful to other organisms [87].

3.3.3.3 Toxicity on Humans

IGR present very low toxicity to humans [81], their very specific mode of action makes them low and even non-toxic to off-target organisms [83, 87].

No clinical effects or overt signs of toxicity are known to have occurred in humans due to methoprene exposure. Some commercial formulations of methoprene can, however, contain additional ingredients that cause skin and eye irritation or allergic reactions after repeated exposures [88].

3.3.4 Neonicotinoids

Neonicotinoids are a new class of synthetic insecticides, chemically associated with nicotine, with extensive use both in veterinary medicine and crop production [89]. The first neonicotinoid (imidacloprid—Fig. 3.18) was introduced in 1991 [90]. These family of insecticides, namely imidacloprid, acetamiprid, dinotefuran, thiamethoxam, and clothianidin [89].

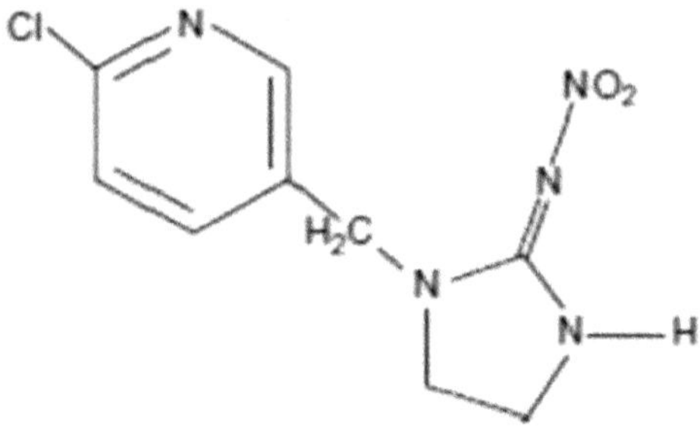

Fig. 3.18 Structure of imidacloprid [89]

3.3.4.1 Chemical Aspects

Nowadays, seven commercial compounds are available and are chemically classified as N-nitroguanidines (imidacloprid, thiamethoxam, clothianidin, and dinotefuran), nitromethylenes (nitenpyram), and N-cyanoamidines (acetamiprid and thiacloprid) [90].

3.3.4.2 Uses and Environmental Persistence

Neonicotinoids besides extensively used in agriculture are also applied to control human and animal parasites such as bed bugs and fleas [90].

Being nowadays one of the most frequently used insecticides, neonicotinoids (and their metabolites), are increasingly detected in several environmental media, namely watercourses nearby agricultural and urban areas [91], soils, drinking water, vegetables and fruits, and even bovine milk [92]. Thus, these insecticides are diffusely in the environment and can accumulate there [92, 93].

One of the main worries since its introduction in 1991 in Europe has been the possible harmful effects on bees. Therefore, in 2013, the Commission severely restricted the use of clothianidin, imidacloprid, and thiamethoxam in plant protection products and treated seeds containing these neonicotinoids to protect honeybees (see Regulation (EU) No 485/2013) [94].

Nicotinic acetylcholine receptors are involved in rapid neurotransmission in both insect and mammalian nervous systems. In Mammalian tissue, several subtypes of receptors can be present that are formed from different combinations of nine α, four β, γ, δ, and ε subunits. However, the neonicotinoids insecticides have much lower activity in vertebrates [89].

3.3.4.3 Toxicity on Human

Neonicotinoids insecticides seem to pose a relatively low risk for the environment and mammals as generally, they are highly specific for the subtypes of nicotinic receptors present in insects [89].

Due to the widespread use of these insecticides, human are generally exposed through dietary sources [92].

In recent reviews concerning the impact of neonicotinoids in human health, authors concluded that the currently available epidemiological, in vivo, and in vitro studies suggest possible detrimental effects to humans and mammals. However, more research is needed to fully understand the impact of these compounds for humans and in particular for vulnerable groups such as children and pregnant women [92, 95].

3.3.5 Organochlorines

Organochlorine pesticides (OCPs) are synthetic pesticides extensively used from the 1940s to 1970s as pests in agriculture and in insect control worldwide [96, 97]. OCPs are highly persistent in the environment compared with other synthetic organic pesticides, being this group classified as persistent organic pollutants [97]. Some examples of compounds included in this group are hexachlorobenzene, hexachlorocyclohexanes, 1,1,1-trichloro-2,2-bis-4-chlorophenyl-ethane (*p,p*-DDT) and its metabolites 1,1-dichloro-2,2-bis-4-chlorophenyl-ethene (*p,p*-DDE) and 1,1-dichloro-2,2-bis-4-chlorophenyl-ethylene (*p,p*-DDD), aldrin, dieldrin, endrin, α-endosulfan, β-endosulfan, and methoxychlor [98, 99].

Stockholm convention banishes the use of OCPs in most countries. However, the World Health Organization permitted the reintroduction of DDT (Fig. 3.19) for control of vector-borne diseases, such as malaria in some tropical countries in 2006 [96].

3.3.5.1 Chemical Aspects

OCPs are a diverse group of chemicals that classically have an aliphatic or aromatic cyclical structure and multiple chlorines. Consequently, OCPs are nonpolar, lipophilic, semi-volatile organic compounds and hydrophobic therefore they tend to adsorb on to particles [98]. The use of most OCPs are now banned around the world, however, due to their chemical properties, OCPs are not easily degraded being persistent in the environment [100].

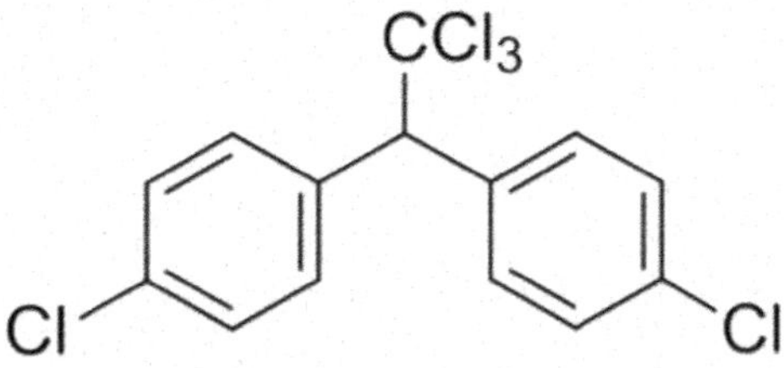

Fig. 3.19 Structure of *p,p*-DDT [98]

3.3.5.2 Uses and Environmental Persistence

OCPs have long-range atmospheric transport, so they can be detected in remote regions far away from the primary site of application [98, 101]. They are used in terrestrial ecosystems for agriculture purposes, however, they can be flushed away to rivers and end up in marine ecosystems [97]. These compounds due to their chemical properties tend to accumulate in the adipose tissues of biota; this characteristic makes these compounds difficult to excrete. They bioaccumulate and are biomagnified through the food chain, as the trophic level increases, the concentrations in living organisms will increase too [102, 103].

OCPs have been found in several different environmental samples [104]. Some of the examples are strawberries [105], carrots [106], rocket salad [107], and seafood such as clams, shrimps, and fish around the world [108].

DDT is one of the most used OCPs since 1945 in agriculture and since 1955 for vector-borne disease control. Humans are still exposed to OCPs, such as DDT, mainly through the consumption of contaminated foods [96] and other routes because they are ubiquity in the environment.

3.3.5.3 Toxicity on Human

The main route of human exposure to OCPs is through diet (more than 90%), mainly attributed to the food of animal origin, particularly seafood [109].

OCPs have been described to accumulate in different biological tissues and fluids, such as adipose tissue, hair, blood, and breast milk [110, 111] These compounds have been reported to be harmful to humans. They are endocrine disrupting chemicals, so they can interfere with several aspects of hormone action. Exposure to this group of insecticides is reported to be associated with numerous adverse health effects. Namely, cancers, neurodegenerative disorders, respiratory disorders, reproductive disorders [112], thyroid dysfunction, and immunological disorders [113], increase the risk for obesity and type 2 diabetes [114].

3.3.6 Pyrethroids

Pyrethroids, the synthetic organic insecticides derived from the natural pyrethrins, have been extensively applied globally since the 1980s [115–118]. These insecticides are considered to have great efficiency, with strong selectivity and considerably low toxicity when compared to other insecticides [117–119].

Nowadays, pyrethroids have diverse applications in agricultural activities but also as active ingredients of many insect-control products such as pet shampoos and lice treatment products [120]. Cypermethrin, deltamethrin, fenpropathrin, fenvalerate, bifenthrin, permethrin, lambda-cyhalothrin, and cyfluthrin are some of the insecticides belonging to this group [117, 118].

3.3.6.1 Chemical Aspects

Pyrethroids are among the insecticides families that possess a higher number of chiral centers. These insecticides usually contain 2–3 chiral centers and each pyrethroid has 2 or 4 enantiomer pairs, or 2 or 4 diastereoisomers [118, 120].

Besides their rich stereochemistry, they also present low water solubility and strong sorptive properties and are fairly photolabile [121]. The presence or absence of the cyanogroup [115] (Fig. 3.20) leads to classification in pyrethroids type I (non-cyano pyrethroids) and type II (with a cyanogroup).

3.3.6.2 Uses and Environmental Persistence

As these insecticides are extremely effective and generally regarded as safer, they are widely used in agriculture, in domestics applications, in veterinary and even in the public health domain [123].

The increased applications of pyrethroids in the last decades make these insecticides detectable in several environmental samples across the globe. In a recent review, comprising data from all continents, it was compiled the determined levels of several pyrethroids and their metabolites in environmental samples and in live organisms. With 31 pyrethroids and 8 metabolites being identified in these media, and cypermethrin being the most frequently detected. Thus as confirmed by this review pyrethroids can reach several environmental compartments including water samples [117].

Therefore, concern over their possible environmental non-target effects has increased [121] particularly for the aquatic organisms as they are highly sensitive even to extremely low concentrations.

Pyrethroids act on the nervous system through several mechanisms which are analogous in insects and mammals and comprise action on the voltage-gated sodium, calcium and chloride channels, GABA-gated chloride channels, nicotinic receptors, membrane depolarization, and intercellular gap junctions [115, 124, 125].

Fig. 3.20 Structure of deltamethrin [122]

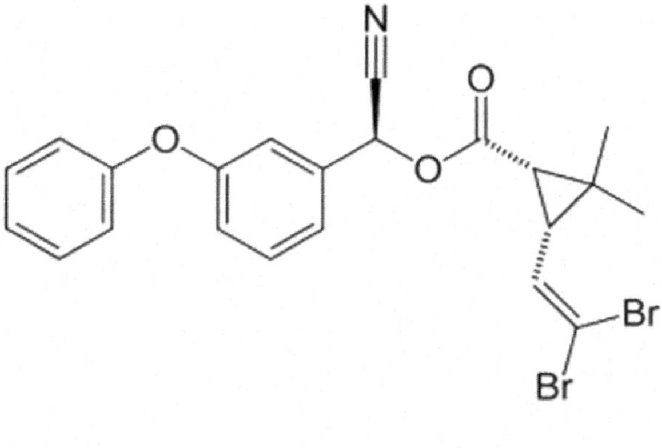

3.3.6.3 Toxicity on Human

Human exposure to pyrethroids is generally linked with dietary intake but also with the application of these insecticides in gardens and homes for pest-control purposes [117, 126].

After uptake in the human body, pyrethroids are rapidly metabolized, with the 3-phenoxybenzoic acid (3-PBA) being the metabolite common to several pyrethroids. In a recent review concerning biomonitoring studies and health effects, from pyrethroids human exposure, the authors concluded that the information regarding the chronic effects from low concentrations is still scarce and controversial. However, authors found the strongest links for several recent epidemiological studies linking adverse effects on the reproductive system and pregnancy outcomes with pyrethroids exposure. Furthermore, other studies addressed and showed a linked between the post-natal neurobehavioral development and in utero exposure [126].

Some pyrethroids are also considered endocrine disruptors [127], and particularly 3-PBA was associated with an increased body mass index [128]. Other studies have linked the exposure to some pyrethroids and cancer, reviewed in Burns et al. [129].

3.3.7 Spinosyns

Spinosyns are a family of insecticides manufactured by *Saccharopolyspora spinosa* and *Saccharopolyspora pogona*. Spinosyns present strong insecticide capacity and wide range of action, have been commercialized and used since 1997 [130]. Due to their biological origin, spinosyns are thought not to be very toxic to non-targets and with low impact on the environment [131].

3.3.7.1 Chemical Aspects

Spinosyns are macrocyclic lactones with a unique characteristic of having a tetracyclic core of a 12-membered macrocyclic lactone fused to 5,6,5-cis-anti-trans tricyclic ring and two sugar moieties, the amino sugar forosamine, and a neutral sugar, 2,3,4-tri-O-methylated rhamnose (Fig. 3.21) [130, 132].

The major spinosyns obtained from the fermentation of *Saccharopolyspora spinosa* are spinosyn A and spinosyn D. This last has an additional methyl group on B-ring [130]. Over 25 different spinosyns can be obtained from *Saccharopolyspora spinosa* differing or in the 2′-, 3′-, 4′-methyl positions of the rhamnose, or in methyl substitution patterns on the forosamine nitrogen and or in the C6, C16 and C21 positions of the tetracycle [132].

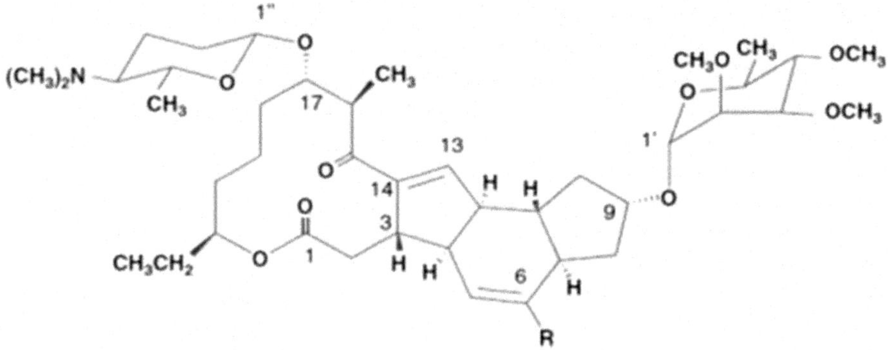

Fig. 3.21 Structure of spinosyns [134]

Spinosad, the first spinosyn-based biopesticide in the market, is a combination of two spinosyns: A and D. Spinetoram is obtained by mixing two synthetically modified spinosyns (spinosyn J and spinosyn L) and was the first semi-synthetic spinosyn known [131, 133].

3.3.7.2 Uses and Environmental Persistence

Spinosyns are neurotoxic when ingested or through contact are able to cause hyper-excitation and rapidly death in numerous insect targets [131, 133, 135]. For this reason, these are used in agriculture and to control sanitary interest insects [131].

Regarded as a biopesticide, spinosad is widely used in agriculture to control several types of pests: caterpillars, leaf miners, horn flies, blowfly strike, lice, maggots thrips, flies, drywood termites, and some beetles [131, 134]. In 2007, the USA authorized the utilization of spinosad in organic farming. Europe follows the same course and in 2008 included spinosad in Annex II of the EU Council Regulation 2092/91, allowing its use in organic farming. Presently, more than 80 counties around the world use this insecticide in agriculture and livestock against numerous target pests, e.g., *Lepidoptera, Diptera, Thysanoptera,* and *Coleoptera* [131, 133].

Spinetoram presents higher efficiency than other spinosad descendants. Owing to similar toxicological properties to spinosad but acting faster and more potently [131, 133].

Spinosyns affect the nervous system of the target pest by passing as allosteric modulators of nicotinic acetylcholine receptors [131, 133], essentially the target insect is killed by the collapse of the firing ability of the neurons [130]. Interestedly, in addition to work as an agonist in nicotinic acetylcholine receptors, spinosad can also interfere with GABA as an antagonistic. The unique mode of action of this class makes resistance very unusually, only a few cases were reported [131].

The high selectivity of spinosyns, specifically spinosad, makes them nearly non-toxic to non-target insects and fish and non-toxic toward mammals [130, 131]. Furthermore, none case of phytotoxicity was reported [131].

Spinosyns are rapid degradable either from the soil, sediments, and water [131] by photolytic processes sunlight-based and biotic degradation and easily metabolized by mammals [134]. All put together makes spinosyns a very low or non-persistence insecticide [130, 134].

3.3.7.3 Toxicity on Humans

Spinosyns are not considered harmful to humans; in chronic toxicity test in mammals, no carcinogenic, teratogenic, mutagenic or neurotoxic effects were found [131, 134]. For that reason, spinosad is used to treat parasitic problems in humans as head lice [134].

However, a case was reported of acute exposure to a mixture of spinosad and flonicamid [136]. The patient underwent neurological manifestation, as consciousness disturbance, shock, respiratory failure, pneumonitis, and urinary retention. Additionally, second-grade corrosive esophageal injury was found in endoscopic examination. Followed resuscitation, detoxification measures, and intensive care, the patient recovered fully without any chronic sequels. It was speculated that the high ingestion of spinosad overpassed the physiological safety does that can be handled by the human body or the clinical toxicities observed were due to the solvent contributions of the conserve insecticide formulation [136].

3.4 Compounds with Herbicide and Insecticide Activity

3.4.1 Carbamates, Thiocarbamates, and Dithiocarbamates

3.4.1.1 Carbamates Herbicides, Thiocarbamate, and Dithiocarbamate

Pesticides that belong to the carbamate class have become progressively important over the years due to their extensive spectrum of activity. They can be employed as insecticides, herbicides, fungicides, nemotocides, miticides, and molluscicides [137]. The compounds in the herbicidal category include derivatives of carbamic acid (e.g., propham, asulam, barban, chlorpropham, chlorbufam, karbutilate, and phenmedipham), derivatives of thiocarbamic acid (e.g., butylate, cycloate, diallate, EPTC, molinate, and triallate) and derivatives of dithiocarbamic acid (such methamsodium) [10]. Dithiocarbamates can also be used as fungicides [138].

Chemical Aspects

The group of carbamate herbicides is in terms of chemistry esters of carbamic acid. They diverge from carbamate fungicides depending on the N-substituted side chain, carbamate herbicides (aromatic moiety), and fungicides (benzimidazole moiety) [139]. They are thermolabile and polar [137]. Herbicidal carbamates structure is constituted usually for an aryl or aromatic group in R and a short-chain alkyl group in R′, as represented in Fig. 3.22 [140].

The toxicity of carbamates can be different between the diverse classes, in the case of carbamate herbicides can be related to the aromatic moiety, the aromatic or aniline-derived structure can be related to a toxic effect in laboratory animals. Carbamate herbicides do not influence the cholinesterase activity as a contrast to the insecticidal carbamates, although they can inhibit cell division [139, 141].

Uses and Environmental Persistence

This group of herbicides is generally pre-emergence, however, some have also used as post-emergence. The first herbicidal carbamate known was propham. Previous data suggest that these herbicides are exclusively distributed via the apoplastic system in plants, they act stopping cell division and plant tissue growth. The main routes of metabolization of herbicidal carbamates by the plants are arylhydroxylation, conjugation, or hydrolytic breakdown [141]. In general, the concentrations used of these herbicides are low and no poisoning problems have been reported. Also relatively to the acute toxicity, that is usually lower when compared to the carbamate insecticides [12, 139].

Toxicity on Human

Carbamate herbicides usually display rapid biotransformation by hydrolysis, oxidation, and conjugation, and the main route of excretion is via urine [139].

For mammals such as rats, these herbicides demonstrated to have low-to-moderate toxicity, not a poising acute danger. For example, chlorpropham giving at high oral doses during organogenesis has shown to cause malformations and various types of developmental toxicity. Studies reported that at high doses by

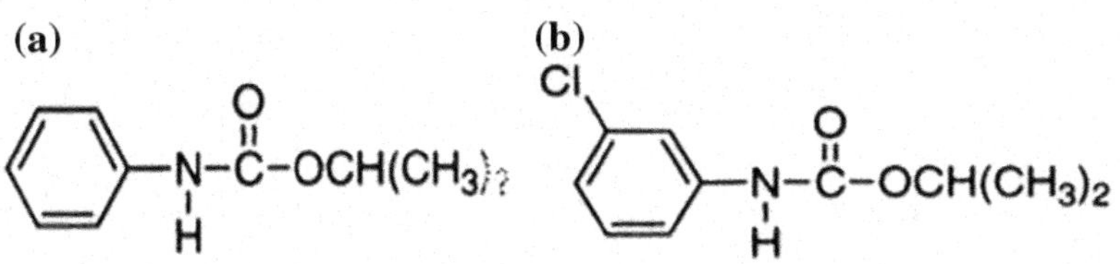

Fig. 3.22 Structure of propham (**a**) and chlorpropham (**b**) [140]

gavage cause testicular effects, such as decreased sperm count, decreased testicular weights, and histopathological changes [10].

Since the acute toxicity of carbamate herbicides are known to be low, acute signs of intoxication in human are improbable to occur under occupational conditions. Studies reported that they do not cause skin or eye irritation, although repeated exposure may cause alopecia for some time after ingestion [12, 139].

3.4.1.2 Carbamate Insecticides

Carbamate insecticides are commonly synthetic derivatives of physostigmine, alkaloid of the *Physostigma venenosum* plant [142]. The main carbamate insecticides include bendiocarb, carbaryl, carbofuran, methomyl, oxamyl, and propoxur [63, 138].

Chemical Aspects

Carbamates are carbamic acid esters (Fig. 3.23) which two R groups could be hydrogen, methyl, ethyl, propyl, or other short-chain alkyls and the other R a phenol, naphthalene, or other cyclic hydrocarbon rings [142, 143].

Uses and Environmental Persistence

Carbamates are a major class of synthetic organic pesticides, introduced in the 1950s, these are used annually on a large scale worldwide [141]. Carbamates act on a broad spectrum of insects, carbaryl and propoxur are commonly applied for residential uses, whereas the remaining carbamates are used in agricultural applications [55, 63].

The action mode of carbamates involves inhibiting the acetylcholinesterase enzyme within the nervous tissue at the neuromuscular junctions. As acetylcholinesterase is inhibited by carbamylation, acetylcholine accumulates at the nerve endings of all cholinergic nerves causing overstimulation of electrical activity [138, 141, 143].

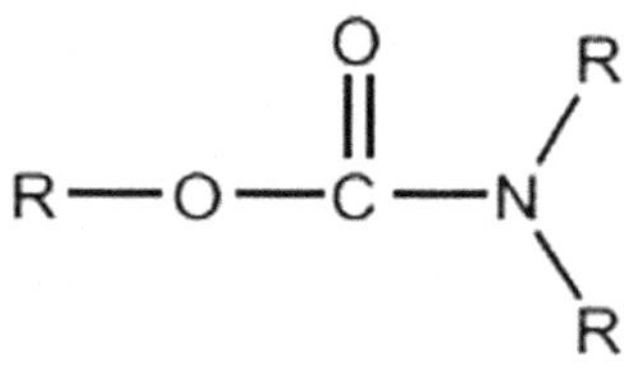

Fig. 3.23 Structure of carbamates insecticides [138]

In general, carbamates are degraded into metabolites of lesser toxicity and they have very little impact in terms of environmental persistence. Carbamates can be degraded by microorganisms, soil, water, light, and animals. These compounds do not bioaccumulate in the food chain or environment [138].

Toxicity on Humans

Risks to humans are significant from overexposure to carbamates insecticides. Symptoms of carbamates acute poisoning, depending on the dose, can include hypersalivation, excessive tracheobronchial secretions, gastrointestinal cramps, lacrimation, dacryorrhea, nausea, excessive sweating, urinary incontinence, diarrhea, miosis, bradycardia, muscle fasciculations, tremors, muscle weakness, flaccid paralysis, blurred vision, vomiting, respiratory muscles paralysis, coma, and even death through cardiac and respiratory failure [138, 142].

Chronic exposure to carbamates causes skin and eye irritation, hemopoietic alterations, degeneration of the liver, kidneys, and testes, functional and histopathologic changes in the nervous system after long-term or high-dose exposures. Moreover, carbamates are also considered embryotoxic, fetotoxic, teratogenic, mutagenic, and carcinogenic [138, 142].

3.4.2 *Organophosphates*

Organophosphorus or organophosphates compounds (OPs) are one of the large groups of chemicals that used over the past 60 years for protecting crops, livestock, human health and as warfare agents. OPs are the most widely used insecticides today. Over 100 OPs representing a variety of chemical, physical, and biological properties are presently in commercial use.

3.4.2.1 Chemical Aspects

OPs are originally produced by the reaction of alcohols and phosphoric acid. Organophosphate pesticides are synthetic in origin and are normally esters, amides, or thiol derivatives of phosphoric, phosphonic, phosphorothioic, or phosphonothioic acids. Figure 3.24 showed a general chemical structure of the OPs.

Fig. 3.24 Structure for OPs [144]

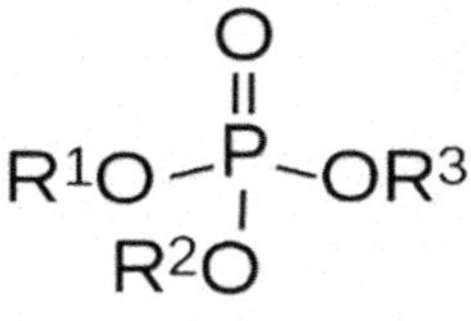

Most of the OPs are slightly soluble in water and have a high oil–water partition coefficient and a low vapor pressure. Most, except for dichlorvos, are of comparatively low volatility and are all degraded by hydrolysis, yielding water-soluble products. Parathion, for example, is freely soluble in alcohols, esters, ethers, ketones, and aromatic hydrocarbons but is practically insoluble in water or in petroleum ether, kerosene, or spray oils [144]. Parathion is stable at a pH below 7.5.

3.4.2.2 Uses and Environmental Persistence

Because OPs are often sprayed on crops and plants, small particles may be carried away from the field or yard before falling to the ground. After OPs are applied, they may be present in the soil, surface waters, and on the surface of the plants. They can move through the soil and contaminate the groundwater. Rain can wash OPs on soil and plant surfaces into surface water. There is a strong possibility that surface and groundwater will become contaminated in areas with high levels of agricultural activity, and this is especially important when the water is intended for use in either irrigation activities or human consumption [145].

OPs exposure pathways overlap for many wildlife species and humans. For example, the spraying of crops with OPs can cause pesticide drift to nearby communities. Similarly, pesticide runoff into water bodies can cause harmful effects on aquatic species, terrestrial species that forage around water bodies, and humans that reside or recreate in the vicinity [146]. Cholinesterases as the site of action of OPs have been identified in a vast array of species, including humans, but not in plants and microorganisms.

3.4.2.3 Toxicity on Humans

OPs can be absorbed by all routes, including inhalation, ingestion, and dermal absorption [147]. Organophosphate insecticides (such as diazinon) are one type of pesticide that works by damaging an enzyme in the body called acetylcholinesterase. This enzyme is critical for controlling nerve signals in the body. The damage to this enzyme kills pests and may cause unwanted side effects in exposed humans. All organophosphates have a common mechanism of toxicity and can cause similar symptoms in humans who have too much exposure [148]. Many OPs are potent nerve agents, functioning by inhibiting the action of acetylcholinesterase in nerve cells [149]. Their inhibitory effects on the acetylcholinesterase enzyme lead to a pathological excess of acetylcholine in the body. Their toxicity is not limited to the acute phase, however, and chronic effects have long been noted. Neurotransmitters such as acetylcholine (which is affected by organophosphate pesticides) are profoundly important in the brain's development, and many organophosphates have neurotoxic effects on developing organisms, even from low levels of exposure [150].

3.4.2.4 OPs Herbicides: Organic Phosphorus and Phosphonomethly

Glyphosate

Glyphosate is a very broad spectrum and the most commonly used herbicide worldwide, and the only one to target the enzyme 5-enolpyruvyl-3-shikimate phosphate synthase. Glyphosate is applied to large cultivation areas as well as small allotments, home, and industrial greeneries, those along with traffic routes and around public areas. Glyphosate is highly valued for its lack of direct risks to humans and animals or no tendency to bioaccumulate in trophic networks. However, controversies over its applications have been growing; therefore, it is increasingly implied that its effects on the environment, living organisms, and people residing in it should be verified.

Chemical Aspects

Ghyphosate is a phosphonic acid resulting from the formal oxidative coupling of the methyl group of methylphosphonic acid with the amino group of glycine. Ghyphosate has the empirical formula $C_3H_8NO_5P$ (Fig. 3.25) and a molecular weight of 169.1. The solubility of glyphosate in water is 1.2% at 25 °C. It is not generally soluble in organic solvents.

The scientific literature reported the movement and residues of glyphosate and its breakdown product Aminomethyl phosphonic acid (AMPA) in soil and water, their toxicity to macro- and microorganisms, their effects on microbial compositions and potential indirect effects on plant, animal, and human health [151].

Glyphosate may contaminate soils in and around treated areas. Glyphosate adsorbs to clay and organic matter, slowing its degradation by soil microorganisms and leading to accumulation in soils over time [152]. An example, in US where genetically modified glyphosate-resistant crops are grown, glyphosate and AMPA occur widely in soil, surface water, and groundwater [153]. Glyphosate has been measured in river water and streams water at levels from 2 to 430 μg L^{-1} [151].

Toxicity on Humans

Toxicologically, effects of glyphosate on humans and animals can be considered from three ways: (a) direct contact with the preparations sprayed in agrocenoses;

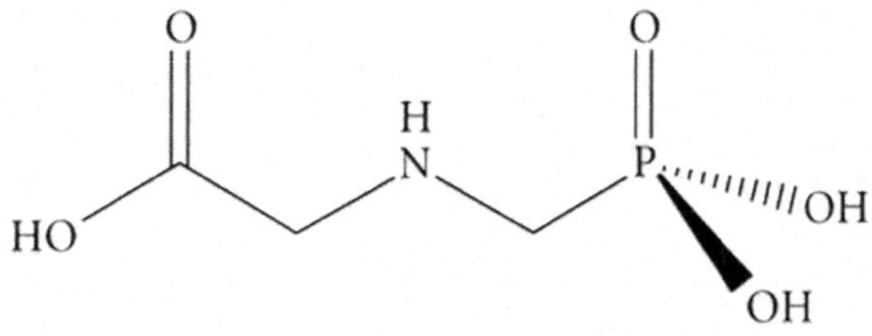

Fig. 3.25 Structure of ghyphosate [154]

(b) transfer of individual components of herbicides to surface waters; and (c) penetration of herbicide residues to the food chain [155]. Besides acute toxicity, the following issues should be considered: various changes occurring at subacute or chronic exposure, abnormalities in individual metabolic pathways, teratogenicity, disorders of the cell cycle, genetic dysfunctions, and endocrine pathophysiologies. As far as endocrine pathophysiologies are concerned, possible inability to reproduce due to dysregulation of sexual functions should be considered. The widespread occurrence of endocrine disrupting compounds has been confirmed in various environmental elements [156].

Glufosinate

One of the most successful natural products used as an herbicide is glufosinate. Glufosinate is the synthetic version of the natural compound phosphinothricin produced by *Streptomyces spp.* It is a potent inhibitor of glutamine synthetase and is the only commercial herbicide with this molecular site of action, although there are numerous other natural compounds that inhibit this enzyme. Glufosinate is sold throughout the world under various trade names.

Chemical Aspects

Glufosinate is soluble in water (>500 g L^{-1} at pH 5–9, 20 °C) and stable to light and hydrolysis at pH 5, 7, and 9 [157]. Glufosinate is a racemic mixture of D- and L-phosphinothricin. L-Phosphinothricin is a naturally-occurring bacterial amino acid (Fig. 3.26). Phosphinothricin was first synthesized by scientists at Hoechst in the 1970s as a racemic mixture; this racemic mixture is called glufosinate and is the commercially relevant version of the chemical [158].

Uses and Environmental Persistence

Glufosinate is a relatively fast acting, broad-spectrum herbicide with reduced environmental impact, making it ideal for use with transgenic herbicide-resistant crops (soybean, corn, and cotton). Plants exposed to glufosinate have reduced glutamine and increased ammonia levels in their tissues, which stops photosynthesis and results in plant death [160]. Glufosinate also has shown to provide some protection against various plant diseases, as it also acts to kill fungi and bacteria on contact [161].

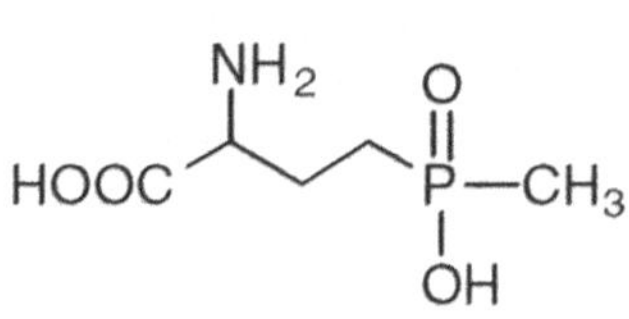

Fig. 3.26 Structure of glufosinate [159]

Toxicity on Human

The acute oral toxicity of glufosinate is low, and glufosinate is slightly more hazardous than glyphosate. Common signs of toxicity include CNS excitation and hypothermia in animals [162, 163]. Glufosinate ammonium formulation has been involved in a number of poisoning cases (cardiovascular and CNS adverse effects) possibly due to surfactant-induced penetration into the CNS [164]. The compound is not considered to be mutagenic, teratogenic, or carcinogenic, except in whole-embryo culture. Teratogenic effects in mice have been observed, resulting in apoptosis in the neuroepithelium of the developing embryo [165].

3.4.3 Ureas and Thioureas

Urea-derivates are considered as one of the most important herbicides together with the triazines and phenoxy derivatives [166]. Generally, the urea-derivates with a good herbicidal action are trisubstituted ureas, containing a free imino-hydrogen, with this hydrogen bonding being significant for their activity [166, 167]. Besides herbicidal action, some of the aliphatic guanidine derivatives have a fungicidal action and some of the aromatic thioureas a rodenticidal action. Some even have pharmaceutical properties as antidiabetic drugs (derivatives of sulfonylurea) [167]. The group of benzoylureas has shown insecticidal activity.

3.4.3.1 Chemical Aspects

Chemically, the urea-type herbicides contain a urea bridge substituted by triazine, benzothiazole, sulfonyl, phenyl, alkyl, or other moieties [166].

In the general structure of a phenylurea herbicide, the phenyl ring is frequently substituted with chlorine or bromine atoms, but methoxy, methyl, trifluoromethyl, or 2-propyl replacement is also possible (Fig. 3.27).

Sulphonylureas are a group of selective herbicides with R1–NH–C(O)–NH–SO2–R2 as general structure (Fig. 3.27). R1 and R2 are usually substituted by heterocyclic rings such as 4,6-dimethylpyrimidin-2-yl, and 2-(benzoic acid methyl ester) [168].

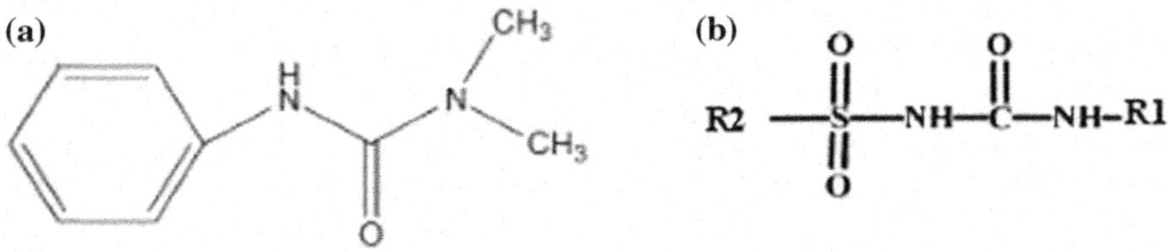

Fig. 3.27 Structure of phenylurea (**a**) and sulphonylurea (**b**) herbicides [169, 170]

3.4.3.2 Uses and Environmental Persistence

The phenylureas include diuron, fluometuron, and isoproturon, and they act as herbicides by inhibiting photosynthesis [169].

The sulfonylureas interfere with plant biosynthesis of certain amino acids by interfering with acetolactate synthase which is a key enzyme in such biosynthesis [170].

The benzoylureas are widely used insect growth regulators acting by inhibiting or blocking the synthesis of chitin [166, 171]. Diflubenzuron, the prototype of all benzoylurea insecticides, was discovered in 1972 [172].

Due to the extensive use of urea pesticides in agricultural practices, their residues have been detected in several environmental samples such as wastewater effluents, surface, and drinking water sources, as well as food items, worldwide [166, 173].

3.4.3.3 Toxicity on Humans

Particular interest is given to these family of compounds as some possible have carcinogenic properties [167].

Although several phenylureas are available in the market, the toxicity information available is still scarce. The acute toxicity potential for all phenylureas appears to be low, however, for high-level acute or chronic exposures, there have been reports on neurobehavioral alterations, body weight reductions, hematotoxicity, and hepatotoxicity [169]. For diuron a study, with the Comet assay, in MCF-7 and placental cells, it was observed an increased DNA fragmentation and several markers for cell stress were also increased in MCF-7. These results suggested a potential cytotoxic and genotoxic effect for diuron [174]. Moreover, it was observed that diuron can cross human placenta and may be metabolized to more toxic metabolites; this may cause fetotoxicity [175].

The sulfonylureas appear to have a quite low toxicity [12]. Concerning, benzoylureas in mammals seem to have a considerable antiproliferative and antitumor effects [176].

3.5 Conclusion

Pesticides are essential to maintaining agricultural and livestock production and the economy. However, some of these chemicals, by targeting systems or enzymes similar or even same present in other living beings, as humans and by persist in the environment or bioaccumulate, are a threat to the ecosystem. Insecticides common mode of action targets the nervous system, though skeletal muscles, digestive, cardiovascular, respiratory, ophthalmic, reproductive, endocrine, dermal, and immune systems can also be affected. Some insecticides have hazard effects in non-target species including humans, something even death. Herbicides can be more selective than insecticides, nonetheless, some of these are toxic to other species including human being.

The improper use and disposal are the primary sources of health problems from exposure. From the moment of its application, pesticides are distributed by the main ecosystems: atmosphere, water, soil, and sediments, as well as plants and animals. Furthermore, the ability of some herbicides and insecticides to bioaccumulate, resistance to degradation and biomagnification are the main aspects that the public should be aware and enforce a more consciously use of pesticides.

Acknowledgements S. Sousa, M. L. Maia, and V. C. Fernandes are grateful to FCT (Fundação para a Ciência e a Tecnologia) for the Ph.D. grant (SFRH/BD/137516/2018), Ph.D. grant (SFRH/BD/128817/2017) and Postdoc grant (SFRH/BPD/109153/2015), respectively. This work received financial support from the Fundação para a Ciência e a Tecnologia (FCT)/the Ministério da Ciência, Tecnologia e Ensino Superior (MCTES) through national funds (UID/QUI/50006/2019). The authors would also like to thank the EU and FCT for funding through the projects: FOODnanoHEALTH—Qualidade e Segurança Alimentar—uma abordagem (nano)tecnológica (Portugal2020, Norte-01-0145-FEDER-000011). The authors would also like to thank the EU and FCT/UEFISCDI, Romania/The Swedish Research Council for Environment, Agricultural Sciences and Spatial Planning (FORMAS), Sweden for funding, in the frame of the collaborative international consortium REWATER financed under the ERA-NET Cofund WaterWorks2015 Call. This ERA-NET is an integral part of the 2016 Joint Activities developed by the Water Challenges for a Changing World Joint Programme Initiative (Water JPI).

References

1. Gavrilescu M (2005) Fate of pesticides in the environment and its bioremediation. Eng Life Sci 5:497–526. https://doi.org/10.1002/elsc.200520098
2. Carson R (1962) Silent spring. Houghton Mifflin Company, Boston
3. Álvarez-Muñoz D, Llorca M, Blasco J, Barceló D (2016) Contaminants in the marine environment. In: Marine ecotoxicology. Elsevier, pp 1–34
4. Claver A, Ormad P, Rodríguez L, Ovelleiro JL (2006) Study of the presence of pesticides in surface waters in the Ebro river basin (Spain). Chemosphere 64:1437–1443. https://doi.org/10.1016/j.chemosphere.2006.02.034
5. Cruzeiro C, Rocha E, Pardal MÂ, Rocha MJ (2016) Seasonal-spatial survey of pesticides in the most significant estuary of the Iberian Peninsula—the Tagus River estuary. J Clean Prod 126:419–427. https://doi.org/10.1016/j.jclepro.2016.03.005
6. Nollet LML, Rathore HS (2010) Handbook of pesticides methods of pesticide residues analysis. CRC Press, Taylor & Francis Group
7. Au AM (2003) Pesticides and herbicides | Types, uses, and determination of herbicides. In: Encyclopedia of food sciences and nutrition. Elsevier, pp 4483–4487
8. Zimdahl RL (2018) Properties and uses of herbicides. In: Fundamentals of weed science. Elsevier, pp 463–499
9. Gupta PK (2014) Herbicides and fungicides. In: Biomarkers in toxicology. Elsevier, pp 409–431
10. Gupta PK (2011) Herbicides and fungicides. In: Reproductive and developmental toxicology. Elsevier, pp 503–521
11. Choudhury PP (2019) Transformation of herbicides in the environment. In: Herbicide residue research in India. Springer, Singapore, pp 415–442

12. Gupta PK (2018) Toxicity of herbicides. In: Veterinary toxicology, 3rd edn. Elsevier, pp 553–567
13. Kanawi E, Van Scoy AR, Budd R, Tjeerdema RS (2016) Environmental fate and ecotoxicology of propanil: a review. Toxicol Environ Chem 98:689–704. https://doi.org/10.1080/02772248.2015.1133816
14. Antonious GF (2010) Mobility and half-life of bensulide in agricultural soil. J Environ Sci Heal Part B Pestic Food Contam Agric Wastes 45:1–10. https://doi.org/10.1080/03601230903404283
15. Santos TCR, Rocha JC, Alonso RM, Marti E (1998) Rapid degradation of propanil in rice crop fields. Environ Sci Technol 32:3479–3484
16. Richardson R, Whaley CM, Wilson HP, Hines TE (2004) Weed control and potato (*Solanum tuberosum*) tolerance with dimethenamid isomers and other herbicides. Am J Pot Res 81:299. https://doi.org/10.1007/BF02870175
17. Glavaški OS, Petrović SD, Rajaković-Ognjanović VN et al (2016) Photodegradation of dimethenamid-P in deionised and ground water. Chem Ind Chem Eng Q 22:101–110. https://doi.org/10.2298/CICEQ150608025G
18. Liu J (2014) Propanil. Encycl Toxicol Third Ed 3:1092–1093. https://doi.org/10.1016/B978-0-12-386454-3.00187-1
19. Harp PR (2010) Dicamba. In: Hayes' handbook of pesticide toxicology. Elsevier, pp 1849–1852
20. Kavrakovski ZS, Rafajlovska VG (2015) Development and validation of thin layer chromatography method for simultaneous determination of seven chlorophenoxy and benzoic acid herbicides in water. J Anal Chem 70:995–1000. https://doi.org/10.1134/S1061934815080122
21. Fogarty AM, Tuovinen OH (1995) Microbiological degradation of the herbicide dicamba. J Ind Microbiol 14:365–370. https://doi.org/10.1007/BF01569952
22. Public Eye Brazil bans paraquat and the agribusiness lobby is gearing up for action. https://www.publiceye.ch/en/news/detail/brazil-bans-paraquat-and-the-agribusiness-lobby-is-gearing-up-for-action/. Accessed 30 Jan 2019
23. Pope CN (2005) Diquat. In: Encyclopedia of toxicology. Elsevier, pp 78–79
24. Baer KN (2005) Paraquat. In: Encyclopedia of toxicology. Elsevier, pp 329–330
25. Janz DM (2014) Dinitrophenols. In: Encyclopedia of toxicology. Elsevier, pp 177–178
26. Raman P (2005) Dinoseb. In: Encyclopedia of toxicology. Elsevier, pp 63–65
27. Badanthadka M, Mehendale HM (2014) Dinoseb. In: Encyclopedia of toxicology. Elsevier, pp 183–185
28. Young RA (2014) Dinitroanilines. Encycl Toxicol Third Ed 2:172–174. https://doi.org/10.1016/B978-0-12-386454-3.00670-9
29. Wallace DR (2014) Trifluralin. Encycl Toxicol Third Ed 4:846–848. https://doi.org/10.1016/B978-0-12-386454-3.00205-0
30. Hess FG, Harris JE, Pendino K, Ponnock K (2010) Imidazolinones. In: Hayes' handbook of pesticide toxicology. Elsevier, pp 1853–1863
31. Lamberth C, Dinges J (2012) Bioactive heterocyclic compound classes: agrochemicals. Wiley-VCH Verlag GmbH & Co. KGaA, Weinheim, Germany
32. Dube S, Lesoli MS, Fatunbi AO (2009) The efficacy and safety of bromacil based herbicide for the control of the invasive bush species in South Africa rangelands. Afr J Biotechnol 8:1776–1781
33. Caverly DJ, Denney RC (1977) Determination of residues of the herbicides bromacil, lenacil and terbacil in soils by gas chromatography. Analyst 102:576–579
34. Boydston RA, Al-Khatib K (2016) Terbacil and bromacil cross-resistance in powell amaranth (*Amaranthus powellii*). Weed Sci 40:513–516
35. Chelme-Ayala P, El-Din MG, Smith DW (2010) Degradation of bromoxynil and trifluralin in natural water by direct photolysis and UV plus H_2O_2 advanced oxidation process. Water Res 44:2221–2228. https://doi.org/10.1016/j.watres.2009.12.045

36. Kennepohl E, Munro IC, Bus JS (2010) Phenoxy herbicides (2,4-D). Hayes' Handb Pestic Toxicol 1829–1847. https://doi.org/10.1016/B978-0-12-374367-1.00084-7
37. Mei X-Y, Hong Y-Q, Chen G-H (2016) Review on analysis methodology of phenoxy acid herbicide residues. Food Anal Methods 9:1532–1561. https://doi.org/10.1007/s12161-015-0327-z
38. Tayeb W, Chaieb I, Hammami M (2011) Environmental fate and effects of 2,4-dichlorophenoxyacetic herbicide. In: Herbicides: properties, crop protection and environmental hazards. Nova Science Publishers
39. Sterling TD, Arundel AV (1986) Health effects of phenoxy herbicides. A review. Scand J Work Environ Health 12:161–173. https://doi.org/10.5271/sjweh.2160
40. Alexander BH, Mandel JS, Baker BA et al (2007) Biomonitoring of 2,4-dichlorophenoxyacetic acid exposure and dose in farm families. Environ Health Perspect 115:370–376. https://doi.org/10.1289/ehp.8869
41. Hatch EE, Nelson JW, Stahlhut RW, Webster TF (2010) Association of endocrine disruptors and obesity: perspectives from epidemiological studies. Int J Androl 33:324–332. https://doi.org/10.1111/j.1365-2605.2009.01035.x
42. Carbone V, Velkov T (2013) Interaction of phthalates and phenoxy acid herbicide environmental pollutants with intestinal intracellular lipid binding proteins. Chem Res Toxicol 26:1240–1250. https://doi.org/10.1021/tx400170t
43. Pochettino AA, Bongiovanni B, Duffard RO, Evangelista de Duffard AM (2013) Oxidative stress in ventral prostate, ovary, and breast by 2,4-dichlorophenoxyacetic acid in pre- and postnatal exposed rats. Environ Toxicol 28:1–10. https://doi.org/10.1002/tox.20690
44. Duke SO, Kenyon WH (1988) Chapter 2. Polycyclic alkanoic acids. In: Herbicides chemistry: degradation and mode of action
45. Hao G-F, Zuo Y, Yang S-G, Yang G-F (2011) Protoporphyrinogen oxidase inhibitor: an ideal target for herbicide discovery. Chim Int J Chem 65:961–969. https://doi.org/10.2533/chimia.2011.961
46. Dayan FE, Duke SO (2010) Protoporphyrinogen oxidase-inhibiting herbicides. In: Hayes' handbook of pesticide toxicology, 3rd edn. Elsevier, pp 1733–1751
47. Sartori SK, Alvarenga ES, Franco CA et al (2018) One-pot synthesis of anilides, herbicidal activity and molecular docking study. Pest Manag Sci 74:1637–1645. https://doi.org/10.1002/ps.4855
48. Chiriac CI, Onciu M, Tanasa F (2004) Synthesis of aromatic amides at room temperature using triphenyl phosphite-4-dimethylaminopyridine as reagent. Des Monomers Polym 7:331–335. https://doi.org/10.1163/1568555041475284
49. Busi R (2014) Resistance to herbicides inhibiting the biosynthesis of very-long-chain fatty acids. Pest Manag Sci 70:1378–1384. https://doi.org/10.1002/ps.3746
50. Christopher DN, Udu IA, Florence O et al (2013) Acute toxicity of the chloroacetanilide herbicide butachlor and its effects on the behavior of the freshwater fish *Tilapia zillii*. Afr J Biotechnol 12:499–503. https://doi.org/10.5897/AJB12.2433
51. Chang J, Liu S, Zhou S et al (2013) Effects of butachlor on reproduction and hormone levels in adult zebrafish (*Danio rerio*). Exp Toxicol Pathol 65:205–209. https://doi.org/10.1016/j.etp.2011.08.007
52. Geng B, Lin L, Zhang Q, Zhong B (2010) Genotoxicity of the pesticide dichlorvos and herbicide butachlor on Rana zhenhaiensis tadpoles. Asian Herpetol Res 1:118–122. https://doi.org/10.3724/SP.J.1245.2010.00118
53. Grube A, Donaldson D, Kiely T, Wu L (2011) Pesticides industry sales and usage: 2006 and 2007 market estimates. United States Environmental Protection Agency, Washington, DC
54. Counihan JL, Duckering M, Dalvie E et al (2017) Chemoproteomic profiling of acetanilide herbicides reveals their role in inhibiting fatty acid oxidation. ACS Chem Biol 12:635–642. https://doi.org/10.1021/acschembio.6b01001
55. Barr DB, Buckley B (2011) In vivo biomarkers and biomonitoring in reproductive and developmental toxicity. In: Reproductive and developmental toxicology. Elsevier, pp 253–265

56. Stevens J, Breckenridge C, Simpkins J, Eldridge J (2001) Symmetrical and asymmetrical triazine herbicides. In: Handbook of pesticide toxicology, 2nd edn. Elsevier, pp 1511–1519
57. Breckenridge CB, Werner C, Stevens JT, Sumner DD (2008) Hazard assessment for selected symmetrical and asymmetrical triazine herbicides. In: The triazine herbicides. Elsevier, pp 387–398
58. Fan AM (2014) Triazines. In: Encyclopedia of toxicology. Elsevier, pp 810–815
59. Armendáriz CR, de la Torre AH, Fernández ÁJG, González GL (2014) Metribuzin. In: Encyclopedia of toxicology. Elsevier, pp 327–329
60. Liu J (2014) Atrazine. In: Encyclopedia of toxicology. Elsevier, pp 336–338
61. Maples RD (2014) Propazine. In: Encyclopedia of toxicology. Elsevier, pp 1096–1098
62. Billington R, Gehen SC, Hanley TR (2010) Toxicology of triazolopyrimidine herbicides. In: Hayes' handbook of pesticide toxicology, 3rd edn. Elsevier, pp 1865–1885
63. Doble M, Kumar A (2005) Biodegradation of pesticides. In: Biotreatment of industrial effluents. Elsevier, pp 89–100
64. Gupta RC, Milatovic D (2014) Insecticides. In: Biomarkers in toxicology. Elsevier, pp 389–407
65. Zhang Y, Wu J, Xu W et al (2017) Cytotoxic effects of avermectin on human HepG2 cells in vitro bioassays. Environ Pollut 220:1127–1137. https://doi.org/10.1016/j.envpol.2016.11.022
66. Bai SH, Ogbourne S (2016) Eco-toxicological effects of the avermectin family with a focus on abamectin and ivermectin. Chemosphere 154:204–214. https://doi.org/10.1016/j.chemosphere.2016.03.113
67. Türkan F, Huyut Z, Atalar MN (2018) The toxicological impact of some avermectins on human erythrocytes glutathione S-transferase enzyme. J Biochem Mol Toxicol 32:e22205. https://doi.org/10.1002/jbt.22205
68. Awasthi A, Razzak M, Al-Kassas R et al (2012) An overview on chemical derivatization and stability aspects of selected avermectin derivatives. Chem Pharm Bull 60:931–944. https://doi.org/10.1248/cpb.c12-00258
69. Boukhrissa A, Ferrag-Siagh F, Rouidi L-M et al (2017) Study of the degradation in aqueous solution of a refractory organic compound: avermectin type used as pesticide in agriculture. Water Sci Technol 76:1966–1980. https://doi.org/10.2166/wst.2017.310
70. Thuan NH, Pandey RP, Sohng JK (2014) Recent advances in biochemistry and biotechnological synthesis of avermectins and their derivatives. Appl Microbiol Biotechnol 98:7747–7759. https://doi.org/10.1007/s00253-014-5926-x
71. Horsberg TE (2012) Avermectin use in aquaculture. Curr Pharm Biotechnol 13:1095–1102. https://doi.org/10.2174/138920112800399158
72. Zhang Y, Luo M, Xu W et al (2016) Avermectin confers its cytotoxic effects by inducing DNA damage and mitochondria-associated apoptosis. J Agric Food Chem 64:6895–6902. https://doi.org/10.1021/acs.jafc.6b02812
73. Yang C-C (2012) Acute human toxicity of macrocyclic lactones. Curr Pharm Biotechnol 13:999–1003. https://doi.org/10.2174/138920112800399059
74. Gupta P (2018) Pesticides (agrochemicals). In: Illustrated toxicology. Elsevier, pp 165–194
75. Shankland DL, Hollingworth RM, Smyth T (1978) Pesticide and venom neurotoxicity. Springer US, Boston, MA
76. Gao X, Tan Y, Guo H (2017) Simultaneous determination of amitraz, chlordimeform, formetanate and their main metabolites in human urine by high performance liquid chromatography–tandem mass spectrometry. J Chromatogr B 1052:27–33. https://doi.org/10.1016/j.jchromb.2017.03.004
77. Costa LG (2014) Chlordimeform. In: Encyclopedia of toxicology. Elsevier, pp 849–850
78. Hollingworth RM (1976) Chemistry, biological activity, and uses of formamidine pesticides. Environ Health Perspect 14:57–69. https://doi.org/10.1289/ehp.761457
79. Filazi A, Yurdakok-Dikmen B (2018) Amitraz. In: Veterinary toxicology, 3rd edn. Elsevier, pp 525–531

80. Pener MP, Dhadialla TS (2012) An overview of insect growth disruptors; applied aspects. In: Advances in insect physiology. Elsevier Science, pp 1–162
81. Karuppuchamy P, Venugopal S (2016) Integrated pest management. In: Ecofriendly pest management for food security. Elsevier, pp 651–684
82. Siddall JB (1976) Insect growth regulators and insect control: a critical appraisal. Environ Health Perspect 14:119. https://doi.org/10.2307/3428369
83. Lee S-H, Oh H-W, Fang Y et al (2015) Identification of plant compounds that disrupt the insect juvenile hormone receptor complex. Proc Natl Acad Sci 112:1733–1738. https://doi.org/10.1073/pnas.1424386112
84. Subramanian S, Shankarganesh K (2016) Insect hormones (as pesticides). In: Ecofriendly pest management for food security. Elsevier, pp 613–650
85. Kennepohl E, Munro IC, Bus JS (2010) Phenoxy herbicides (2,4-D). In: Hayes' handbook of pesticide toxicology. Elsevier, pp 1829–1847
86. Merzendorfer H (2013) Chitin synthesis inhibitors: old molecules and new developments. Insect Sci 20:121–138. https://doi.org/10.1111/j.1744-7917.2012.01535.x
87. Sun R, Liu C, Zhang H, Wang Q (2015) Benzoylurea chitin synthesis inhibitors. J Agric Food Chem 63:6847–6865. https://doi.org/10.1021/acs.jafc.5b02460
88. Monteiro JP, Jurado AS (2014) Methoprene. In: Encyclopedia of toxicology. Elsevier, pp 246–249
89. Ensley SM (2018) Neonicotinoids. In: Veterinary toxicology. Elsevier, pp 521–524
90. Bass C, Field LM (2018) Neonicotinoids. Curr Biol 28:R772–R773. https://doi.org/10.1016/j.cub.2018.05.061
91. Butcherine P, Benkendorff K, Kelaher B, Barkla BJ (2019) The risk of neonicotinoid exposure to shrimp aquaculture. Chemosphere 217:329–348. https://doi.org/10.1016/j.chemosphere.2018.10.197
92. Han W, Tian Y, Shen X (2018) Human exposure to neonicotinoid insecticides and the evaluation of their potential toxicity: an overview. Chemosphere 192:59–65. https://doi.org/10.1016/j.chemosphere.2017.10.149
93. Goulson D (2013) REVIEW: an overview of the environmental risks posed by neonicotinoid insecticides. J Appl Ecol 50:977–987. https://doi.org/10.1111/1365-2664.12111
94. Auteri D, Arena M, Barmaz S et al (2017) Neonicotinoids and bees: the case of the European regulatory risk assessment. Sci Total Environ 579:966–971. https://doi.org/10.1016/j.scitotenv.2016.10.158
95. Wang Z, Brooks B, Zeng EY, You J (2019) Comparative mammalian hazards of neonicotinoid insecticides among exposure durations. Environ Int 125:9–24. https://doi.org/10.1016/j.envint.2019.01.040
96. Mansouri A, Cregut M, Abbes C et al (2017) The environmental issues of DDT pollution and bioremediation: a multidisciplinary review. Appl Biochem Biotechnol 181:309–339. https://doi.org/10.1007/s12010-016-2214-5
97. Tsai WT (2010) Current status and regulatory aspects of pesticides considered to be persistent organic pollutants (POPs) in Taiwan. Int J Environ Res Public Health 7:3615–3627. https://doi.org/10.3390/ijerph7103615
98. Shen L, Wania F (2005) Compilation, evaluation, and selection of physical-chemical property data for organochlorine pesticides. J Chem Eng Data 50:742–768. https://doi.org/10.1021/je049693f
99. Jayaraj R, Megha P, Sreedev P (2016) Review article. Organochlorine pesticides, their toxic effects on living organisms and their fate in the environment. Interdiscip Toxicol 9:90–100. https://doi.org/10.1515/intox-2016-0012
100. Al-Malki AL, Moselhy SS (2011) Impact of pesticides residue and heavy metals on lipids and fatty acids composition of some seafoods of Red Sea (KSA). Hum Exp Toxicol 30:1666–1673. https://doi.org/10.1177/0960327110396535
101. Ali U, Syed JH, Malik RN et al (2014) Organochlorine pesticides (OCPs) in South Asian region: a review. Sci Total Environ 476–477:705–717. https://doi.org/10.1016/j.scitotenv.2013.12.107

102. Aksoy A, Guvenc D, Yavuz O et al (2012) Seasonal variation of polychlorinated biphenyls and organochlorine pesticide levels of sea and cultured farm fish in the Samsun Region of Turkey. Bull Environ Contam Toxicol 88:842–849. https://doi.org/10.1007/s00128-012-0561-7
103. Choi M, Lee I-S, Jung R-H (2016) Rapid determination of organochlorine pesticides in fish using selective pressurized liquid extraction and gas chromatography–mass spectrometry. Food Chem 205:1–8. https://doi.org/10.1016/j.foodchem.2016.02.156
104. Correia-Sá L, Fernandes VC, Carvalho M et al (2012) Optimization of QuEChERS method for the analysis of organochlorine pesticides in soils with diverse organic matter. J Sep Sci 35:1521–1530. https://doi.org/10.1002/jssc.201200087
105. Fernandes VC, Domingues VF, Mateus N, Delerue-Matos C (2011) Organochlorine pesticide residues in strawberries from integrated pest management and organic farming. J Agric Food Chem 59:7582–7591. https://doi.org/10.1021/jf103899r
106. Correia-Sá L, Fernandes VC, Calhau C et al (2013) Optimization of QuEChERS procedure coupled to GC-ECD for organochlorine pesticide determination in carrot samples. Food Anal Methods 6:587–597. https://doi.org/10.1007/s12161-012-9463-x
107. Maia ML, Correia-Sá L, Coelho A et al (2015) *Eruca sativa*: benefits as antioxidants source versus risks of already banned pesticides. J Environ Sci Heal Part B Pestic Food Contam Agric Wastes 50. https://doi.org/10.1080/03601234.2015.1000178
108. Maia ML, Correia-Sá L, Sousa S et al (2017) Organochlorine pesticides in seafood: occurrence and risk assessment. In: Advances in medicine and biology, pp 175–222
109. Ferrante MC, Cirillo T, Naso B et al (2007) Polychlorinated biphenyls and organochlorine pesticides in seafood from the Gulf of Naples (Italy). J Food Prot 70:706–715
110. Kafkas E, Daglioglu N, Yarpuz-Bozdogan N, Zarifikhosroshahi M (2019) Pesticide analysis techniques, limitations, and applications. In: Wani KA, Mamta (eds) Handbook of research on the adverse effects of pesticide pollution in aquatic ecosystems. IGI Global, Hershey, PA, USA, pp 301–317
111. Fernandes VC, Pestana D, Monteiro R et al (2012) Optimization and validation of organochlorine compounds in adipose tissue by SPE-gas chromatography. Biomed Chromatogr 26:1494–1501. https://doi.org/10.1002/bmc.2723
112. Mostafalou S, Abdollahi M (2017) Pesticides: an update of human exposure and toxicity. Arch Toxicol 91:549–599. https://doi.org/10.1007/s00204-016-1849-x
113. Starek-Świechowicz B, Budziszewska B, Starek A (2017) Hexachlorobenzene as a persistent organic pollutant: toxicity and molecular mechanism of action. Pharmacol Rep 69:1232–1239. https://doi.org/10.1016/j.pharep.2017.06.013
114. Xiao X, Clark JM, Park Y (2017) Potential contribution of insecticide exposure and development of obesity and type 2 diabetes. Food Chem Toxicol 105:456–474. https://doi.org/10.1016/j.fct.2017.05.003
115. Bragança I, Domingues V, Lemos P, Delerue-Matos C (2016) Biodegradation of pyrethroid pesticides. In: Soil remediation. CRC Press, pp 59–74
116. Bragança I, Lemos PC, Delerue-Matos C, Domingues VF (2019) Pyrethroid pesticide metabolite, 3-PBA, in soils: method development and application to real agricultural soils. Environ Sci Pollut Res 26:2987–2997. https://doi.org/10.1007/s11356-018-3690-7
117. Tang W, Wang D, Wang J et al (2018) Pyrethroid pesticide residues in the global environment: an overview. Chemosphere 191:990–1007. https://doi.org/10.1016/j.chemosphere.2017.10.115
118. Ye J, Zhao M, Liu J, Liu W (2010) Enantioselectivity in environmental risk assessment of modern chiral pesticides. Environ Pollut 158:2371–2383. https://doi.org/10.1016/j.envpol.2010.03.014
119. Palmquist K, Salatas J, Fairbrother A (2012) Pyrethroid insecticides: use, environmental fate, and ecotoxicology. In: Insecticides—advances in integrated pest management. InTech
120. Feo ML, Eljarrat E, Barceló D, Barceló D (2010) Determination of pyrethroid insecticides in environmental samples. TrAC Trends Anal Chem 29:692–705. https://doi.org/10.1016/j.trac.2010.03.011

121. Antwi FB, Reddy GVP (2015) Toxicological effects of pyrethroids on non-target aquatic insects. Environ Toxicol Pharmacol 40:915–923. https://doi.org/10.1016/j.etap.2015.09.023
122. Domingues V, Cabral M, Alves A, Delerue-Matos C (2009) Use and reuse of SPE disks for the determination of pyrethroids in water by GC-ECD. Anal Lett 42:706–726. https://doi.org/10.1080/00032710902721949
123. Ensley SM (2012) Pyrethrins and pyrethroids. In: Veterinary toxicology. Elsevier, pp 591–595
124. Ensley SM (2018) Pyrethrins and pyrethroids. In: Veterinary toxicology. Elsevier, pp 515–520
125. Koureas M, Tsakalof A, Tsatsakis A, Hadjichristodoulou C (2012) Systematic review of biomonitoring studies to determine the association between exposure to organophosphorus and pyrethroid insecticides and human health outcomes. Toxicol Lett 210:155–168. https://doi.org/10.1016/j.toxlet.2011.10.007
126. Saillenfait A-M, Ndiaye D, Sabaté J-P (2015) Pyrethroids: exposure and health effects—an update. Int J Hyg Environ Health 218:281–292. https://doi.org/10.1016/j.ijheh.2015.01.002
127. Tyler CR, Beresford N, van der Woning M et al (2000) Metabolism and environmental degradation of pyrethroid insecticides produce compounds with endocrine activities. Environ Toxicol Chem 19:801–809. https://doi.org/10.1002/etc.5620190404
128. Yoo M, Lim Y-H, Kim T et al (2016) Association between urinary 3-phenoxybenzoic acid and body mass index in Korean adults: 1st Korean National Environmental Health Survey. Ann Occup Environ Med 28:2. https://doi.org/10.1186/s40557-015-0079-7
129. Burns CJ, Pastoor TP (2018) Pyrethroid epidemiology: a quality-based review. Crit Rev Toxicol 48:297–311. https://doi.org/10.1080/10408444.2017.1423463
130. Tietze LF, Schützenmeister N, Grube A et al (2012) Synthesis of spinosyn analogues for modern crop protection. Eur J Org Chem 2012:5748–5756. https://doi.org/10.1002/ejoc.201200600
131. Bacci L, Lupi D, Savoldelli S, Rossaro B (2016) A review of spinosyns, a derivative of biological acting substances as a class of insecticides with a broad range of action against many insect pests. J Entomol Acarol Res 48:40. https://doi.org/10.4081/jear.2016.5653
132. Huang K, Xia L, Zhang Y et al (2009) Recent advances in the biochemistry of spinosyns. Appl Microbiol Biotechnol 82:13–23. https://doi.org/10.1007/s00253-008-1784-8
133. Biondi A, Mommaerts V, Smagghe G et al (2012) The non-target impact of spinosyns on beneficial arthropods. Pest Manag Sci 68:1523–1536. https://doi.org/10.1002/ps.3396
134. Kirst HA (2010) The spinosyn family of insecticides: realizing the potential of natural products research. J Antibiot (Tokyo) 63:101–111. https://doi.org/10.1038/ja.2010.5
135. Zhao C, Huang Y, Guo C et al (2017) Heterologous expression of spinosyn biosynthetic gene cluster in *Streptomyces* species is dependent on the expression of rhamnose biosynthesis genes. J Mol Microbiol Biotechnol 27:190–198. https://doi.org/10.1159/000477543
136. Su T, Lin J, Lin-Tan D-T et al (2011) Human poisoning with spinosad and flonicamid insecticides. Hum Exp Toxicol 30:1878–1881. https://doi.org/10.1177/0960327111401639
137. Rossi M, Rotilio D (1997) Analysis of carbamate pesticides by micellar electrokinetic chromatography. HRC J High Resolut Chromatogr 20:265–269. https://doi.org/10.1002/jhrc.1240200505
138. Gupta RC (2014) Carbamate pesticides, 3rd edn. Elsevier
139. Machemer LH, Pickel M (1994) Chapter 16. Carbamate herbicide and fungicides. Toxicology 91:105–109. https://doi.org/10.1016/0300-483X(94)90249-6
140. Moreland DE (1999) Biochemical mechanisms of action of herbicides and the impact of biotechnology on the development of herbicides. J Pestic Sci 24:299–307. https://doi.org/10.1584/jpestics.24.299
141. Morais S, Dias E, Pereira MDL (2012) Carbamates: human exposure and health effects. Impact Pestic 21–38
142. Matsumura F (1985) Toxicology of insecticides. Springer US, Boston, MA

143. Gupta RC, Sachana M, Mukherjee IM et al (2018) Organophosphates and carbamates, 3rd edn. Elsevier Inc.
144. Gallo MA, Lawryk NJ (1991) Organic phosphorus pesticides. In: Hayes WJ, Laws ER (eds) Handbook of pesticide toxicology. Academic Press, New York
145. Herrero-Hernández E, Andrades MS, Álvarez-Martín A et al (2013) Occurrence of pesticides and some of their degradation products in waters in a Spanish wine region. J Hydrol 486:234–245. https://doi.org/10.1016/j.jhydrol.2013.01.025
146. Vermeire T, McPhail R, Waters M. Organophosphorous pesticides in the environment. Toxicology advisory centre national institute of public health and the environment, the Netherlands and national health and environmental effects research laboratory office of research and development environmental protection agency, USA
147. Beyond Pesticides (2015) Health effects of 30 commonly used lawn pesticides. Archived (PDF) from the original on 2011-11-12
148. Centers for disease control and prevention organophosphates. https://www.cdc.gov/nceh/clusters/Fallon/organophosfaq.htm. Accessed 23 Jan 2019
149. Eddleston M, Buckley NA, Eyer P, Dawson AH (2008) Management of acute organophosphorus pesticide poisoning. Lancet 371:597–607. https://doi.org/10.1016/S0140-6736(07)61202-1
150. Jokanović M, Kosanović M (2010) Neurotoxic effects in patients poisoned with organophosphorus pesticides. Environ Toxicol Pharmacol 29:195–201. https://doi.org/10.1016/j.etap.2010.01.006
151. Van Bruggen AHC, He MM, Shin K et al (2018) Environmental and health effects of the herbicide glyphosate. Sci Total Environ 616–617:255–268. https://doi.org/10.1016/j.scitotenv.2017.10.309
152. Sidoli P, Baran N, Angulo-Jaramillo R (2016) Glyphosate and AMPA adsorption in soils: laboratory experiments and pedotransfer rules. Environ Sci Pollut Res 23:5733–5742. https://doi.org/10.1007/s11356-015-5796-5
153. Battaglin WA, Meyer MT, Kuivila KM, Dietze JE (2014) Glyphosate and its degradation product AMPA occur frequently and widely in U.S. soils, surface water, groundwater, and precipitation. JAWRA J Am Water Resour Assoc 50:275–290. https://doi.org/10.1111/jawr.12159
154. Prasad S, Srivastava S, Singh M, Shukla Y (2009) Clastogenic effects of glyphosate in bone marrow cells of Swiss albino mice. J Toxicol 2009:1–6. https://doi.org/10.1155/2009/308985
155. Baran MJ, Cibiński M (2014) Glyphosate-based phospho-organic herbicides—an outline of action, metabolism and the selected effects on humans and other organisms. Arch Physiother Glob Res 18:35–45. https://doi.org/10.15442/apgr.19.2.10
156. Zarzycki PK, Włodarczyk E, Baran MJ (2009) Determination of endocrine disrupting compounds using temperature-dependent inclusion chromatography. J Chromatogr A 1216:7612–7622. https://doi.org/10.1016/j.chroma.2009.03.066
157. Worthing CR, Walker SB, British Crop Protection Council (1983) The pesticide manual: a world compendium, 7th edn. British Crop Protection Council, Croydon
158. Donn G, Köcher H (2002) Inhibitors of glutamine synthetase. In: Herbicide classes in development. Springer, Berlin, Heidelberg, pp 87–101
159. Nagatomi Y, Yoshioka T, Yanagisawa M et al (2013) Simultaneous LC-MS/MS analysis of glyphosate, glufosinate, and their metabolic products in beer, barley tea, and their ingredients. Biosci Biotechnol Biochem 77:2218–2221. https://doi.org/10.1271/bbb.130433
160. Duke SO (2005) Taking stock of herbicide-resistant crops ten years after introduction. Pest Manag Sci 61:211–218. https://doi.org/10.1002/ps.1024
161. Duke SO (2014) Biotechnology: herbicide-resistant crops. In: Encyclopedia of agriculture and food systems. Elsevier, pp 94–116
162. Ebert E, Leist K-H, Mayer D (1990) Summary of safety evaluation toxicity studies of glufosinate ammonium. Food Chem Toxicol 28:339–349. https://doi.org/10.1016/0278-6915(90)90108-Y

163. Hack R, Ebert E, Ehling G, Leist K-H (1994) Glufosinate ammonium—some aspects of its mode of action in mammals. Food Chem Toxicol 32:461–470. https://doi.org/10.1016/0278-6915(94)90043-4
164. Watanabe T, Sano T (1998) Neurological effects of glufosinate poisoning with a brief review. Hum Exp Toxicol 17:35–39. https://doi.org/10.1177/096032719801700106
165. Watanabe T (1997) Apoptosis induced by glufosinate ammonium in the neuroepithelium of developing mouse embryos in culture. Neurosci Lett 222:17–20. https://doi.org/10.1016/S0304-3940(97)13330-4
166. Morais S, Correia M, Domingues V, Delerue-Matos C (2011) Urea pesticides. In: Pesticides—strategies for pesticides analysis. InTech
167. Lányi K, Dinya Z (2005) Photodegradation study for assessing the environmental fate of some triazine-, urea- and thiolcarbamate-type herbicides. Microchem J 80:79–87. https://doi.org/10.1016/j.microc.2004.12.001
168. Niessen WMA (2010) Group-specific fragmentation of pesticides and related compounds in liquid chromatography–tandem mass spectrometry. J Chromatogr A 1217:4061–4070. https://doi.org/10.1016/j.chroma.2009.09.058
169. Liu J (2010) Phenylurea herbicides. In: Hayes' handbook of pesticide toxicology. Elsevier, pp 1725–1731
170. Duggleby RG, McCourt JA, Guddat LW (2008) Structure and mechanism of inhibition of plant acetohydroxyacid synthase. Plant Physiol Biochem 46:309–324. https://doi.org/10.1016/j.plaphy.2007.12.004
171. Yang M, Hong K, Li X et al (2017) Freezing temperature controlled deep eutectic solvent dispersive liquid–liquid microextraction based on solidification of floating organic droplets for rapid determination of benzoylureas residual in water samples with assistance of metallic salt. RSC Adv 7:56528–56536. https://doi.org/10.1039/C7RA11030H
172. Matsumura F (2010) Studies on the action mechanism of benzoylurea insecticides to inhibit the process of chitin synthesis in insects: a review on the status of research activities in the past, the present and the future prospects. Pestic Biochem Physiol 97:133–139. https://doi.org/10.1016/j.pestbp.2009.10.001
173. Bolte M (2004) Environmental fate of pollutants: example of phenylurea pesticides. Actual Chim 33–39
174. Huovinen M, Loikkanen J, Naarala J, Vähäkangas K (2015) Toxicity of diuron in human cancer cells. Toxicol In Vitro 29:1577–1586. https://doi.org/10.1016/j.tiv.2015.06.013
175. Mohammed AM, Karttunen V, Huuskonen P et al (2018) Transplacental transfer and metabolism of diuron in human placenta. Toxicol Lett 295:307–313. https://doi.org/10.1016/j.toxlet.2018.07.012
176. Huang Q, Wu X, Yu X et al (2015) Comparison of the cytotoxic impact of chlorfluazuron on selected insect and human cell lines. Environ Toxicol Chem 34:1675–1682. https://doi.org/10.1002/etc.2969

Chapter 4
Controlled Release Pesticides as a Route to Sustainable Crop Production

Maya Rajan, Vinaya Chandran, S. Shahena and Linu Mathew

Abstract Pesticides are substances which prevent the growth of unwanted organisms. They are formulated in the form of powders, aqueous solutions, aerosols or spray formulations. But these have shown varying degrees of health hazards and environmental problems when used in excess. Controlled release of pesticide can be developed as a method to reduce the pesticide overuse. Controlled release of pesticides helps in reducing the pesticide concentrations and prolonging the half-life of pesticides. Controlled release formulations contain pesticides or their active ingredients incorporated into a polymer matrix by chemical or physical methods. The compound is released either by cleavage of chemical bond or by physical degeneration of the polymer matrix. Microencapsulation, microcapsules and hydrogels polymers are extensively used for controlled release. Biodegradable microbial polyesters like poly hydroxyl alkanoates are also used as a carrier for pesticides. This technology provides a viable alternative to pesticide overdose by improving the efficiency of already existing pesticides and thereby reducing the environmental problems. This application has a major role in maintaining the sustainability of agricultural practices.

Keywords Pesticides · Biological control · 2,4-D · DDT · PHA

4.1 Introduction

Pesticide is a broader term including all those substances or mixtures of compounds which are used to kill or controls pests includes insects, fungi, bacteria and weeds. Use of pesticides has its origins in ancient Rome, but their extensive use started after the Second World War by the introduction of dichloro-diphenyl-trichloroethane (DDT), beta-hexachlorocyclohexane (BHC), Aldrin and 2,4-D. FAO defines that pesticides are "any substances or mixture of substances for

M. Rajan · V. Chandran · S. Shahena · L. Mathew (✉)
School of Biosciences, Mahatma Gandhi University, Kottayam, Kerala, India
e-mail: linumathew@mgu.ac.in

Rakhimol K. R. et al. (eds.), *Controlled Release of Pesticides for Sustainable Agriculture*, https://doi.org/10.1007/978-3-030-23396-9_4

preventing, repelling, destroying pests including vectors of humans, animal disease, unwanted species of plants or animals causing harms during production, processing, storage, transport or marketing of food and agricultural commodities". Pesticides can be differentiated into different classes based on their targeting organisms (pesticides, fungicides and herbicides), chemical structure (organochlorines, organophosphates, carbamates and phenoxy acids) and physical states (solid, liquid, aerosol). On the basis of their chemical structure, pesticides can again be classified into many groups.

They are as follows:

1. Organochlorines
2. Organophosphates
3. Carbamic and thiocarbamic derivatives
4. Carboxylic acids and their derivatives
5. Urea derivatives
6. Heterocyclic compounds
7. Phenol and nitrophenol derivatives
8. Hydrocarbons, ketones, aldehydes and their derivatives
9. Fluorine-containing compounds
10. Copper-containing compounds
11. Metal organic and inorganic compounds
12. Natural and synthetic pyrethroids and others.

4.2 Biological Pesticides

In insects, certain compounds called semio-chemicals causes behavioural changes. These chemicals are secreted from the exocrine glands of insects. These secreted compounds are used by insects for finding their mates, alarm signalling, defence mechanisms, etc. [1]. One example of such pheromones is antifeedants which inhibited food intake activities. Normally GABA receptors are the receptor which may responsible for neurotransmitter gamma aminobutyric acid. Insect taste sensitivity is controlled by this receptor. So, this antifeedant works against these receptors [2]. Antifeedants are useful for controlling various pest replacing toxic pesticides [3, 4]. *Hylobiusabietis* (pine weevil) causes a lot of damage in conifer forests in Europe and North America. A molecule called BHT (butylated hydroxytoluene) protects pine stem against pine weevil [5–7]. Earlier pyrethroids like permethrin have been applied to reduce the pest attacks. But its usage has detrimental effects on aquatic organisms and hence the use of this compound was aborted. Cypermethrin is a synthetic analogue used as an alternative pesticide. But this has toxicity to aquatic and terrestrial insects as well as mammals [8]. Plant-derived carboxylic acids are found highly active against pine weevil.

Examples of these plant-derived carboxylic acids are hexanoic acid, esters of 2,4- and 3, 5-dimethoxy benzoic acids [9]. Simmondsin and simmondsin 2-ferulate are the two glucosides, derived from jojoba seed with antifeedant activity against *Spodopteralittoralis* an agricultural pest [10].

4.3 Benefits and Hazards of Pesticides

There are primary and secondary benefits of pesticides: primary benefits include protection of human, animal and crop health. Secondary benefits arise from primary benefits, and these are less immediate or have longer-term consequences. The relation between negative effects of pesticides and their benefit is complex and not easy to follow.

Primary benefits of pesticides include (1) Controlling pest and plant disease vectors, (2) Controlling disease vectors and noxious organisms and (3) Preventing or controlling of organisms that harm human activities. Secondary benefits are the long-term benefits to the community, nation and world as a whole.

Pesticide use causes a number of hazards including environmental problems and human and animal health hazards. Food can also be contaminated with these toxic pesticides causing many diseases. 98% of pesticides and 95% of herbicides reach a destination other than their target species including air, water and soil [11]. Prolonged use of these chemicals results in various diseases such as neurological changes, alterations in psychology, hormonal imbalance, immunity-related diseases, infertility problems, cancers, etc. The usage of pesticides can also contaminate water, soil and other non-targeted vegetation. Pesticides cannot only kill pests, weeds and other insects but also kill beneficial organisms including birds, fish, beneficial insects and non-targeted plants. Recent reviews report toxicological and epidemiological evidences for various health effects associated with pesticides to which general population may be chronically exposed. They are potential carcinogens, neurotoxins and immunotoxins [12–14]. According to González [15], the exposure of 2,4-D showed DNA damage in Chinese hamster ovary cells (CHO). Pesticide contamination on both ground and surface water can affect both aquatic fauna and flora as well as affect human health when this water is consumed by public [16].

Hence, search for safer analogues of pesticides of natural origin is in the rise. Safer alternatives to pesticides include use of biological pest controls, plant genetic engineering, maintaining good crop hygiene, cultivation practices interfering with insect breeding, release of organisms that fight pest, soil steaming, etc. These alternatives are usually safer than traditional chemical pesticides.

4.4 Chemical Pesticides Versus Biological Control in Farming Practices

Pesticides in general are highly effective to kill or control pests. Nevertheless, even by applying pesticides at very high dose, not all pests are killed. Some insects with a slight variation in their genetic make-up will become resistant and survive. Gradually, a population with pesticide resistance will develop and farmers will be forced to apply higher doses and more toxic alternatives to combat the resistant pest species. The insecticide resistance will spread throughout the ecosystem at very fast rate. Pesticide resistance and pest resurgence to epidemic proportion are often a problem faced by farming communities.

Persistent organic pollutants are a serious threat to the environment, under which pesticides feature a major role. They continue to poison non-target organisms in the environment and to increase the human health hazards. The main advantage of pesticides is that they are readily available and easy to use, unlike the alternative methods such as biological control which may produce no immediate effect on pests. Unlike chemicals, biological control reduces but does not eradicate pest populations completely. Therefore, a successful programme relies on finding the correct balance between the level of suppression of pest and also level of how much a user is prepared to tolerate the pest infection. Biocontrol is most effective against pests with high tolerance threshold, i.e. a pest that defoliates plants but does not much reduce crop yield. Chemical pesticides are very effective, if used safely, and they are useful for those pests with low threshold and those which directly affect the crop field. Biological control is more effective against pests that are tolerable enough so that it only needs to be suppressed and not destroyed entirely (high threshold pests). Hence, chemical pesticides can be termed as necessary evils in agriculture.

4.5 Controlled Release of Pesticides as a Route to Sustainable Crop Production

Controlled release technology ultimately aims at effective pesticide treatment, avoidance of side effects, prolonging efficiency and boosting the safety and reliability [17, 18]. When applied in the field, the chemical pesticides are invariably subjected to leaching, evaporation and degradation which will remove them before they reach the target and perform their intended actions. This prompts the farmer to increase the pesticide dosage leading to heavier damage to the ecosystem and development of pesticide resistance. A safer alternative is the use of a persistent analogue. But this again acts as a "Persistent organic pollutant", and therefore, key invention to reduce pesticide use is the design and development of sustainable delivery devices and formulations.

Controlled release technology is an alternative method to avoid problems accompanying the use of pesticides that causes side effects [19]. Controlled release technology was developed to protect store grains against insects and rodent pests [20]. Overall aim is to protect the active ingredient and to allow its automatic delivery to the target at controlled rates for maintaining optimum concentration over a long period of time [21]. Controlled release is a method in which active chemicals are made to act at a specified target at a rate and duration designed to accomplish the intended effect [22].

There are many potent novel pesticides and agrochemicals designed to control pests. The usage of these agents produces inefficient biological response because of their inability to deliver these agents to their targets at precise time and optimum quantity. The term controlled release means "a combination of biologically active agents and excipient normally a polymer which allows the agent delivered to the target at controlled rates over a specified period". The main advantages of controlled release of pesticides are that it allows slow dosing of pesticides without affecting its duration of activity. It also comes into aid when normal half-life of a potent pesticide is short when applied by conventional methods. In conventional methods, pesticides are subject to evaporation, degradation, etc., all of which remove their active material from target before they can perform their functions. Controlled release of polymer system can be divided into two classes: first case is when the agent is dissolved, dispersed and encapsulated within in the matrix. The release of these components takes place through diffusion or after biological or chemical breakdown of the polymer. Secondly, this active compound either constitutes a part of the macromolecular backbone or it may be attached to it so that the release is obtained by the cleavage of chemical or biological bond between polymer and the biological agent [23].

4.6 Bioengineering Aspects of Controlled Release of Pesticides

Agriculture is highly depended on synthetic pesticides. From the agrochemicals that are used in crop protection, by a rough estimation, only 0.1% reach target pests. The rest remains and enters the environment and causes certain hazards to non-target organisms [1] and are associated with groundwater contamination, vitalization and degradation [24]. In order to overcome these limitations, controlled release technology can be applied. Folkman and Long invented the membrane diffusion device with silicone rubber in 1964 to control the release of drug. Alex Zaffaroni continued their research with novel control release drug delivery system in the 1960s. In the 1970s, controlled release technology was extended to agrochemical, cosmetic and food industries [25–27].

On the basis of the application method, controlled release mechanisms are classified as follows: (1) chemically controlled system (2) diffusion controlled system. In chemically controlled system, release normally occurs either by a bio-erodible system or by a drug-polymer conjugate system. In case of diffusion controlled system, the release may occur from membrane reservoir systems and matrix systems by matrix diffusion, polymer erosion, polymer swelling, etc. [28]. This technology helps to deliver the active components to the target without releasing too little or too much so that over release can be suppressed.

The release of pesticides in controlled systems happens by releasing their biological constituents into their environment over a defined period of time. In this system, substances or active material are incorporated into a polymer matrix which is chemically bound or physically incorporated by means of different techniques. So, the release of compound occurs either by bond cleavage by chemical reactions or by physical transport process in and through polymers which can be modelled mathematically [29].

Controlled release technology provides a promising effort for improving the efficiency of some already existing pesticides and reducing the environmental problems. In this system, the crucial step is identifying the correct matrix which combines with pesticides capable of altering its availability to the target organisms. The greatest challenge of this technique is changing the environmental behaviour of pesticides. Certain environmental parameters are volatility, leaching, microbial metabolisms, etc. Laboratory and field techniques are developed for measuring some of these parameters, affecting pesticide stability [30].

Many pesticides such as chlorinated hydrocarbons face some environmental and toxicological issues. Many a times, less persistent and thus environmentally safer alternative pesticides result in acute toxicity and many other side effects. There are greater chances of high exposure of these pesticides to operators and farm workers which can be solved by controlled release formulation of these pesticides. Another aspect is that one of the insect growth regulators, for example, methoprene is unstable in aquatic environment and its application could only be effected as a controlled release formulation [31]. This technique is based on the concept of combining biologically active substances with polymeric materials either by physical mixing or by chemical attachment that act as a carrier with subsequent chemical cleavage.

a. Physical mixtures: There are two different approaches in physically combining biologically active materials with polymeric materials. Firstly, the biological agent can be encapsulated in polymeric materials in which active agent that can be released on the basis of Fick's law of diffusion through micropores in the capsule wall. In the second method, the biologically active agent which is dissolved in solid polymeric matrix can be either degradable or biodegradable to release the active ingredient [32]. The release of active agent is normally controlled by diffusion phenomena through the matrix by either chemical or biological erosion. In most cases, the polymer should have properties such as compatibility with the active agent so that it can be readily dissolved to avoid physical interactions and low softening point so as to avoid thermal degradation of active agent. Crystallisation of the polymer should be avoided and also polymer should be stable and of low cost.

b. Chemical attachment: In this type, the active agent is chemically bound by natural or synthetic polymeric material by a chemical bond. Biologically active agent having at least one reactive functional group can be used as a link to the functionalized polymer which is used in this technique [32]. Two synthetic methods are used for the preparation of polymeric chemically bonded active agent. In the first method, preformed polymer with desired active agent can be chemically modified through chemical bonds leading to a polymer having active species linked to the main chain as a pendant group. The second method involves the polymerization of biologically active monomers so that polymers having active group exist as repeat units in the main backbone. For example, copolymers containing pendant pentachlorophenol prepared by free radical co-polymerization of corresponding monomer. The advantage of polymerization is the ability to control the molecular weight of the polymer, hydrological and hydrophilic balance of the polymer and the group distribution along the backbone.

When compared to chemical pesticides, biological pesticides help to increase the efficiency and also are eco-friendly. For example, avermectin is a biopesticide possesses certain advantages over chemical pesticides [33]. Nowadays, the development of nanotechnology helps in improving the development of conventional pesticides by constructing nano-delivery system using nanoparticles as a carrier molecule [33, 34]. Many studies suggested that nanoparticles used as a carrier molecule improve the release time of drug molecules and improve the drug bioavailability. Pesticide nano-delivery system provides with great potential to control the release of pesticides and lowering the dosage by maintaining the effective concentration in the target for long period of time [35, 36]. This helps to increase the agricultural production and also avoid the environmental pollution. Pesticides delivered through nano-delivery systems lead to a better distribution on leaf surfaces of crops by the form of nanosize molecule. Moreover, this system allows good penetration thereby allowing slow and controlled release of ingredients on the target [32, 37].

Another alternate approach is of releasing pesticides through silica nanoparticles. Silica nanoparticles can be used as a carrier due to their low cost, non-toxicity, high reactivity and also high surface area [38, 39]. But currently used silica nanoparticle as pesticide carrier did not show porous surface properties.

4.6.1 Microencapsulation of Pesticides

Controlled release of pesticides can be achieved by microencapsulation. Pesticide microcapsule systems can be designed to reduce mammalian toxicity and extended activity, reduced evaporative losses, reduced phytotoxicity, protecting pesticides from environmental degradation, reduced leaching, reduced pesticide levels in the environment, etc. [40]. Encapsulation in a polymeric matrix would maintain active ingredient levels adequate for pest control as well as reducing degradation losses and evaporative losses [41, 42]. An aqueous dispersion of pesticide microcapsules is a particularly useful controlled release formulation because it is composed of discrete microcapsules. It can be diluted with water or liquid fertilizers and sprayed using conventional equipment. It requires less polymeric component per pound of pesticide than monolithic devices; it is capable of establishing a constant pesticide release rate; pesticide release rate can be varied over wide limits by varying microcapsule particle size distribution, wall thickness and wall permeability; additives such as film forming agents can be added directly to the formulation.

Microencapsulation processes are grouped into three categories (phase separation, interfacial reactions and physical methods). A good pesticide should show an efficient biological activity and should be eco-friendly [43, 44]. Normally pesticides are in the form of wettable powders, water solutions, emulsifiable concentrates, powder or granules, spray formulation or aerosols. These have different types of hazards. Powder or aerosol forms that penetrate through the skin are hazardous. Considering all this, microencapsulation can be considered as an effective way of delivering pesticides [45].

Releasing of active substances within the controlled release system helps to release the active compound slowly. As a result, the protection of the seedlings is prolonged until they are physically strong enough to withstand insect attack [46]. Microencapsulations help to prolong the residual activity [47], to overcome environmental degradation [48, 49] and to reduce the aquatic toxicity [43, 50].

Microencapsulation of the pesticides provides an efficient and economically beneficial pest control [51]. For prolonging time of protection, microcapsule must be adhered to the seedling surface. An efficient way to disperse microcapsules in a matrix which is compatible with the microcapsule shell material is the key in achieving this [52]. Water-based latex will ensure the strong attachment of microcapsules on the seedling surface [53].

Biodegradable microbial polyesters called polyhydroxylalkonates are used as carriers for pesticides, and alginates are used for the controlled release of herbicides [28, 50, 54–59]. Intelimer is a temperature sensitive polymer which is used to encapsulate pesticides. These provide a slow release under specific temperature. As a result, we can protect the active pesticides from leaching and degradation [24]. Insecticides and also pesticides can also be used in a granular system [60]. The encapsulation of pesticides is done by releasing the pesticides into an aqueous dispersion of gelatinized starch thereby crosslinking the starch by calcium chloride. This improves the workers' safety in handling of pesticides and also helps to decrease the crop phytotoxicity [61]. One such example is liquid neem oil which is dispersed into a matrix where urea formaldehyde polymer is crosslinked with natural polymers like starch [62].

4.6.2 Microcapsule as a Source for Releasing Pesticides

There are different types of controlled release devices, like microsphere coated granules, microcapsules, etc. Much industrial research during the last 20 years was devoted to the utilization and specialization of microcapsules as a source for releasing pesticides [63]. Microcapsule is a system in which the chemical present within the core is surrounded by a shell or a membrane. It contains one or several shells and one or several cores. The inside core may be made up of solid, liquid, gaseous or combination of these phases. The matrix is organic or an inorganic polymer or a metal oxide [64]. The active material within the microcapsule is protected by the wall material from adverse reactions and restrict the direct exposure to outside environment [65, 66]. Naturally occurring and synthetic polymers are employed in microcapsule as wall material. The naturally occurring polymer employed in microcapsule is biodegradable or non-biodegradable. Natural polymers normally used are polysaccharides including cellulose, starch and chitosan. Commonly used synthetic polymers include polysterene, polymethylacrylates, polyesters and amino resins [64, 67, 68]. For the preparation of microspheres, inorganic materials can be used. They are made up of silica, zeolites, inorganic oxides, glass beads and ceramics [68–72].

4.6.3 Hydrogel as a Delivery Vehicle for Controlled Release of Pesticides

According to Rahim et al. [73], controlled relese technology supplies an advanced pesticide delivering technology coupled with reduction in toxicity and transportation hazards and prompts the development of newer and better alternative pesicides. Hydrogels are hydrophilic macromolecular three-dimensional networks that can absorb water many times their dry mass expands in their volume [74, 75].

Hydrogels are physical or chemical cross-links of homo- or copolymers which give three-dimensional structures. Hydrogels can be grouped into different groups based on their properties [76].

1. Physical structure: amorphous, semi-crystalline and hydrogen bonded
2. Electric charge: ionic or neutral
3. Crosslink: physically or chemically
4. Response to external effects: stimulus sensitive and insensitive
5. Origin: natural and synthetic.

Hydrogel could be used as potential delivery vehicles for the controlled release of pesticide. Examples of biodegradable hydrogel include polyacrylamide and methyl cellulose which play an important role in usage of hydrogels in controlled release systems [77]. The dispersal of chemicals which is entrapped in a hydrogel occurs only after soaking in water resulting in swelling up of the polymer. Polymer swelling causes the chemicals to become dissolved and diffuse to the surface of the device. Releasing chemicals is closely related to swelling features of hydrogels which in turn depends on the chemical architecture of the hydrogels [78].

4.6.4 Nanocomposites

Nanocomposites can be prepared by the combinations of organic polymer matrix and nano-dimension inorganic filter. As a result, a hybrid material can be developed and these hybrids exhibit high strength; it has high durability and process flexibility which can be used for agriculture, transportation, defence mechanisms, etc. [79, 80]. For example, the polymer and clay-based coats containing azadirachtin-A were used for maintaining soya bean seed quality during storage time [81]. Ethyl cellulose, methyl cellulose, polyvinyl chloride, polyvinyl acetate, hydroxyl ethyl cellulose, and agrimer VA 6 polymers and bentonite clay were used as carriers. The release of azadirachtin-A into water ranged from 8.02 to 21.36 h. Half-life in the seed coats ranged from 4.37 to 11.22 months when compared to 3.45 months in azadirachtin-A WP. These coats acted as a moisture barrier reducing the degradation and preventing fungal proliferation. The release of active components can be achieved by several mechanisms such as diffusion through a membrane, osmosis, ion exchange and matrix degradation.

4.7 Polymer Active Ingredient Conjugates

In this system, active compound is covalently bound to the polymer. The release of the compound depends on the chemical or biological cleavage of bond [21]. Amidated polyacrylamide gel derivatives with pendant herbicide (2,4-D) residues were reported [82]. Biodegradable fungicide polymers containing diniconazole were also devised. The combinations of pesticide and polymer were hydrolysed or depolymerized at controlled rate so as to protect the plants.

4.7.1 Matrix-Based Systems

In this system, the active components are dispersed in a monolithic polymer. When this polymer is not biodegradable, the release occurs by diffusion through the matrix. Matrix includes rubber, polyester, cellulose, etc. A matrix is immersed in a concentrated matrix swelling solution of active agent. After this, the solvent is removed.

4.7.2 Membrane-Based System

In this system, the active ingredient is in a membrane and the releases of the active ingredients are of two main types: the release through diffusion across the membrane and an osmotic system through a semipermeable membrane [83].

4.8 Environmental Safety of the Controlled Release Formulations

Application of pesticides directly in agriculture causes severe damage to the ecosystem. But application of these pesticides through controlled release significantly reduces risks to environment and human health. The synthetic polymer and their degradable products may also cause toxicity when compared to natural polymer [84]. When the pesticides are applied in the field, it moves from one zone to another, and it gets degraded or moves into other biological systems. Through the processes of evaporation, leaching and plant uptake, about 95% of pesticides disappear from soil. Plant losses pesticides through evaporation, in root exudates and in stubbles after harvest. Residues remaining in the plant or soil are metabolized. Persistent pesticides constitute only a small proportion of the whole [85, 86]. Artificial polymers which degrade over several days or months may affect the environment and also soil fertility. But in case of natural polymers, the degradation products are non-toxic and they do not damage the plant and ecosystems. Hence, in these controlled release systems, usually bio-polymers or biodegradable polymers are preferred. In controlled release systems, naturally available biodegradable polymers are favoured because their decomposition may enhance the soil fertility. Cheaper, easily available and effective carrier materials are required. Naturally occurring pesticides of reduced toxicity should be considered as safer alternatives. Encapsulations further reduce their toxicity. Application of pesticides into a single formulation would make it effective and economic and may contribute to green technology in agriculture production.

4.9 Conclusion

Pesticides are chemicals used to prevent growth of unwanted living organisms or pests. But these have certain health hazards and environmental problems. Normally pesticides are subjected to evaporation, degradation, etc., as a result of which the active ingredient is removed from the target. Controlled release formulations are combinations of biologically active agent with a carrier like a polymer, nano-composite, etc. and allow the release of the active ingredients to the target at a controlled rate. Both synthetic and natural polymers can be used in this system. The degradation of synthetic polymer may cause ecosystem toxicity and hence biodegradable non-toxic natural polymer is used. Certain nanoparticles can also be used for the release of pesticides. Silica nanoparticle can be used as a carrier for release of pesticides. Microencapsulation, hydrogels and microcapsules can also be used as a delivery vehicle for the release of pesticides. The release of active ingredients from polymers occurs mainly through diffusion a membrane, osmosis and matrix degradation.

References

1. Varma J, Dubey NK (1999) Prospectives of botanical and microbial products as pesticides of tomorrow. Curr Sci 172–179
2. Duchowicz PR, Goodarzi M, Ocsachoque MA, Romanelli GP, Ortiz EDV, Autino JC, Bennardi DO, Ruiz DM, Castro EA (2009) QSAR analysis on Spodopteralitura antifeedant activities for flavone derivatives. Sci Total Environ 408(2):277–285
3. Pickett JA, Tatchell GM, Glen DM, Raccah B, Singer S, Veierov D, Hart J (1995) Summaries of lectures presented at a binational UK—Israel conference agriculture in a cleaner environment—new technologies from Britain and Israel. Phytoparasitica 23(4):359–366
4. Witzgall P, Kirsch P, Cork A (2010) Sex pheromones and their impact on pest management. J Chem Ecol 36(1):80–100
5. Månsson PE, Eriksson C, Sjödin K (2005) Antifeedants against *Hylobius abietis* pine weevils: an active compound in extract of bark of *Tilia cordata* linden. J Chem Ecol 31(5):989–1001
6. Månsson PE, Schlyter F, Eriksson C, Sjödin K (2006) Nonanoic acid, other alkanoic acids, and related compounds as antifeedants in *Hylobius abietis* pine weevils. Entomol Exp Appl 121(3):191–201
7. Shtykova L, Masuda M, Eriksson C, Sjödin K, Marling E, Schlyter F, Nydén M (2008) Latex coatings containing antifeedants: formulation, characterization, and application for protection of conifer seedlings against pine weevil feeding. Prog Org Coat 63(2):160–166
8. Månsson PE, Eriksson C, Sjödin K, Schlyter F (2008) Antifeedants and feeding stimulants in bark extracts of ten woody non-host species of the pine weevil *Hylobius abietis*. J Chem Ecol 34(10):1290–1297
9. Månsson PE (2005) Host selection and antifeedants in *Hylobius abietis* pine weevils, vol 2005, no 16
10. Van Boven M, Toppet S, Cokelaere MM, Daenens P (1994) Isolation and structural identification of a new simmondsinferulate from jojoba meal. J Agric Food Chem 42 (5):1118–1121
11. Miller GT (2004) Sustaining the earth, vol 9. Thompson Learning, Inc., Pacific Grove, California, pp 211–216
12. Kubrak OI, Atamaniuk TM, Husak VV, Drohomyretska IZ, Storey JM, Storey KB, Lushchak VI (2012) Oxidative stress responses in blood and gills of Carassius auratus exposed to the mancozeb-containing carbamate fungicide tattoo. Ecotoxicol Environ Saf 85:37–43
13. Lushchak V, Kubrak OI, Storey JM, Storey KB, Lushchak VI (2009) Low toxic herbicide roundup induces mild oxidative stress in goldfish tissues. Chemosphere 76(7):932–937
14. Maksymiv IV, Husak VV, Mosiichuk NM, Matviishyn TM, Sluchyk IY, Storey JM, Storey KB, Lushchak VI (2015) Hepatotoxicity of herbicide Sencor in goldfish may result from induction of mild oxidative stress. Pestic Biochem Physiol 122:67–75
15. González M, Soloneski S, Reigosa MA, Larramendy ML (2005) Genotoxicity of the herbicide 2,4-dichlorophenoxyacetic and a commercial formulation, 2,4-dichlorophenoxyacetic acid dimethylamine salt. I. Evaluation of DNA damage and cytogenetic endpoints in Chinese Hamster ovary (CHO) cells. Toxicol In Vitro 19(2):289–297

16. Cerejeira MJ, Viana P, Batista S, Pereira T, Silva E, Valério MJ, Silva A, Ferreira M, Silva-Fernandes AM (2003) Pesticides in Portuguese surface and ground waters. Water Res 37(5):1055–1063
17. Celis R, Hermosín MC, Carrizosa MJ, Cornejo J (2002) Inorganic and organic clays as carriers for controlled release of the herbicide hexazinone. J Agric Food Chem 50(8):2324–2330
18. Roy A, Bajpai J, Bajpai AK (2009) Development of calcium alginate–gelatin based microspheres for controlled release of endosulfan as a model pesticide
19. Akelah A, Selim A (1987) Applications of functionalized polymers for controlled release formulations of agrochemicals and related biocides. Chimica & L Industria 69(1–2):62–67
20. Cardarelli NF (1970) U.S. patent no. 3,489,591. U.S. Patent and Trademark Office, Washington, DC
21. Akelah A (1984) Biological applications of functionalised polymers—a review. J Chem Technol Biotechnol Chem Technol 34(6):263–286
22. Kydonieus AF (1980) Fundamental concepts of controlled release. In: Controlled release technologies: methods, theory, and applications, vol 1, pp 1–19
23. Mitrus M, Wojtowicz A, Moscicki L (2009) Biodegradable polymers and their practical utility. In: Thermoplastic starch: a green material for various industries, pp 1–33
24. Deborah H, Meyers PA, Greene CL (1992) Temperature-activated release of trifluralin and diazinon. In: Pesticide formulations and application systems: eleventh volume. ASTM International
25. Baker RW (1987) Controlled release of biologically active agents. Wiley, London
26. Brannon-Peppas L (1993) Controlled release in the food and cosmetics industries. In: ACS Symposium series-American Chemical Society (USA)
27. Hoffman AS (2008) The origins and evolution of "controlled" drug delivery systems. J Control Release 132(3):153–163
28. Lee PI, Good WR (1987) Controlled-release technology: pharmaceutical applications, vol 348. Amer Chemical Society
29. Bhadi M, Fister GP (1990) Controlled release formulations of pesticides. In: Chemistry of plant protection book series, vol 6, pp 1–64
30. Kearney PC (1977) A challenge for controlled release technology. Symposium series, chapter 3, vol 53, pp 30–36
31. Zweig G (1977) Environmental aspects of controlled release pesticide formulations. In: ACS Symposium Series-American Chemical Society (USA)
32. Zeoli LT, Kydonieus AF (1983) Physical methods of controlled release. Control Release Technol
33. Dass CR, Su T (2001) Particle-mediated intravascular delivery of oligonucleotides to tumors: associated biology and lessons from genotherapy. Drug Deliv 8(4):191–213
34. Xu H, Song T, Bao X, Hu L (2005) Site-directed research of magnetic nanoparticles in magnetic drug targeting. J Magn Magn Mater 293(1):514–519
35. Marc GM, Akin H, Hasirci N, Trantolo DJ, Gresser JD, Wise DL (1996) Controlled release of biologically active agents for purposes of agricultural crop management. Resour Conserv Recycl 16(1–4):289–320
36. Li ZZ, Xu SA, Wen LX, Liu F, Liu AQ, Wang Q, Chen JF (2006) Controlled release of avermectin from porous hollow silica nanoparticles: influence of shell thickness on loading efficiency, UV-shielding property and release. J Control Release 111(1–2):81–88
37. Martin CR, Kohli P (2003) The emerging field of nanotube biotechnology. Nat Rev Drug Discov 2(1):29
38. Cao B, Yang M, Zhu Y, Qu X, Mao C (2014) Stem cells loaded with nanoparticles as a drug carrier for in vivo breast cancer therapy. Adv Mater 26(27):4627–4631
39. Zhang Q, Zhang T, Ge J, Yin Y (2008) Permeable silica shell through surface-protected etching. Nano Lett 8(9):2867–2871
40. Scher HB, Rodson M, Lee KS (1998) Microencapsulation of pesticides by interfacial polymerization utilizing isocyanate or aminoplast chemistry. Pestic Sci 54(4):394–400

41. Shaviv A (2001) Advances in controlled-release fertilizers. 71:1–49
42. Rudzinski WE, Dave AM, Vaishnav UH, Kumbar SG, Kulkarni AR, Aminabhavi TM (2002) Hydrogels as controlled release devices in agriculture. Des Monomers Polym 5(1):39–65
43. Scher HB (1999) Controlled-release delivery systems for pesticides. CRC Press. SB 951.145:65–66
44. Weatherston I, Miller D, Dohse L (1985) Capillaries as controlled release devices for insect pheromones and other volatile substances—a reevaluation. J Chem Ecol 11(8):953–965
45. Siewierski M (2011) Determination and assessment of pesticide exposure, vol 24. In: Proceedings of a working conference, Hershey, PA, 29–31 Oct 1980. Elsevier, Amsterdam
46. McGuire MR, Shasha BS, Lewis LC, Nelsen TC (1994) Residual activity of granular starch-encapsulated *Bacillus thuringiensis*. J Econ Entomol 87(3):631–637
47. Guessan NR, Knols BG, Pennetier C, Rowland M (2008) DEET microencapsulation: a slow-release formulation enhancing the residual efficacy of bed nets against malaria vectors. Trans R Soc Trop Med Hyg 102(3):259–262
48. Fries E, Püttmann W (2002) Analysis of the antioxidant butylated hydroxytoluene (BHT) in water by means of solid phase extraction combined with GC/MS. Water Res 36(9):2319–2327
49. Greene LC, Meyers PA, Springer JT, Banks PA (1992) Biological evaluation of pesticides released from temperature-responsive microcapsules. J Agric Food Chem 40(11):2274–2278
50. Pepperman AB, Kuan JCW, McCombs C (1991) Alginate controlled release formulations of metribuzin. J Control Release 17(1):105–111
51. Wege PJ, Hoppe MA, Bywater AF, Weeks SD, Gallo TS (1999, July) A microencapsulated formulation of lambda-cyhalothrin. In: Proceedings of the 3rd international conference on urban pests (pp 19–22)
52. Kim EY, Kim HD (2005) Preparation and properties of microencapsulated octadecane with waterborne polyurethane. J Appl Polym Sci 96(5):1596–1604
53. Peng YX, Zheng ZH, Ding XB, Zhang WC, Ye ZH (2003) Nanometer polymer latex dispersion and its application in water-based coating. Prog Org Coat 48(2–4):161–163
54. Voinova ON, Kalacheva GS, Grodnitskaya ID, Volova TG (2009) Microbial polymers as a degradable carrier for pesticide delivery. Appl Biochem Microbiol 45(4):384–388
55. Romero-Cano MS, Vincent B (2002) Controlled release of 4-nitroanisole from poly (lactic acid) nanoparticles. J Control Release 82(1):127–135
56. Wang Y, Lapitsky Y, Kang CE, Shoichet MS (2009) Accelerated release of a sparingly soluble drug from an injectable hyaluronan–methylcellulose hydrogel. J Control Release 140 (3):218–223
57. Byrne ME, Salian V (2008) Molecular imprinting within hydrogels II: progress and analysis of the field. Int J Pharm 364(2):188–212
58. Colombo I, Grassi M, Lapasin R, Pricl S (1997) Determination of the drug diffusion coefficient in swollen hydrogel polymeric matrices by means of the inverse sectioning method. J Control Release 47(3):305–314
59. Fundueanu G, Constantin M, Ascenzi P (2008) Preparation and characterization of pH-and temperature-sensitive pullulan microspheres for controlled release of drugs. Biomaterials 29 (18):2767–2775
60. Watanabe A, Takebayashi Y, Ohtsubo T, Furukawa M (2009) Permeation of urea through various polyurethane membranes. Pest Manag Sci 65(11):1233–1240
61. Wing RE, Carr ME, Doane WM, Schreiber MM (1992) Starch encapsulated herbicide formulations: scale-up and laboratory evaluations. In: Pesticide formulations and application systems: eleventh volume. ASTM International
62. Kulkarni AR, Soppimath KS, Aminabhavi TM, Dave AM, Mehta MH (1999) Urea–formaldehyde crosslinked starch and guar gum matrices for encapsulation of natural liquid pesticide [Azadirachta Indica A. Juss. (neem) seed oil]: swelling and release kinetics. J Appl Polym Sci 73(12):2437–2446
63. Lemetter CYG, Meeuse FM, Zuidam NJ (2009) Control of the morphology and the size of complex coacervate microcapsules during scale-up. AIChE J 55(6):1487–1496

64. Arshady R (1999) Preparation of biodegradable microspheres and microcapsules: 2. Polyactides and related polyesters. J Controlled Release 17(1):1–21
65. Sheorey DS, Kshirsagar MD, Dorle AK (1991) Study of some improved shellac derivatives as microencapsulating materials. J Microencapsul 8(3):375–380
66. Riyajan SA, Sakdapipanich JT (2009) Development of a controlled release neem capsule with a sodium alginate matrix, crosslinked by glutaraldehyde and coated with natural rubber. Polym Bull 63(4):609–622
67. Chang CP, Kimura M, Yamamoto T, Nobe M, Dobashi T (2003) Effect of dispersing medium on permeability of microcapsule membrane. Colloids Surf B 30(1–2):123–127
68. Chang CP, Chang JC, Ichikawa K, Dobashi T (2005) Permeability of dye through poly (urea-urethane) microcapsule membrane prepared from mixtures of di-and tri-isocyanate. Colloids Surf B 44(4):187–190
69. Chang CP, Dobashi T (2003) Preparation of alginate complex capsules containing eucalyptus essential oil and its controlled release. Colloids Surf B 32(3):257–262
70. Hsieh WC, Chang CP, Gao YL (2006) Controlled release properties of chitosan encapsulated volatile citronella oil microcapsules by thermal treatments. Colloids Surf B 53(2):209–214
71. Shukla PG, Sivaram S, Mohanty B (1992) Structure of carbofuran in crosslinked starch matrix by 13C nmr: correlation of release and swelling kinetics with the dynamic behaviour of polymer chains. Polymer 33(17):3611–3615
72. Chang CP, Leung TK, Lin SM, Hsu CC (2006) Release properties on gelatin-gum arabic microcapsules containing camphor oil with added polystyrene. Colloids Surf B 50(2): 136–140
73. Rahim M, Hakim MR, Haris HM (2005) Application of advanced polymeric materials for controlled release pesticides, vol 146
74. Aouada FA, Guilherme MR, Campese GM, Girotto EM, Rubira AF, Muniz EC (2006) Electrochemical and mechanical properties of hydrogels based on conductive poly (3, 4-ethylene dioxythiophene)/poly (styrenesulfonate) and PAAm. Polym Test 25(2):158–165
75. Aouada FA, Pan Z, Orts WJ, Mattoso LH (2009) Removal of paraquat pesticide from aqueous solutions using a novel adsorbent material based on polyacrylamide and methylcellulose hydrogels. J Appl Polym Sci 114(4):2139–2148
76. Deligkaris K, Tadele TS, Olthuis W, van den Berg A (2010) Hydrogel-based devices for biomedical applications. Sens Actuators B Chem 147(2):765–774
77. Stepto RF (1998) Polymer networks: principles of their formation, structure, and properties. Van Nostrand Reinhold
78. Singh B, Sharma DK, Gupta A (2008) In vitro release dynamics of thiram fungicide from starch and poly (methacrylic acid)-based hydrogels. J Hazard Mater 154(1–3):278–286
79. Kabiri K, Omidian H, Zohuriaan-Mehr MJ, Doroudiani S (2011) Superabsorbent hydrogel composites and nanocomposites: a review. Polym Compos 32(2):277–289
80. Kuzma J (2007) Moving forward responsibly: oversight for the nanotechnology-biology interface. J Nanopart Res 9(1):165–182
81. Nisar K, Kumar J, Arun Kumar MB, Walia S, Shakil NA, Parsad R, Parmar BS (2009) Pesticidal seed coats based on azadirachtin-A: release kinetics, storage life and performance. Pest Manag Sci 65(2):175–182
82. Kenawy ER (1998) Biologically active polymers: controlled-release formulations based on crosslinked acrylamide gel derivatives. React Funct Polym 36(1):31–39
83. Reddy JRK, Jyothsna MV, Saleem TM, Chetty CMS (2009) Review on: pulsatile drug delivery systems. J Pharm Sci Res 1(4):109
84. Kumar LP, Panneerselvam N (2008) Toxic effects of pesticides: a review on cytogenetic biomonitoring studies. Med Biol 15(2):46–50
85. Chinnamuthu CR, Boopathi PM (2009) Nanotechnology and agroecosystem. Madras Agric J 96(1/6):17–31
86. Moraru CI, Panchapakesan CP, Huang Q, Takhistov P, Liu S, Kokini JL (2003) Nanotechnology: a new frontier in food science. Food Technol 57(12):24–29

Chapter 5
Manufacturing of Slow- and Controlled-Release Pesticides

V. R. Remya, Jesiya Susan George and Sabu Thomas

Abstract To ensuring the food safety, pesticides are used as the fundamental material against the major biological disasters. In recent years, pesticides have obtained enormous attention in the field of agriculture and food industry. Therefore to overcome the limitations of using traditional pesticide formulation, a new facile and innovative method of pesticide formulation has been addressed. This approach designs an intelligent smart process like 'slow and controlled release of pesticides'. This concept may be greatly improving the use of pesticides by reducing waste and pollution. In this chapter, a brief description about pesticides, raw materials, manufacturing methods of slow and controlled release of pesticides and preparation methods of pesticide microcapsule like nanopesticides is discussed. This novel and intelligent slow- and controlled-release technology could enhance the target ability and improve the dispersibility and stability of active ingredients, pesticide loading and environmental viability.

Keywords Pesticide · Synthesis · Formulation · Application

5.1 Introduction

5.1.1 Pesticides

Pesticides are indispensable agents, which kill plants or animals that compete for humanity's food supply. It is mainly used for improving the crop yield by destroying the pests and disease [1]. Pesticides are classified into herbicides, insecticides, nematicides, molluscicides, piscicide, avicide, rodenticide, bactericide, insect repellent, animal repellent, antimicrobials, fungicides and disinfectants [2]. Because of the organic solvent contents, poor dispersibility and dust drift of traditional

V. R. Remya · J. S. George · S. Thomas (✉)
International and Inter University Centre for Nanoscience and Nanotechnology, Mahatma Gandhi University, Kottayam, India
e-mail: sabuthomas@mgu.ac.in

Rakhimol K. R. et al. (eds.), *Controlled Release of Pesticides for Sustainable Agriculture*, https://doi.org/10.1007/978-3-030-23396-9_5

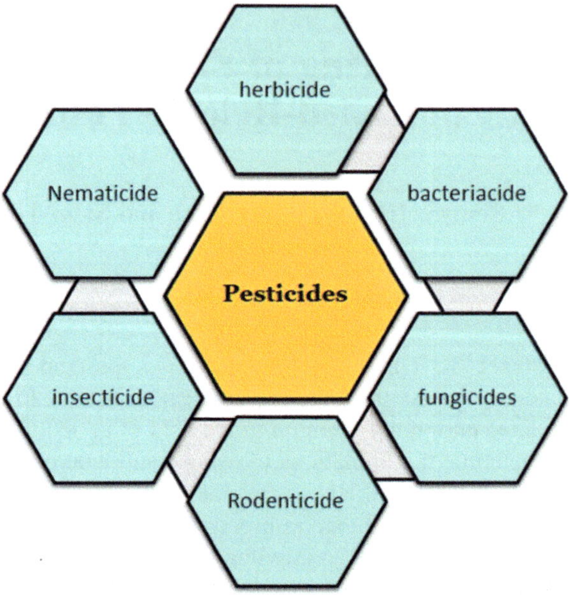

Fig. 5.1 Classification of pesticides

pesticides, its applicability has some limitations hence 99% of pesticides are lost to the environment rather than attaining the target. This will create a serious environmental pollution. This low effectiveness contributes to serious environmental pollution [3]. Therefore necessary action should be taken to reduce waste, environmental pollution and production costs associated with pesticides through slow and controlled release of pesticides on crops [4] (Fig. 5.1).

5.1.2 Raw Materials

Pesticides mainly consist of two parts; one active and inert part. The active ingredients, which kill pests, coupled with inert ingredients, which help the coating and spraying. In past decades, active ingredients were extracted from natural sources such as plants (e.g. azadirachtin, pyrethrum and rotenone) and microbes/insects (e.g. bacillus thuringiensis), but at present it is largely synthesized via laboratory works. Most common active ingredients are hydrocarbons which are derived from petroleum source. In pure form, the active ingredients are not suitable for application. This is because of the high toxic effect of concentrated form, since many of them do not mix effectively with water, it also difficult to handle, store, transport and unstable. To solve this problem, manufacturers are adding inert ingredients along with active ingredients for the production of pesticides. The composition of inert ingredients depends on the type of pesticides. In fact, inert ingredients have no pesticidal activity; it simply acts as a carrier or diluents in the

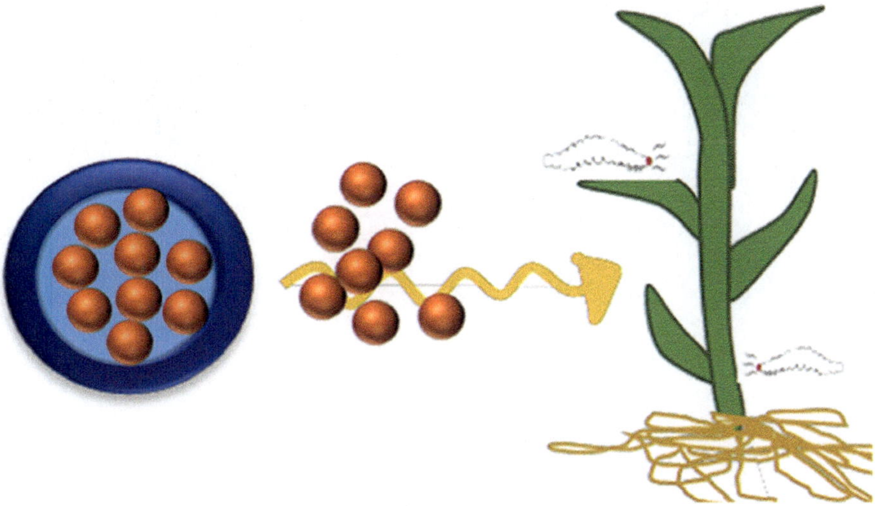

Scheme 5.1 Targeted delivery of pesticide nanoformulation [11]

formulation of pesticides. In addition to the active part, formulated pesticides contain surface active ingredients, stabilizers, dyes, chemicals and carriers or diluents. Kerosene and other petroleum distillates are used as carriers in liquid pesticides while vegetable matters like ground up nut shells, clays and powdered minerals are used in dust pesticides [5–10] (Scheme 5.1).

5.2 Manufacturing Process of Pesticides

The three separate activities are involved in the manufacturing of pesticides. The first step is the synthesis of active ingredient in a chemical laboratory, and later it can be formulated in the same place or to another formulator in the form of liquid or powder form. The users diluted it to for apply in fields.

5.2.1 Synthesizing the Pesticide

The pesticide synthesis is a complex chemical method; it essentially needs sophisticated laboratories, trained chemists. In the preliminary steps of synthesizing pesticide, small quantity of pesticides only developed. Later the viability studies only the large-scale production of pesticides initiates. The basic unit that required for the synthesis of pesticide is an organic molecule; it can be converted to the form

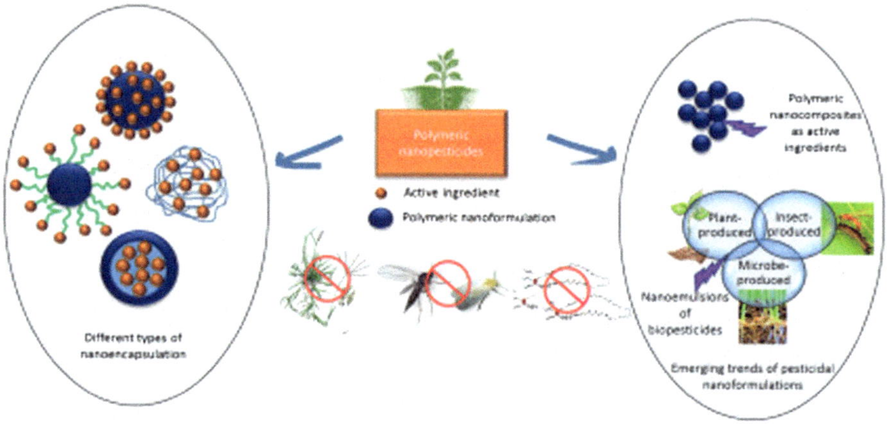

Fig. 5.2 Nano-based smart pesticide formation [11]

of pesticides via series of reactions with the help of certain reagents, catalysts, specific temperature in a controlled atmosphere [7–10, 12–14] (Fig. 5.2).

5.2.2 Formulating the Pesticide

The first step of the formulation of pesticide is the selection of active ingredient with proper amount followed by mixing with inert carriers. The quantity of pesticide formulation depends on the need of anticipated farmers [6–10, 12, 13].

5.2.3 Diluting the Pesticide

The final product need to apply in field is 0.5–1% of the original active ingredient. For that, pesticide is stored in a short time and then transported to the anticipated agent: they diluted the concentrated pesticide for applying in fields [8–10, 12–14].

5.2.4 Applying the Pesticide

There are a lot of techniques used for the application of pesticides. For example, Americans are the export of crop dusting and its use is limited to large flat areas. Others used tanks and booms to a tractor and spray closer to the ground. Farmers used the most common method of applying pesticides like hand-held sprayers attached to small tanks [11, 15–20] (Fig. 5.3).

Fig. 5.3 Image of pesticide application [21]

5.2.5 *Nanopesticides*

The production and discovery of intelligent pesticides paves a way for future generation, especially nanopesticides, since nanomaterials are used in all areas of life including agricultural, medical, industrial etc. In the field of pesticides, applications of nanomaterials and technologies have made substantial changes. The slow and controlled release and targeted delivery of nanopesticides can improve the pesticide utilization and also reduce the residue and pollution [22–26] (Fig. 5.4).

The smaller size, wettability, improvable pesticide droplet ductility and target adsorption of nanopesticides formulation result in the slow and controlled release and higher protection performance of pesticides: this is highly beneficial for non-hazards and environmentally friendly advantages [27, 28] (Fig. 5.5).

Popular pesticide microcapsule preparation methods are described in this part. There are seven methods used for the preparation of pesticide microcapsule. They are: interfacial polymerization, in situ polymerization, emulsion polymerization, membrane emulsification, solvent evaporation, nano-precipitation and double coacervation [29–40] (Fig. 5.6).

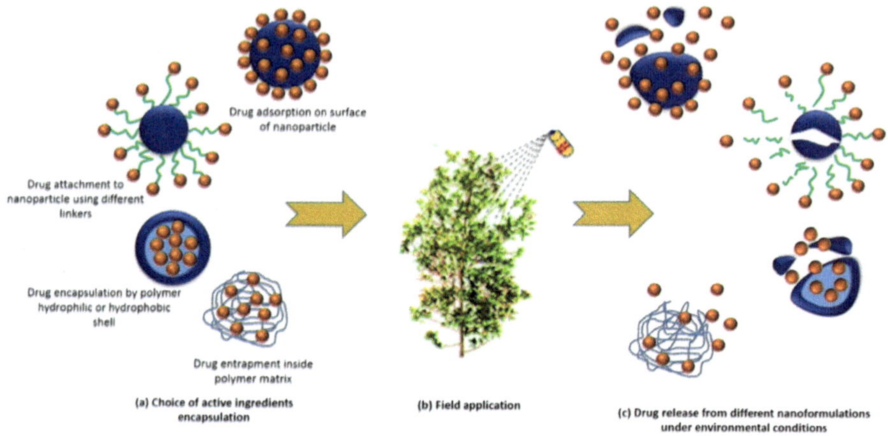

Fig. 5.4 Nanoencapsulation of pesticides [11]

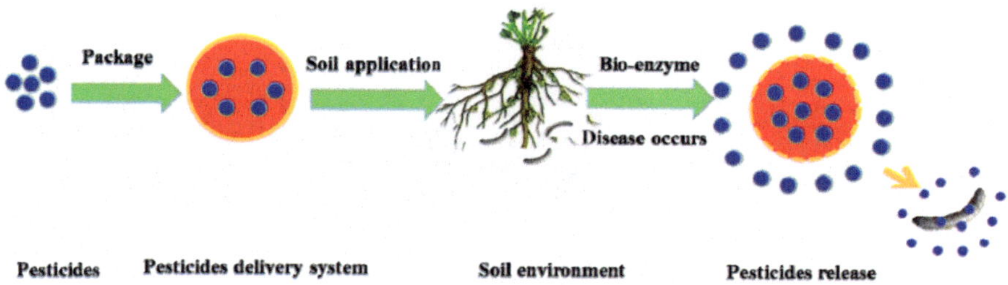

Fig. 5.5 Slow and controlled release of pesticides [29]

5.2.6 Interfacial Polymerization

Interfacial polymerization is the widely used method for encapsulation of pesticides for their slow and controlled release of pesticides. In this technique, two reactive monomers are dissolved in different solvents and mixed together. During the mixing procedure, two monomers undergo a polycondensation reaction between the interface of two monomer solution and form microcapsules. It is a type step-growth polymerization, and it utilizes the Schotten–Baumann reaction. There are a lot of core shell type nanopesticides prepared via this method. Natural pyrethrin nanocapsules are the examples of nanopesticide which synthesis through interfacial polymerization [30, 31] (Fig. 5.7).

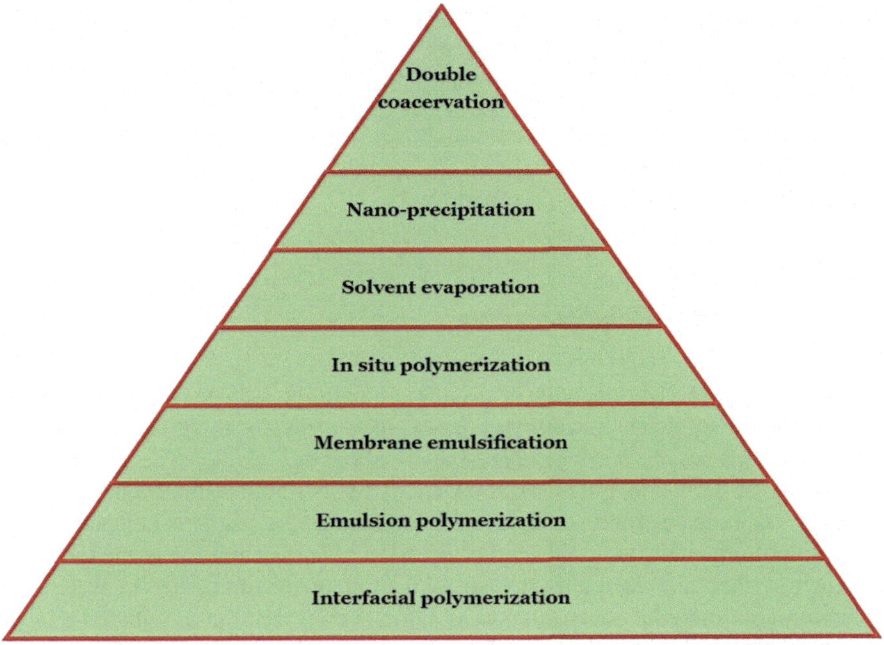

Fig. 5.6 Preparation methods of pesticide microcapsules

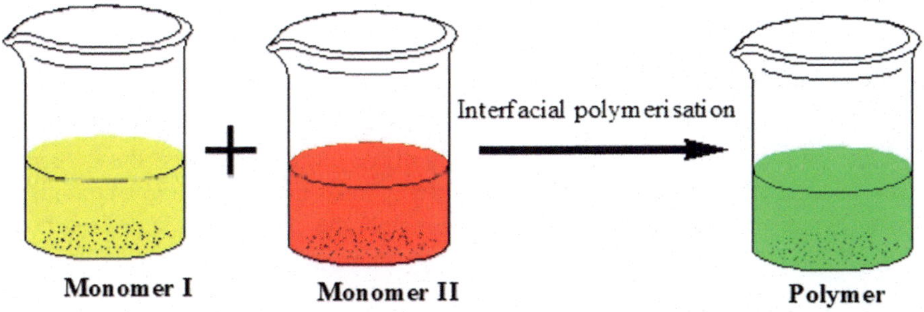

Fig. 5.7 Image of interfacial polymerization

5.2.7 *In situ Polymerization*

The word 'in situ' polymerization means 'polymerization mixture'. In some cases, in situ polymerization is similar to interfacial coating. In this method, two or more water-soluble monomers are undergoing polymerization reaction and form water-insoluble polymer. This product is accumulated on the surface of the core material. To minimize the disadvantages and maximize the effectiveness of pesticide microencapsulation, target concentration is the essential criteria adapted. The factors influencing the encapsulation of pesticides are (a) they must not react with the pesticide; (b) the molecular weight, vitreous transition temperature and structure must allow sufficient liberation; (c) the polymer and its degradation products must not cause environmental contamination; (d) they must be stable during use and storage; and (e) they must be easily produced at low cost. S-ethyl dipropylthiocarbamate, acetochlor, atrazine and methotrexate are the examples of pesticides formed by in situ polymerization [29, 32, 33].

5.2.8 Emulsion Polymerization

In emulsion polymerization, solvent-insoluble monomer is dispersed in a solvent to form uniform emulsion various processes like mechanical agitation, vigorous shaking, mechanical stirring, high-speed shearing along with surfactants. Surfactants in emulsion polymerization are used to stabilize the emulsion form. Generally anionic, cationic or non-ionic surfactants are used in emulsion polymerization. Surfactants are used for the reduction of surface tension at the monomer solvent interface and facilitate the emulsification of the monomer in water. For achieving core material encapsulation, polymerization reaction is initiated to form the polymer. Abamectin nanocapsule suspension and natural pyrethrin nanocapsules are the examples of pesticides formed through emulsion polymerization [34–36] (Fig. 5.8).

5.2.9 Membrane Emulsification

This is a novel technique for single and multiple emulsions. Here the dispersed phase is compelled through the pores of a microporous membrane into a continuous phase. The drop–by-drop mechanism in the membrane emulsification process helps

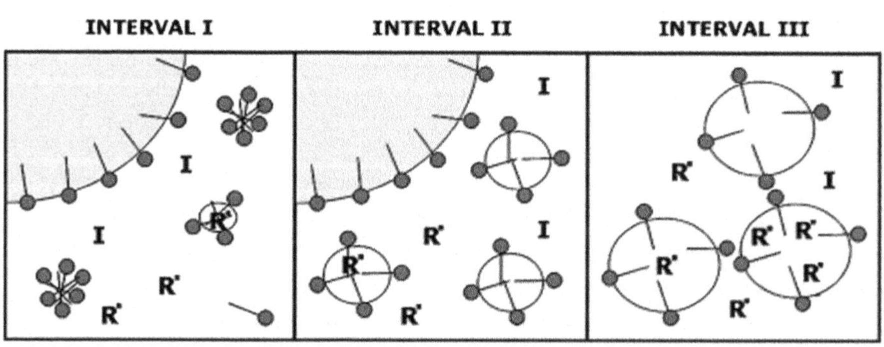

Fig. 5.8 Image of emulsion polymerization [41]

to detach the emulsified droplets formed. In the case of pesticides, the dispersed phase is forced through a shirasu porous glass membrane under inert gas pressure into the continuous phase. Chlorantraniliprole nanocapsules and avermectin nanocapsules are the examples of membrane-emulsified pesticides [37, 38, 42] (Fig. 5.9).

5.2.10 Solvent Evaporation

In this method, the organic phase, the wall material and core material are dispersed, is added to the solution immiscible with the wall material, and the wall material is precipitated to form the microcapsule by heating and evaporating the solvent. Spinosad nanocapsules are the example of solvent-evaporated pesticide [29, 39] (Fig. 5.10).

5.3 Nano-Precipitation

The presence of interfacial interaction between solvent and non-solvent helps to disperse the polymer and drug from the oil phase into the aqueous phase. Here the nanocapsules are formed through precipitation via wrapping of pesticide to the

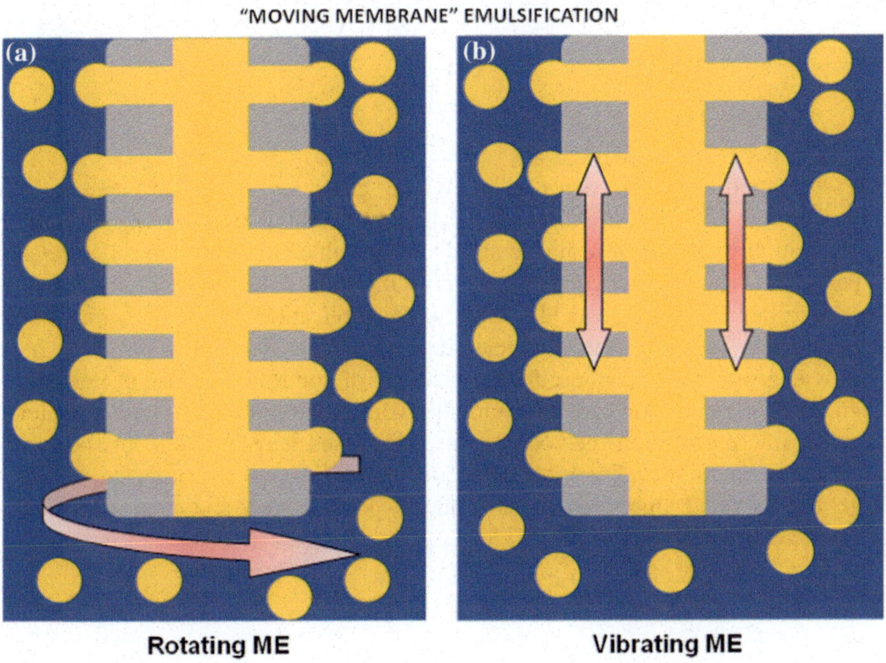

Fig. 5.9 Image of membrane emulsification [11, 43]

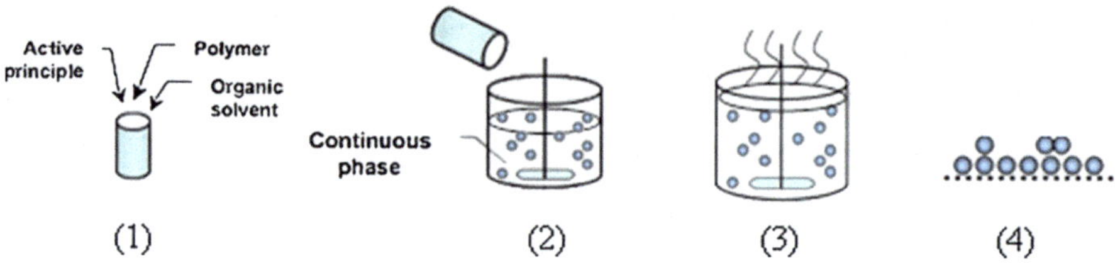

Fig. 5.10 Image of solvent evaporation [44]

material. Pyrazole azoxystrobin nanocapsules and azoxystrobin microcapsules are the examples of nano-precipitated pesticides [29, 40].

5.4 Double Coacervation

This process is mainly used for water-soluble polymers; here two water-soluble polymers with opposite charges form a wall around water-insoluble pesticide active ingredient. The double coacervation involves a spontaneous liquid to liquid separation due to electrostatic interactions. Azoxystrobin microcapsule is the example of this type of pesticide. The following figure shows the nano-based pesticide formulation which increases the bioavailability and efficiency [29] (Fig. 5.11).

5.5 Conclusions

Nanostructured materials are chosen as the emerging intelligent material for responding to external stimuli such as light, humidity, pH and temperature to achieve an environmental response and via to feasible slow and controlled release of pesticides. This chapter comprehensively summarized the manufacturing of slow and controlled release of pesticides and also the application of polymeric material to enhance their slow and controlled release of pesticides. Nowadays, polymers have been widely used in agriculture for applications of controlled release of pesticides. Their slow and controlled release helps to avoid environmental hazards. Pesticide microcapsule preparation methods are also discussed here in detail. In short, this chapter can enlighten and promote the readers to the field of environmental friendly, target concentrated, improved wettable and dispersible slow and controlled release of intelligent nanopesticides.

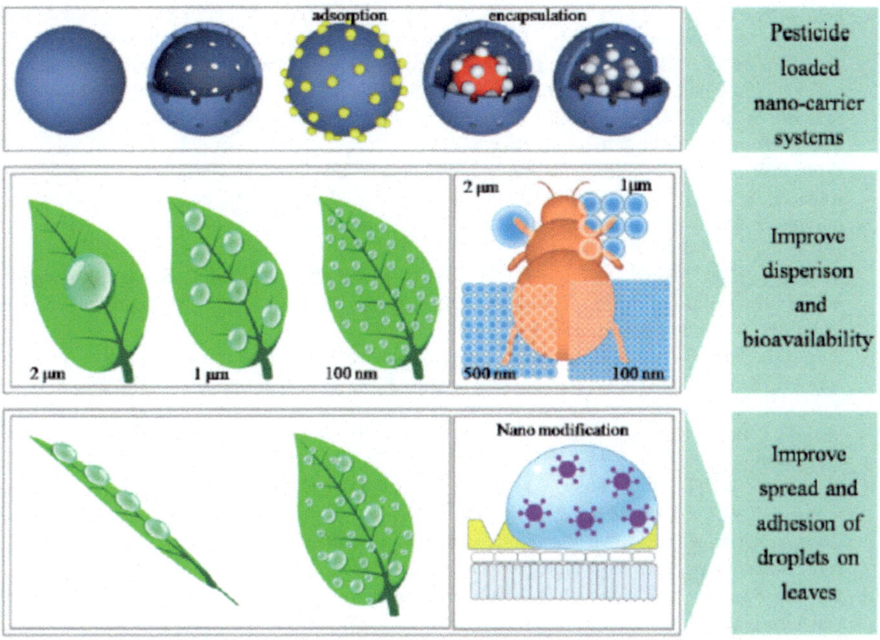

Fig. 5.11 Image of nano-based pesticide formulation [29]

References

1. Sun C, Cui H, Wang Y, Zeng Z, Zhao X, Cui B (2016) Studies on applications of nanomaterial and nanotechnology in agriculture. J Agric Sci Technol (Beijing) 18(1):18–25
2. Carolyn R, Winand H, Edward C, Colleen H (2013) National pesticide applicator certification core manual. National Association of State Departments of Agriculture Research Foundation, Washington, DC
3. Jin P, Yao R, Qin D, Chen Q, Du Q (2019) Enhancement in antibacterial activities of eugenol-entrapped ethosome nanoparticles via strengthening its permeability and sustained release. J Agric Food Chem 67(5):1371–1380
4. Reichenberger S, Bach M, Skitschak A, Frede HG (2007) Mitigation strategies to reduce pesticide inputs into ground-and surface water and their effectiveness; a review. Sci Total Environ 384(1–3):1–35
5. Gibbons A (1990) Overkilling the insect enemy; an OTA report says massive spraying of pesticides was misguided-and faults aid agencies for operating in a crisis mode. Science 249 (4969):621–622
6. Holmes B (1992) U.S. News and World Report, 73–74
7. Jacobsen RM, Youngs GA Jr, Goreham GA, Watt DL, Dahl BL, Sell RS, Stearns LD (1991) Selected characteristics of North Dakota farm families engaged in sustainable agricultural practices (no 1189-2016-94239)
8. Richmond S (1990) Changing Times, 102
9. Satchell M (1990) News and World Report, 31–32
10. Carson R (2009) Silent spring, 1962

11. Kumar S, Nehra M, Dilbaghi N, Marrazza G, Hassan AA, Kim KH (2018) Nano-based smart pesticide formulations: emerging opportunities for agriculture. J Control Release
12. GARBAGE DO (2010) Glausiusz, Josie. Buzz: the intimate bond between humans and insects. San Francisco, Calif.: Chronicle Books, 2004. Lee, Sally. Pesticides. New York: Franklin Watts, 1991. Macfarlane, Katherine. Pesticides. Chicago, Ill.: KidHaven Press, 2007. The Truth about Environmental Hazards, 92
13. Ware GW (1982) Pesticides: theory and application. Ed. Freeman, WH, 308
14. Ruan G, Feng SS, Li QT (2002) Effects of material hydrophobicity on physical properties of polymeric microspheres formed by double emulsion process. J Control Release 84(3):151–160
15. Sun N, Wei X, Wu B, Chen J, Lu Y, Wu W (2008) Enhanced dissolution of silymarin/polyvinylpyrrolidone solid dispersion pellets prepared by a one-step fluid-bed coating technique. Powder Technol 182(1):72–80
16. Kim DJ, Jung JY (2007) Granule performance of zirconia/alumina composite powders spray-dried using polyvinyl pyrrolidone binder. J Eur Ceram Soc 27(10):3177–3182
17. Kulkarni AR, Soppimath KS, Aminabhavi TM (1999) Controlled release of diclofenac sodium from sodium alginate beads crosslinked with glutaraldehyde. Pharm Acta Helv 74 (1):29–36
18. Devassine M, Henry F, Guerin P, Briand X (2002) Coating of fertilizers by degradable polymers. Int J Pharm 242(1–2):399–404
19. Rosenkranz K, Kasper MM, Werther J, Brunner G (2008) Encapsulation of irregularly shaped solid forms of proteins in a high-pressure fluidized bed. J Supercrit Fluids 46(3):351–357
20. Dhumal SS, Wagh SJ, Suresh AK (2008) Interfacial polycondensation—modeling of kinetics and film properties. J Membr Sci 325(2):758–771
21. Konradsen F (2007) Acute pesticide poisoning–a global public health problem. Dan Med Bull 54(1):58–59
22. Kang M, Myung SJ, Jin HJ (2006) Nylon 610 and carbon nanotube composite by in situ interfacial polymerization. Polymer 47(11):3961–3966
23. Nair R, Varghese SH, Nair BG, Maekawa T, Yoshida Y, Kumar DS (2010) Nanoparticulate material delivery to plants. Plant Sci 179(3):154–163
24. Khot LR, Sankaran S, Maja JM, Ehsani R, Schuster EW (2012) Applications of nanomaterials in agricultural production and crop protection: a review. Crop prot 35:64–70
25. Mahjub R, Dorkoosh FA, Amini M, Khoshayand MR, Rafiee-Tehrani M (2011) Preparation, statistical optimization, and in vitro characterization of insulin nanoparticles composed of quaternized aromatic derivatives of chitosan. AAPS Pharmsci Tech 12(4):1407–1419
26. Tebaldi ML, Belardi RM, Montoro SR (2016) Polymers with nano-encapsulated functional polymers. In: Design and applications of nanostructured polymer blends and nanocomposite systems. Elsevier, Amsterdam, The Netherlands
27. Han LL, Bi LW, Zhao ZD, Xing YL (2012) Microencapsulation exploration of squalene by ultrasonic spraying and freeze-drying. In Advanced materials research, vol 554. Trans Tech Publications, pp 1835–1840
28. Wang N, Qi L, Wang Y, Li XG (2017) Preparation and performance of thermo-sensitive pyraclostrobin microcapsules. Chin J Pestic Sci 19:381–387
29. Huang B, Chen F, Shen Y, Qian K, Wang Y, Sun C, Zhao X, Cui B, Gao F, Zeng Z, Cui H (2018) Advances in targeted pesticides with environmentally responsive controlled release by nanotechnology. Nanomaterials 8(2):102
30. Guan W, Tang L, Wang Y, Cui H (2018) Fabrication of an effective avermectin nanoemulsion using a cleavable succinic ester emulsifier. J Agric Food Chem 66(29):7568–7576
31. Zhou YF, Wu J, Chen J, Nie WY (2007) Applied research of APEP-type polymerizable emulsifier in the preparation of pesticide nanocapsule [J]. J Anhui Univ (Nat Sci) 3
32. Pan G, Bao YJ, Xu J, Liu T, Liu C, Qiu YY, Shi XJ, Yu H, Jia TT, Yuan X, Yuan ZT (2016) Esterase-responsive polymeric prodrug-based tumor targeting nanoparticles for improved anti-tumor performance against colon cancer. RSC Adv 6(48):42109–42119

33. Roy A, Singh SK, Bajpai J, Bajpai AK (2014) Controlled pesticide release from biodegradable polymers. Cent Eur J Chem 12(4):453–469
34. Song Q, Mei XD, Huang QL, Wang ZY, Ning J (2009) Preparation of abamectin microcapsules by means of emulsion polymerization and it bioactivity. Chin J Pestic Sci 11:392–394
35. Shang Q, Shang ZH, Ci SY (2007) HPLC analysis of abamectin-nanocapsules suspension-concentrate. Pestic-Shenyang 46(3):185
36. Wu J, et al (2008) Preparation of natural pyrethrum nanocapsule by means of microemulsion polymerization. Polym Mater Sci Eng 24(2):38
37. Liu B, Zhou X, Yang F, Shen H, Wang S, Zhang B, Zhi G, Wu D (2014) Fabrication of uniform sized polylactone microcapsules by premix membrane emulsification for ultrasound imaging. Polym Chem 5(5):1693–1701
38. Feng JG, Xu Y, Luo XR, Yan H, Wu XM (2011) Discussion on the solvent evaporation method for preparation of microcapsules and the development of the pesticides microcapsules. Chin J Pestic Sci 13:568–575
39. Cao M, Guan P, Hu L, Chen XP (2009) The research progress and influential factors on preparation of microcapsules by W/O/W multiple emulsion-solvent evaporation method. Chem Bioeng 9(002)
40. Zhou XQ, Cao LD, Liu YJ, Li FM (2014) Preparation and performance characteristics of azoxystrobin microcapsules. Chin J Pestic Sci 16:213–219
41. Thickett SC, Gilbert RG (2007) Emulsion polymerization: state of the art in kinetics and mechanisms. Polymer 48(24):6965–6991
42. Göpferich A, Luschmann C (2017) Solutions of lipophilic substances, especially medicinal solutions, Universitaet Regensburg. US patent 9,700,509
43. Piacentini E, Drioli E, Giorno L (2014) Membrane emulsification technology: twenty-five years of inventions and research through patent survey. J Membr Sci 468:410–422
44. Li M, Rouaud O, Poncelet D (2008) Microencapsulation by solvent evaporation: state of the art for process engineering approaches. Int J Pharm 363(1–2):26–39

Chapter 6
Characteristics and Types of Slow/Controlled Release of Pesticides

Ayilalath Ashitha and Jyothis Mathew

Abstract Control release formulations of pesticides are designed to improve the delivery and effectiveness of pesticides for a long period of time. These formulations also can prevent loss by physical, chemical, and microbial processes. Non-persistent pesticides are mainly benefited by control release formulations. Nanopesticides are new trends in forming controlled and slow release formulations. It can enhance stability, decrease volatility, control the release of core material, and reduce or obscure odors in the pesticide formulations. Different types of nanopesticides such as nanocapsulated formulations, nanogels, nanospheres, nanoemulsions and metal and metal oxide nanoparticles have been developed by researchers. Nanopesticide formulations are also environment friendly by improved pesticide utilization and reduced residue and pollution.

Keywords Controlled release · Nanopesticides · Slow release · Characteristics · Mechanism of release

6.1 Introduction

Pesticides are substances that are intended to kill pests. The term "**-cide**" comes from the Latin word which means "**to kill**." In general, they are chemical or biological agents such as virus, bacterium, antimicrobials, or disinfectant that prevent, injure, or kill pests. In the different type, the intended pests include insects, plant pathogens, weeds, mollusks, birds, mammals, fish, roundworms, and microbes that destroy goods, cause trouble, spread disease, or disease vectors. Pesticides are crucial to modern agricultural production as they can effectively reduce or control a variety of agricultural pests that can damage crops and livestock and decrease farm productivity, to improve crop yields [1].

A. Ashitha (✉) · J. Mathew
School of Biosciences, Mahatma Gandhi University, Priyadarshini Hills P.O, Kottayam, Kerala, India
e-mail: ash.9199999@gmail.com

Rakhimol K. R. et al. (eds.), *Controlled Release of Pesticides for Sustainable Agriculture*, https://doi.org/10.1007/978-3-030-23396-9_6

A good pesticide should have the following characteristics:

(i) It should be biodegradable.
(ii) It should not kill the friendly insects. In other words, it should selectively kill the harmful pests only.
(iii) It should not be toxic to man.
(iv) It should be cheap and easy to use.

6.2 Classification of Pesticides

Pesticides can be classified based on their target, function, and chemistry.

6.2.1 Target Classification

See Table 6.1.

6.2.2 Function Classification

See Table 6.2.

Table 6.1 Classification of pesticide based on target effected

Pesticide type	Target
Bactericide	Bacteria
Defoliant	Crop foliage
Desiccant	Promote crop drying as a harvest aid
Fungicide	Fungi
Herbicide	Plants
Insecticide	Insects
Miticide	Mites and ticks
Molluscicide	Slugs and snails
Nematicide	Nematodes
Plant growth regulator	Growth processes of crop plants
Plant-incorporated protectant (PIP)	Transgenic event for insecticidal protein
Rodenticide	Rodents
Wood preservative	Wood destroying organisms

Table 6.2 Classification of pesticide based on its function

Type	Function
Contact	Kill by contact with pest
Eradicant	Effective after infection by pathogen
Fumigants	Enter pest in gas state
Non-selective	Poisonous to both crop and weed
Post-emergence	Effective when applied after crop or weed emergence
Pre-emergence	Effective when applied after planting and before crop or weed emergence
Preplant	Effective when applied prior to planting
Protectants	Effective when applied before pathogen infects plant
Selective	Poisonous only to weed
Soil sterilant	Poisonous to all vegetation
Stomach poison	Kills animal pests after ingestion
Systemic	Adsorbed and moved through crop or pest

6.2.3 Chemical Classification

Pesticides can be either organic (carbon containing) or inorganic. Generally used inorganic pesticides are copper fungicides, lime sulfur used to control fungi and mites, boric acid used for cockroach control and ammonium sulfamate herbicides. Organic insecticides include both synthetic and natural (extracted from plants or bacteria). Synthetic organic pesticides are mostly used nowadays, and they can be grouped into chemical families based on their structure and mode of action in the pest.

Based on the toxicity, pesticides that are used in the landscapes are placed into categories. The inherent poisonous nature or how hazardous a pesticide is under experimental circumstances determines its toxicity. Toxicity is normally expressed as LD_{50}, the dose required to kill 50% of a pest population. It can be considered as either acute or chronic toxicity. Immediate effects of a single, short-term exposure to a pesticide are acute toxicity whereas repeated exposure to a pesticide refers to chronic toxicity. The mode of entry of pesticides to the body, such as orally, dermally or by inhalation, can determine the LD_{50}. This detail is estimated to humans. The lower value of LD_{50} means the pesticide is more toxic to humans [2]. Based on the level of toxicity, the pesticides are categorized and indicated by appropriate words (Table 6.3).

6.3 Chemical and Physical Characteristics of a Pesticide

The knowledge on the chemical and physical characteristics of pesticide helps the applicator to make better conclusion about which pesticide active ingredient or formulation to apply for particular situation.

Table 6.3 Classification based on the toxicity of pesticides

Category I	Contain signal words—Danger/poison In addition, a skull and cross-bones symbol is required on labels for all Category I pesticides, which are described as highly toxic. These pesticides have an acute oral LD_{50} range of 0–50 mg/kg
Category II	Contain signal word—Warning Pesticides in this category are described as moderately toxic and have an acute oral LD_{50} range of 50–500 mg/kg
Category III	Contain signal word—Caution These are slightly toxic pesticides with an acute oral LD_{50} range of 500–5000 mg/kg
Category IV	Contain signal word—Caution These are very low toxicity pesticides that have an acute oral LD_{50} greater than 5000 mg/kg

Chemical characteristics

Significant chemical characteristics are solubility, adsorption, persistence, and volatility.

6.3.1 Solubility

The ability of a pesticide to dissolve in a solvent, usually water, is termed as its solubility. Solubility measurements are subjected to temperature, pH, polarity of the substance, molecular size, hydrogen bonding, and the method used [3]. The increased water solubility of pesticide means greater the potential for runoff and leaching. The highly water-soluble pesticide remains dissolved in the water and moves around in the environment. They may most probably contaminate nearby rivers, lakes, streams, wells, and storm sewers, especially following heavy rainfall or excessive irrigation. If groundwater sources are close to the surface or if soils are sandy, highly water-soluble pesticides are not a good choice for use in these areas as the dissolved pesticide has a greater possibility of moving into groundwater. But less water-soluble pesticides are less likely to move around in the environment as they deposit in the soil or on plants more quickly. The pesticides with values greater than 1000 ppm are considered as soluble whereas those with a value of 100 ppm or less are considered comparatively insoluble.

6.3.2 Adsorption

Adsorption is the process by which a pesticide binds to soil colloids. Water solubility, pH, polarity, size, and configuration mostly establish a molecule's adsorption

[4]. Soil colloids are microscopic inorganic and organic particles in the soil, which have an extremely large surface area in proportion to given volume. Adsorption occurs as a result of an attraction between the soil particles and the chemical. Usually the oil-soluble pesticides are more attracted to the organic matter and clay particles in the soil than water-soluble pesticides. Since the soil particles are negatively charged, pesticides with positive charge adsorbs tightly to them and thus, less likely to move from application site.

6.3.3 Persistence

Persistence is the capability of a pesticide to stay present and active in its original form throughout an extended period before degrading and is described in terms of its half-life. Persistence of the pesticide relates to rate of degradation. Persistent pesticide residues provide long-term pest control and reduce the requisite for frequent application. But prevention of persistent pesticides from moving off-site through improper handling, application, drift, leaching, or runoff is essential as they can harm sensitive plants and animals including humans. Application of persistent pesticides is hazardous to humans and non-target animals entering the treated area and the presence of illegal residues may be detected on rotational food or feed crops.

6.3.4 Volatility

The tendency of a pesticide to turn into gas or vapor is termed as its volatility. The potential of pesticide to volatilize is calculated by its vapor pressure. More volatile pesticides have high vapor pressure. The volatile pesticides have greater possibility for drift as the vapors can move off-site and cause injury to non-targeted plants.

The pesticide formulation or physical form of a pesticide can also travel throughout the environment. For example, pesticides in the form of granules can be easily carried by wind or water into undesirable regions (Table 6.4).

6.4 Controlled/Slow Release Formulations

In 1960s, scientists who were working on pesticides recognized the cost and limitations in the design of new pesticides and began to work on improving the delivery of pesticides. Thus resulted in the use of controlled-release (CR) technology during design of pesticide formulations, which is effective for a long period of time. CR formulations of pesticides are based on the amount, the time, and the space of release has an important potential in reducing the consumption and

Table 6.4 Characteristics of major pesticides

Type	Example	Characteristics
Insecticides inorganic chemicals	Mercury, lead, arsenic, copper, sulfate	Highly toxic to many organisms, persistent, bioaccumulates
Organochlorines	DDT, methoxychlor, hepatachlor, HCH, pentachlorophenol, chlordane, toxaphene, aldrin, endrin, dieldrin, lindane	Mostly neurotoxins, cheap, persistent, fast acting, easy to apply, broad spectrum, bioaccumulates, biomagnific
Organophosphates	Parathion, malathion, diazinon, dichlorvos, phosdrin, disulfoton, TEPP, DDVP	More soluble, extremely toxic nerve poisons, fast acting, quickly degrades, do not bioaccumulate, toxic to many organisms, very dangerous to farm workers
Carbamates and urethanes	Carbaryl (Sevin), aldicarb, carbofuran, methomyl, Temik, mancozeb	Quickly degraded, do not bioaccumulate, toxic to broad spectrum of organisms, fast acting, very toxic to honey bees
Formamidines	Amitraz, chlordimeform (fungal and galecron)	Neurotoxins specific for certain stages of insect development, acts synergistically with other insecticides
Microbes	Bacillus thuringiensis, Bacillus papillae, viral diseases	Kills caterpillars, kills beetles, and attacks a variety of moths and caterpillars
Plant products and synthetic analogs	Nicotine, rotenone, pyrethrum, allethrin, decamethrin, resmethrin, fenvalerate, permethrin, tetramethrin	Natural botanical products and synthetic analogs, fast-acting broad insecticide action, low toxicity to mammals, expensive
Fungicides	Captan, maneb, zeneb, dinocap, foipet, pentachlorophenol, methyl bromide, carbon bisulfide, chlorothalonil (Bravo)	Most prevent fungal spore germination and stop plant disease: among most widely used pesticides in USA
Fumigants	Ethylene dibromide, dichloropropane, carbon tetrachloride, carbon disulfide, methyl bromide	Used to kill nematodes, fungi, insects, and other pests in soil, grains, fruits; highly toxic, causes nerve damage, sterility, cancer, birth defects

effect of pesticides on the environment [5]. In the conventional formulations, the pesticide is applied at a higher rate than necessary initially in order to maintain activity for a reasonable period and to balance for dissipation looses. But in CR formulations the pesticide is release over time at a predetermined rate from the matrix where the pesticide is trapped. It also protects against loses by physical, chemical, and microbial processes. CR formulations are mainly useful for non-persistent pesticides.

Advantages of CR formulations over conventional formulations are:

1. Extended effective period of non-persistent pesticides.
2. Pesticide efficacy improved due to better targeting.
3. Reduced toxicity to non-targeted species.
4. Reduced losses due to environmental factors thereby savings in the cost of the active ingredient.
5. Reduced environmental contamination (reduced contamination of surface water and groundwater).
6. Greater safety for users who come into contact with pesticide formulations.
7. Less pesticide use for the same period of activity.

6.4.1 Mechanisms of Release

The principle difference between sustained release formulations and controlled release formulations lies completely not in design but on the mechanism of release [6].

a. **Diffusion Controlled Release**

The degradable and non-degradable material in the matrix confines the agent. The discharge mechanism involved the migration and diffusion inside the reservoir matrix surface and then delivered the active compound in the space between the matrix and surrounding medium and at last to the medium from the surface.

b. **Solvent-Activated Release**

The changes in hydrophobicity and altered interactions between the organic groups may affect some materials and exhibit swelling/shrinkage in water. These actions enhance the movement of the structural chains supporting the pesticide molecules to diffusion and release.

c. **Osmotically controlled Release**

Osmotic systems are promising when it comes to the delivery of poorly soluble compounds. Osmotic pressure facilitates the ejection of the active ingredient dispersed in a polymeric matrix and which cannot diffuse through it.

d. **Stimulus-responsive Release**

Materials those are smart and capable of adapting their response to environmental parameter changes show promise for controlled release.

e. **Chemically activated Release**

In biodegradable polymers, drug release takes place when a non-soluble polymer hydrolyzes into tinier non-toxic molecules.

6.4.2 Classification of Controlled Release System

Controlled release delivery systems of pesticides for agrochemicals can be generally classified as physical or chemical.

A. Physical controlled release system

In physical category, the bioactive substance is mixed with some substances which will retard its loss by volatilization and leaching, by chemical breakdown due to the action of water, air and sunlight, or by microbial degradation. The primitive type of controlled release delivery system is the use of petroleum oils with dissolved herbicides. Later on, the use of polymers as solid equivalent to oil was available. For example, plastic insecticidal strip for flying insect control it consists of a piece of polyvinyl chloride (about $0.1 \times 3 \times 30$ cm) containing dissolved organophosphorus insecticide and while using the strip was placed inside a cardboard tube with open windows to permit insect entry and hung in confined spaces.

Principles of physical controlled release systems are

1. Choice of insecticide

The chemical should have a significant vapor pressure at room temperature to maintain a level of vapor toxic to incoming insects within the open cardboard.

2. Insecticide should be good plasticizer for the polymer

The polymer should be good solvent for the insecticide at room temperature; otherwise, the strip would exude visible drops of liquid insecticide.

3. Decrease non-uniformity of low diffusive insecticides by selection of suitable geometric shapes for plastic

The concentration of insecticide throughout the plastic strip is not uniform when the material on the surface is consumed and is replaced by biocide diffuses from the interior. So in the center of the strip, the concentration will be as it is but considerably depleted at surfaces. This is due to low diffusion of insecticide in the matrix. For example, for flat slab-like geometry, spherical plastic ball is not preferred.

A significant shortcoming and serious economic problem of this initial design were relatively large amount of insecticide (nearly one-third of the original load) still left in the plastic when the strip first becomes ineffective. To avoid this formulation was designed by such a way that the pure chemical was enclosed within some reservoir and released through a thin film of plastic. A simple manifestation of this idea was microencapsulation. But the various processes of encapsulation available were difficult and therefore require large amount of money even when the pesticide is low in cost. Thus, new designs for controlled release systems are developing by considering some principles, they are:

a. Selection of polymeric matrix

The solvent power, its melting point as well as availability, coast, and biodegradability of the polymer are carefully considered.

b. Preparation of the agrochemical plastic combination

The blending of agrochemical, if thermally stable, with polymer is possible in the molten state. This solvent less system can avoid complication of differential removal of solvents.

c. Biodegradability of plastic matrices

Even though polymers are ultimately biodegradable, only some will deteriorate corresponding to the growth cycle and agrochemical needs of agricultural crops. In the order of decreasing susceptibility, the biodegradable polymers are probably proteins, polysaccharides, natural polyphenolics, polyamides, polyurethanes, polyesters, polyethers, polyvinyls, and polyolefins.

B. Chemical controlled release system

Here, the agrochemical will be chemically linked to a substrate by a distinct and identifiable chemical bond. This would entirely deactivate the agrochemical until it was released from its chemical prison by processes which occur in the soil such as hydrolysis.

6.4.3 Types of Controlled Release System

a. Matrix system

This is the most suitable and widely used method. In this the agent/pesticide is evenly dissolved or dispersed all over the material processed to requisite shape and geometry.

b. Reservoir system

For controlled release, the reservoir type carriers are commonly obtained by the method of microencapsulation. Here, the small solid particles, liquid droplets, or gas bubbles are coated with thin coating. A board variety of active materials like adhesives, live cells, active enzymes, agrochemicals, pharmaceuticals, fragrances, flavones, and inks has been encapsulated in this way.

Controlled release pesticide formulations are usually prepared by using natural polymer materials as the capping agents. Due to the low cost, free availability, non-toxicity, and biodegradability of natural polymers, they are preferred as pesticidal carriers. One of the mainly used polymers in controlled release formulations of pesticide is alginate.

Alginate is polysaccharides-based biopolymer derived mainly from brown seaweeds and bacteria. It is employed in many applications due to its low toxicity and readily availability. But the inherent disadvantages of the single alginate as matrices in controlled release preparations are poor mechanical strength, uncontrolled degradation, and extensive water uptake properties which effect in uncontrolled and unpredicted release rates of active pesticides. Alginate can be easily converted into gel by inotropic gelation process with the interaction of divalent cations such as Ca^{2+}, Ba^{2+}, or Sr^{2+} and undergo slow, uncontrolled dissolution. The leaching of pesticides while preparation of alginate beads has been enhanced by blending bi-polymeric beads of alginate with other polymers or with clays. Alginate-clay complexes have been used for various pesticides in slow release formulations such as: diquat bromide [5] 2,6-dichlorobenzonitrile [7], carbofuran, thiobencarb [8], metribuzin and alachlor [9], diuron. Several scientists have been working on this to come up with better controlled release formulations of pesticides.

Incorporation of several adsorbents into alginate-clay formulations reduces the release rate of the herbicide metribuzin. Of the adsorbents used only charcoal reduced the rate of release but caused irreversible adsorption of part of the metribuzin. Thus 1% charcoal used in place of part of clay moderated the release of herbicide [10]. Some work was reported on petroleum ether solutions of linseed oil used as binding agent for pesticides impregnation on various inorganic carriers. Then another slow release formulation of the pesticide metribuzin using alginate–kaolin and linseed oil was prepared, which can reduce the release rate of the herbicide in comparison with conventional formulations [9]. Here, the release rates were reduced due to the combination of polymeric coating formed by linseed oil on the surface of the bead and the partitioning of the herbicide between the oil in the bead and the water surrounding the bead. A work on soil mobility of controlled release formulations of alginate encapsulated metribuzin showed CR-linseed oil formulation can increase the amount of metribuzin retained in soil surface horizon and also decrease the leaching of herbicide [11].

Chitosan is a cationic polysaccharide, a water insoluble polymer mainly present in insect and crustacean exoskeletons and fungal cell walls. It is the deacetylated derivative of chitin (*N*-acetyl-d-glucosamine) [12]. Due to its non-toxic, odorless nature, biodegradability and biocompatibility in animal tissues [13], chitosan has vast applications in medical and pharmaceutical areas. Combining of chitosan with other polymers is an effective method to improve its physical and mechanical properties for applications.

Controlled release formulations of the pesticide were prepared using of alginate, bentonite, and chitosan as hydrogels by extrusion into calcium chloride solutions [14]. Bentonite can improve the hydrogel bead size and sphericity and can increase the drug loading rate and encapsulation efficiency of pesticide. Bentonite also improves the performance of the release by forming hydrogen bonds with sodium alginate and chitosan thereby effectively inhibit calcium alginate large swelling.

Alginate modified starch formulation of trifluralin is another veritable way to decrease the negative environmental effects. It was not feasible to form beads with plain starch, so blending with alginate helps in bead formation and reduce leakage of encapsulated herbicide. The cross-linking of starch with alginate through intermolecular hydrogen bonding modifies the structure of starch [15]. In another study, starch and chitosan were reinforced with alginate for slow release formulation of the herbicide imazaquin. The mechanical strength of chitosan bead is low. But it acquires higher level of mechanical strength through intermolecular hydrogen bonding which is the mechanism of starch–alginate cross-linking and can prevent leakage of encapsulated herbicide [12].

Sinclair [16] developed a novel system in which thermoplastic polymers and copolymers of glycolic and lactic acid was employed as encapsulating agents. The investigations also indicated that polyesters of lactic and glycolic acids biodegrade through hydrolysis, and the products obtained were lactic and glycolic acids, CO^2, and water.

Meng et al. [17] synthesized two different interlayer structured glyphosate (GLY) intercalated MgAI layered double hydroxides as pesticide slow/controlled release materials. They created these materials at different pH by co-precipitation and concluded that at pH 9, GLY-MgAI-LDH-pH9 possessed higher interlayer gallery height and vertical monolayer arrangement of guest anions in the interlayer with larger packing density and also exhibits better slow release property than that accumulated at higher pH 10.

A polyacrylic acid lignin hydrogel (LBPAA) composite was made by grafting lignin on the surface of polyacrylic acid network. These composite gels confirmed high loading and sustained release of a range of pesticides and also found to be sensitive to pH, temperature, and ionic strength [18].

6.5 Nanopesticides

The nanopesticides are the new trends in creating controlled and slow release pesticides after the development of nanomaterials and nanotechnology. The physical and chemical properties of nanomaterials vary from those of macroscopic materials. For pesticide delivery, nano-microcapsules or nanospheres of polymeric materials with light-sensitive, thermo-sensitive, humidity-sensitive, enzyme, and soil pH sensitive were prepared through processes such as adsorption, coupling, encapsulation, and embedding. Thus, the active ingredients of the pesticide can be protected by such formulations and also enhance stability, decrease volatility, control the release of core material, and reduce or obscure odors. Due to the small size, improved pesticide droplet ductility, wettability, and target adsorption when spraying in fields the nanocapsule formulations have environment friendly advantages such as improved pesticide utilization and reduced residue and pollution.

Nanopesticides are minute engineered structures that deliver pesticidal properties or formulation of active ingredient of pesticide in nanoform. They show long time controlled release of active ingredients and slow degradation, which make them less toxic and environmental friendly while comparing with chemical pesticides. Different types of nanopesticides have been developed by researchers such as nanocapsulated formulations, nanogels, nanospheres, nanoemulsions and metal and metal oxide nanoparticles.

Different types of polysaccharide materials have been integrated into the production of nanocapsules, i.e., polyethylene glycol, chitosan, starch, cellulose, and polyester substance. Nanopesticides active against larvae of Culex quinquefasciatus were developed by PEG encapsulated by temephos and imidacloprid with melt-dispersion method. The controlled release of atrazine with PCL nanocapsule enhanced herbicidal activity and also minimized the harmful effect of herbicide than commercial formulations. For fungicide applications, solid lipid and polymeric nanocapsules loaded with carbendazim and tebuconazole were developed.

To improve solubility and spreading capacity of pesticides, nanoemulsions were developed by dispersion into the liquid phase. Nanoemulsions are useful in less deprivation and volatilization of active ingredient and improve the bioavailability for long period of time [19]. Nanoformulation such as formulation of pyridalyl, oil-core silica-shell nanocapsule for fipronil, insecticide coated liposome, Artemisia arborescens essential oil, neem oil, garlic essential oil, imidacloprid, thiamethoxam, carbofuran, thiram and β-cyfluthrin, carbofuran, mancozeb, atrazine and simazine, lansiumamide B were developed to improve efficiency, stability and reduction of effective pesticide concentration.

In 2011, Qian et al. [20] prepared nanocalcium carbonate by reversed-phase microemulsion method and loaded with pesticide validamycin. By changing the water/surfactant molar ratio, the size of nano-CC was adjusted between 50 and 200 nm. This material exhibited enhanced germicidal activity against Rhizoctonia solani compared to usual technical validamycin after about seven days, and the release time of validamycin was expanded to two weeks. Patel et al. [21] developed cypermethrin loaded calcium alginate nanocarriers. The release of cypermethrin from nanocarriers was evaluated, and the mechanism of release of cypermethrin was found to be anomalous transport directed by diffusion and relaxation of alginate chains. In this study, the pesticide molecule occupies the inner side of the nanocarriers which resulted in the decreased size of nanocarriers from the range 108–127 to 115–119 nm.

6.6 Conclusion

The modern technology used for the development of controlled and slow release formulations of pesticides are nanotechnology, thereby nanopesticides are formulated. Different types of nanopesticides developed are currently available in the market and are effective and less harmful than conventional formulations.

References

1. Sun C, Shen Y, Huang B, Zeng Z, Qian K, Gao F, Cui B, Cui H, Chen F, Zhao X, Wang Y (2018) Advances in targeted pesticides with environmentally responsive controlled release by nanotechnology. Nanomaterials 8:102

2. Cloyd RA (2013) IPM of midwest landscapes: tatics and tools for IPM, chapter 8 pesticide characteristics and use. University of Minnesota, pp 40–41
3. Zacharia TJ (2012) Identity, physical and chemical properties of pesticides. Pestic Mod World Trends Pestic Anal 1873
4. da Cunha JPAR, de Oliveira JPR, Pereira VJ, de Morais TP, de Morais JB (2016) Physical-chemical properties of pesticides: concepts, applications, and interactions with the environment. Biosci J 32:627–641
5. Fernández-Pérez M, Villafranca-Sánchez M, Flores-Céspedes F, Garrido-Herrera FJ, Pérez-García S (2005) Use of bentonite and activated carbon in controlled release formulations of carbofuran. J Agric Food Chem 53:6697–6703
6. Lewis DH, Cowsar DR (1977) Principles of controlled release pesticides. In: ACS symposium series American Chemical Soceity, pp 1–16
7. Connick WJ Jr (1982) Controlled release of the herbicides 2,4-D and dichlobenil from alginate gels. J Appl Polym Sci 27:3341–3348
8. Gan J, Hussain M, Rathor NM (1994) Behaviour of an alginate-kaolin based controlled-release formulation of the herbicide thiobencarb in simulated ecosystems. Pestic Sci 42:265–272
9. Pepperman AB, Kuan JCW (1993) Slow release formulations of metribuzin based on alginate-kaolin-linseed oil. J Control Release 26:21–30
10. Pepperman AB, Kuan JCW, McCombs C (1991) Alginate controlled release formulations of metribuzin. J Control release 17:105–112
11. Johnson RM, Pepperman AB (1995) Mobility of atrazine from alginate controlled release formulations. J Environ Sci Heal Part B 30:27–47
12. Nnamonu LA, Sha'Ato R, Onyido I (2012) Alginate reinforced chitosan and starch beads in slow release formulation of imazaquin herbicide—preparation and characterization. Mater Sci Appl 03:566–574
13. Borzacchiello A, Ambrosio L, Netti PA, Nicolais L, Peniche C, Gallardo A, Roman JSAN (2001) Chitosan-based hydrogels : synthesis and characterization 2:861–864
14. Yan H, Feng Y, Hu W, Cheng C, Liu R, Wang C, Li J, Lin Q (2013) Preparation and evaluation of alginate-chitosan-bentonite based beads for the delivery of pesticides in controlled-release formulation. Asian J Chem 25:9936–9940
15. Onyido I, Sha'Ato R, Nnamonu LA (2012) Environmentally friendly formulations of trifluralin based on alginate modified starch. J Environ Prot (Irvine, Calif) 03:1085–1093
16. Sinclair RG (1973) Slow-release pesticide system polymers of lactic and glycolic acids as ecologically beneficial, cost-effective encapsulating materials. Environ Sci Technol 7: 955–956
17. Meng J, Zhang H, Evans DG, Duan X (2005) Novel layered pesticide slow/controlled release materials- supramolecular structure and slow release property of glyphosate intercalated layered double hydroxides. Chinese Sci Bull 50:745–751
18. Sun Y, Ma Y, Ren S, Fang G, Fu Y (2016) Controlled pesticide release from porous composite hydrogels based on lignin and polyacrylic acid. BioResources 11:2361–2371
19. Chhipa H (2017) Nanopesticide: current status and future possibilities. Agric Res Technol Open Access J 5:10–13
20. Qian K, Shi T, Tang T, Zhang S, Liu X, Cao Y (2011) Preparation and characterization of nano-sized calcium carbonate as controlled release pesticide carrier for validamycin against Rhizoctonia solani. Microchim Acta 173:51–57
21. Patel S, Bajpai J, Saini R, Acharya S, Bajpai AK (2018) Sustained release of pesticide (cypermethrin) from nanocarriers: an effective technique for environmental and crop protection. Process Saf Environ Prot 117:315–325

Chapter 7
Advanced Controlled Nanopesticide Delivery Systems for Managing Insect Pests

R. K. Saini, S. Patel, J. Bajpai and A. K. Bajpai

Abstract The need to produce food in sufficient quantities for supplying to the growing population has resulted in the development of a wide range of new technologies. There is always a loss of crops due to pests and, therefore, pesticides have become an essential part of the agricultural system to prevent crop loss. In recent past, a variety of pesticides as organochlorines, organophosphates, carbamates, and pyrethroids have been developed to combat the insect pests, however, conventional methods of using pesticides have severe adverse effects such as reduction in the agricultural yield and undesirable health effects on human and the environment by various processes of degradation of pesticides which results in less than 0.1% of the applied pesticides to be used as active part to kill the target pests. In recent years, advanced nano-controlled delivery systems have been developed for agrochemicals based on biodegradable polymers to maintain optimal pesticide concentrations, protect and stabilize the pesticides and reduce the frequency of administration. This chapter presents an overview of the nanoformulations used as systems for controlled release of pesticides and concentrates on the modern strategies adopted for the delivery of pesticides and their limitations and potential in achieving sustainable agriculture management.

Keywords Nanopesticide · Controlled release · Pest control

7.1 Introduction

The emergence of green revolution has resulted in intensive agricultural practices which have eventually led to the progressive depletion of soil nutrients. The abnormal growth in agriculture production has only been achieved due to the excessive use of agrochemicals which have adversely affected the soil and water

R. K. Saini · S. Patel · J. Bajpai · A. K. Bajpai (✉)
Bose Memorial Research Laboratory, Government Model Science College, Jabalpur 482001, Madhya Pradesh, India
e-mail: akbmrl@yahoo.co.in

Rakhimol K. R. et al. (eds.), *Controlled Release of Pesticides for Sustainable Agriculture*, https://doi.org/10.1007/978-3-030-23396-9_7

quality of the area. The contribution of agriculture to the requirements of health, nutrition, as well as economical developments is globally recognized [1]. There has been obtained a higher yield of crops which has been possible only due to the development of newer high-quality seeds, applications of fertilizers and pesticides in excess amounts then the required ones, and using advanced and high-performance agriculture machinery. Agricultural improvements have been the major sources of boon in meeting out the global requirements of food. However, numerous environmental pollutions issues such as deforestation, soil erosion, salination, and damages to ecosystems have encountered criticisms due to the excessive use of pesticides and other agrochemicals [2–4].

Agriculture contributes positive and negative externalities to society. A typical feature of the negative externalities is the damage caused by nutrients pollution and by pesticide [5]. To fulfill the food requirements of the world, it is important to use high-technology agriculture practices including pesticides and micronutrients to maximize the production of high-quality crops [6, 7]. Today, agrochemicals are used in numerous crops against a large spectrum of weeds, pests, and diseases [8]. Pesticides are commonly used for the control of pests of agriculture and vectors of human and animal diseases [9]. However, their excess use is badly affecting human health and the environment. The term pesticide includes herbicide, insecticide, insect growth regulator, nematicide, termiticide, molluscicide, piscicide, avicide, rodenticide, predacide, bactericide, insect repellent, animal repellent, antimicrobial, fungicide, disinfectant (antimicrobial), and sanitizer [10]. Presently, herbicides have

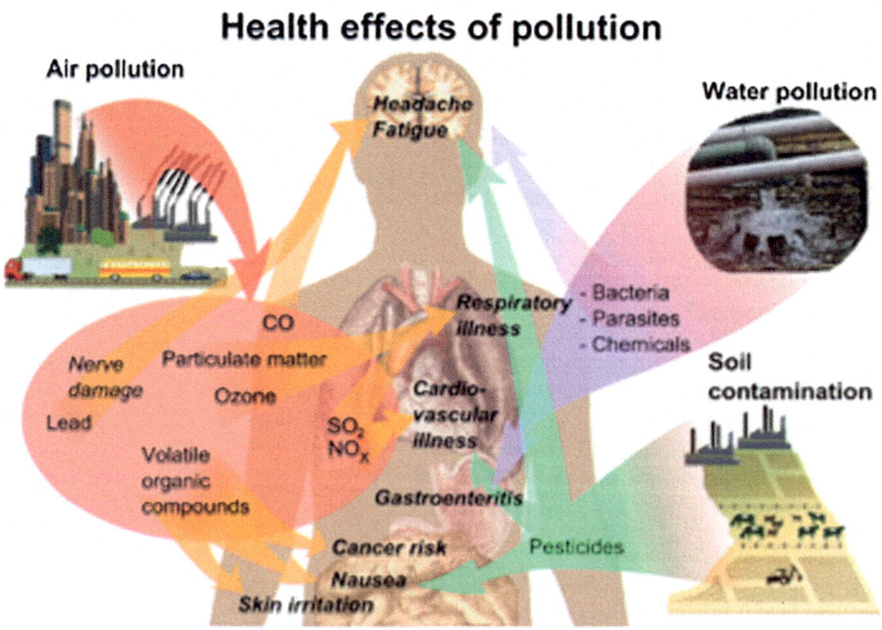

Fig. 7.1 Health effects of pollution due to pesticides

approximately 80% worldwide consumption of total pesticides [11] used as crop-protection materials for plants from weeds, fungi, or insects.

A survey has shown that agricultural producers have used different types of agrochemicals including pesticide for the enhanced production and protection of crops [12]. Among the surveyed farms, 55% used insecticide, 23% used fungicide, 22% farms used herbicides, and 3% used others [13]. However, the main limitation in the use of such chemicals in agriculture is their harmful effects on the environment and human health because most of the pesticide released into the environment is lost before reaching its target. Losses occur as a result of physical, chemical and biological factors. A report states that every year millions of agriculture worker and farmers of developing countries suffer from poisoning due to pesticides and other agrochemicals [14] (Fig. 7.1). Hence, there is an urgent requirement to develop technology for safe use and application of these chemicals [15].

7.2 Application of Insecticides Nanoformulations

In conventional agriculture practices, granular (GR) formulations, solution concentrates, emulsifiable concentrates, wettable powder (WP) formulations, and suspension concentrates (SC) technology use excess active ingredient such as mineral fertilizers and unsafe pesticides as a mixture with other inert materials for their effective application in the field. However, the conventional formulation methods have major drawbacks of burst and immediate release of active ingredients that are quickly lost by volatilization, evaporation, and leaching as well as various degradation processes takes place in the presence of light, water, and microbes. This consequently results in the quick loss of concentration and effectiveness of active ingredients, and, therefore, more amount of such active ingredients are required to maintain their biological efficiency to kill pests that leads to various environmental and health problems. Traditional strategies such as crop rotation, healthy crop variety, manipulations in sowing dates, and integrated pest management (IPM) (combination of techniques such as biological control, habitat manipulation, and modification of agronomic practices) have been commonly used by farmers for the management of insect pest in agriculture. It is, however, to note that if an excess amount of pesticide is used, it may sometimes cause phytotoxicity.

Government and regulatory authorities are giving directions to agrochemical producers to develop formulations which must be safe in manufacturing and use, reduce the amount of applied pesticides, produce less waste and effluents, as well as develop multiple active ingredient formulations for control and target release of pesticides to enhance biological activity and reduce pesticide dosage.

To overcome these problems and less efficiency of traditional methods, advanced technology approaches for the management of insect pest must be used as safer tools for agriculture. Nanotechnology has gained much attention as new technology not only to increase crop yields with quality but to detect and treat the plant diseases, deliver active ingredients to target sites, and enhance the absorption

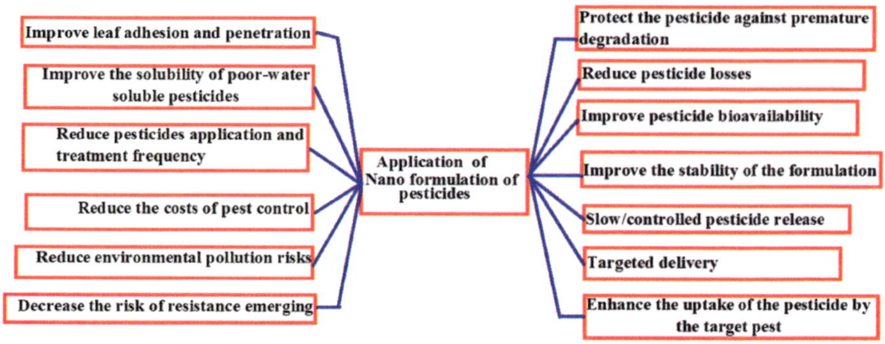

Fig. 7.2 Various applications of nanoformulation in agriculture

of nutrients in plants. Advanced high-performance eco-friendly nanomaterials have been developed not only to improve the efficiency of active ingredients in the field but also reduce the damage to non-target plant tissues and minimize harmful environmental pollutions [16]. There is a pressing need to apply nanotechnology, and this warrants a detailed study in this area. The applications of nanoformulation in agriculture are listed as shown in Fig. 7.2.

7.2.1 Nanoemulsions

Nanoemulsions are kinetically stable transparent non-toxic and non-irritant colloidal particulate systems which are mini-emulsion of either oil in water or water in oil and surfactant (for the stability of systems that decreases the interfacial tension) in the size 10–1000 nm and cannot be formed spontaneously (Fig. 7.3). They have gained much interest as carriers of drug molecules; improve the bioavailability of drug and physical stability due to their greater surface area, robust stability, tunable rheology, as well as they can be easily formulated in the form of foams, creams, liquids, and sprays by dispersion or high-energy emulsification methods and low-energy methods [17]. A phase inversion temperature (PIT) was developed by Shinoda and Saito that works as a low-energy emulsification industrial method for the formation of nanoemulsion. In order to enhance the uptake of agricultural agents (AIs), nanoemulsions have been used [18, 19].

7.2.2 Classical Micro- and Nanoparticles

Nanomaterials may be spherical, tubular, or irregular in shapes and have size less than 100 nm in one dimension (e.g., surface films or surface coatings in circuitry of

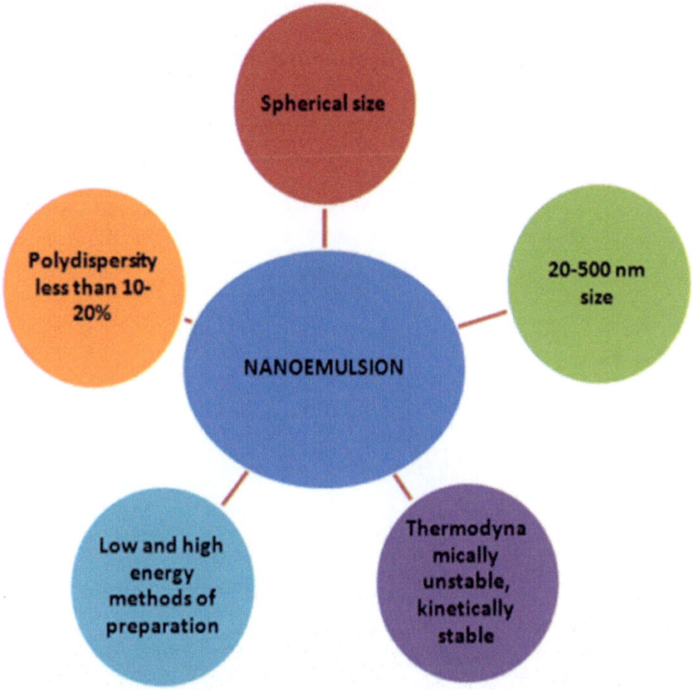

Fig. 7.3 An illustrative presentation of various properties of nanoemulsion

computer chips), two dimensions (e.g., strands or fibres, or nanopore filters used for small particle separation and filtration), or three dimensions (e.g., nanoparticles with various morphologies) as single, dispersed aerosols, as suspensions/colloids or aggregated forms and show properties different from those of the bulk particles due to their greater surface area [20–22].

Nanoparticles can be classified as natural environment nanomaterials and synthesized or engineered nanomaterials on the basis of their source or their manufacturing, respectively. Natural environmental nanoparticles are mostly metal oxides and metal sulfides found in minerals. They can bind with contaminants, responsible for causing chemical reactions in soil and water, synthesized by various biological and chemical processes and occur in nature. At present, many environmental nanoparticles have been discovered by scientist in the last 20 years, but their properties and significance to the environment are poorly understood and require more research [23].

Synthesized or engineered nanoparticles (e.g., silicon dioxide, titanium dioxide, cerium oxide, and iron oxides) are designed in laboratory by using synthetic or natural polymers, metals, etc. with specific properties or compositions (e.g., shape, size, surface properties, and chemistry) and widely used in numerous biomedical

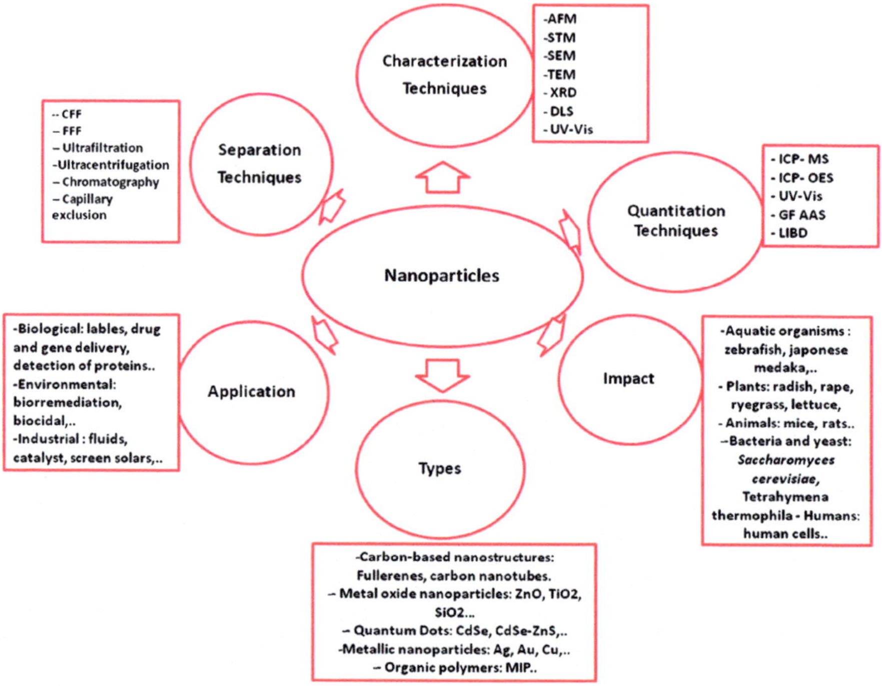

Fig. 7.4 Diagrammatic depiction of properties and application of nanoparticles

and environment applications and provide opportunities to scientists across many disciplines to manipulate properties that might otherwise be incompatible on a single device (Fig. 7.4).

7.2.3 Entomopathogenic Microorganisms Encapsulated Novel Micro- and Nanoparticles for Bioinsecticides

In recent years, due to worldwide pesticide pollutions, there are demands for pesticide-free agriculture products, and biological control-agent-based materials for crop production. Microorganisms such as *Pseudomonas deceptionensis*, *Brevibacterium frigoritolerans*, *Weissella oryzae*, *Bacillus methylotrophicus*, and *Bhargavaea indica* are used as nano-factors to synthesize metal nanoparticles because they are eco-friendly, less toxic, produced reductase enzymes that accumulate and detoxify heavy metals, and cost-effective [24].

Recently, entomopathogenic fungi such as *Metarhizium brunneum*, *Metarhizium pingshaense*, *Metarhizium anisopliae*, and *Metarhizium robertsii* have gained attention as biological control agents for controlling root-feeding herbivore in potato plants, and soil-borne insect pests like wireworms in the field as they do not affect beneficial non-target organisms such as bees and earthworms. Ca-alginate-based hollow beads and coated beads are used as gentle and non-toxic materials for the encapsulation of fungal conidia as they provide better handling, reduced dusting, high flowability, high shelf life, reduced abrasion, and enhanced establishment in soil [25–27].

Ureña-Saborío et al. [28] used green chemistry to encapsulate bacterial metabolic infiltrates isolated from different bacillus strains as biological control agents in biopolymer (chitosan) nanoparticles against *Radopholus similis* in order to improve the persistence and spread of these biocontrol agents in the soil.

7.2.4 Commercial Products

Since 1930, synthetic pesticides have been synthesized and widely used as active ingredient to protect crops from pests but their excessive use to fulfill the food requirements of world has created environmental pollution and damages to human health as well as non-target sites because only 0.1% of the active ingredient reaches at the target pests by conventional methods. Therefore, controlled release technology has been developed as a versatile tool to address the problems originating from the excessive use of pesticides.

Carvalho [29] reported that for safe food production and sustainable development of agriculture, fisheries, and aquaculture, it is necessary to advance the agricultural practices, and educate the farmers and users about careful use of pesticides, as well as prior testing and careful risk assessment of pesticides, measures must be taken for the better protection of environmental ecosystems.

Melo et al. [30] conducted adsorption experiments for evaluating the release of agriculture active ingredients at different pH, using an aqueous medium as well as soil and plants and found that use of tropical peat could be used as materials to improve agriculture productivity.

Tebuthiuron (TBH) is the herbicides that are mostly used to kill weed in cotton and sugar crops, and highly water soluble and degrade by metabolites thus causing underground as well as surface water contaminations. Therefore, there is a need for safe delivery systems for TBH to reduce the harmful effects of it and protect the environment. Moreto et al. [31] developed and investigated the calcium alginate microparticles prepared from the crosslinking of sodium alginate by Ca^{2+} ions containing varying amounts of TBH as a safer delivery system for TBH delivery for the effective control of weeds.

Kumar and coworkers [32] concluded that nanoplatforms and processes such as nano-encapsulation could be used as controlled, safer, and sustainable releases systems for agrochemicals to deliver various active ingredients to target sites, deliver macromolecules needed for improved plant disease resistance, for genetic modification of plants, efficient utilization of nutrients and enhance plant growth as well as provide new ways for crop protection.

Kanhed et al. [33] synthesized copper-based compounds using copper nitrate and isopropyl alcohol with surfactant, cetyl trimethylammonium bromide, for the release of pesticides or agricultural fungicides and found them effective against four pathogenic fungi: *Phoma destructiva*, *Curvularia lunata*, *Alternaria alternata* and *Fusarium oxysporum*. However, such nanopesticide formulations are harmful to human beings also.

7.3 Development of New Nanopesticides

Nanoparticle are nanosize materials that can be classified as inorganic, organic, or hybrid materials and possess many unique properties such as small size and greater surface area, and are capable of delivering pesticides to target sites, minimizing pesticides loss and toxicity, thus offering higher loading capacity, biodegradability, stability and long shelf life in comparison with the bulk materials (Fig. 7.5).

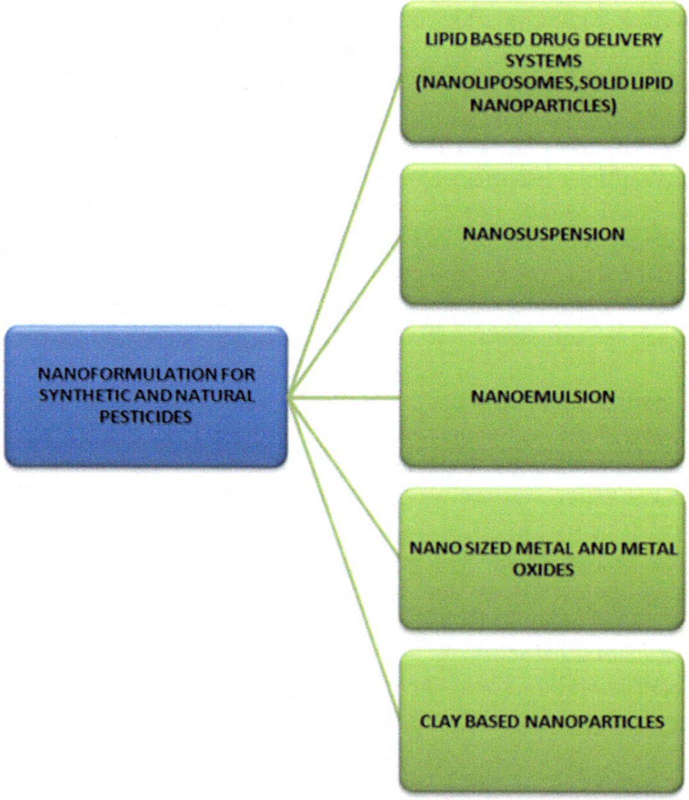

Fig. 7.5 Nanoformulation used for the delivery of synthetic and natural pesticides

There are three types of nanoparticles that can be distinguished as isodimensional nanoparticles (three dimensions in nanometers range), elongated structure (two dimensions in the nanometer range and one in larger), and polymer-layered crystal nanoparticles (only one dimension in nano range). Isodimensional nanoparticles such as silica nanoparticles can be synthesized by in situ sol-gel methods or surface polymerization. Elongated structure like nanoparticles such as carbon nanotubes or whiskers extensively studied as high-performance materials for various applications. Polymer-layered crystal nanoparticles are in the form of sheets, and they are developed by the polymerization of monomers and intercalation of the prepared polymer inside the galleries of layered host crystals.

Nanopesticides are formulations as particles or micelles that consist of organic (polymers), inorganic (metal oxide) ingredients in nm size and able to increase the solubility of poorly soluble active ingredients, release pesticides in a slow and controlled targeted manner to protect it from premature degradation.

Metal-based nanoparticles such as silver (SNP), aluminum oxide (ANP), zinc oxide, and titanium dioxide show antibacterial and antifungal properties and are effective in killing pests, plants, pathogens, and insects. Therefore, they can be used as nanoformulations for the preparation of pesticides, insecticides, and insect repellants as well as deliver DNA into plant tissues for the protection of the host plants against insect pests.

Many researchers' studies showed that porous hollow silica nanoparticles (PHSNs), nanoemulsions and essential oil-loaded solid lipid nanoparticles could be used as safer and sustainable controlled release delivery system of pesticides as insect pests use cuticular lipids for their protection. When these particles adsorb into the cuticular lipids by physisorption, they cause the death of insects.

7.3.1 Biopolymers Nanoparticles

Natural occurring polymers such as polysaccharides and proteins are cheap, readily available, biocompatible, and have low toxicity. They can also be chemically, physically, or enzymatically modified to modulate their functional behavior. Biopolymers such as starch, cellulose, chitosan, alginate, pectin, casein, and gelatin have been used for preparing super-absorbents and widely used in many environmental and biomedical applications such as wastewater treatment, drug delivery, developing hygienic product due to their unique properties such as low cost, renewable, non-carcinogenic nature, abundance, biocompatible nature, stability, and hydrophilic nature [34–55] (Fig. 7.6).

Natural nanoparticles (composed of biopolymer with a dense crosslinked matrix which has the ability to incorporate a limited amount of liquid) can play an important role as efficient delivery systems for sustained, controlled, and targeted release aiming to improve the therapeutic effects and also minimize the side effects of the pesticides by reducing the degradation rate of pesticides by delaying the release and increased residual activity [56].

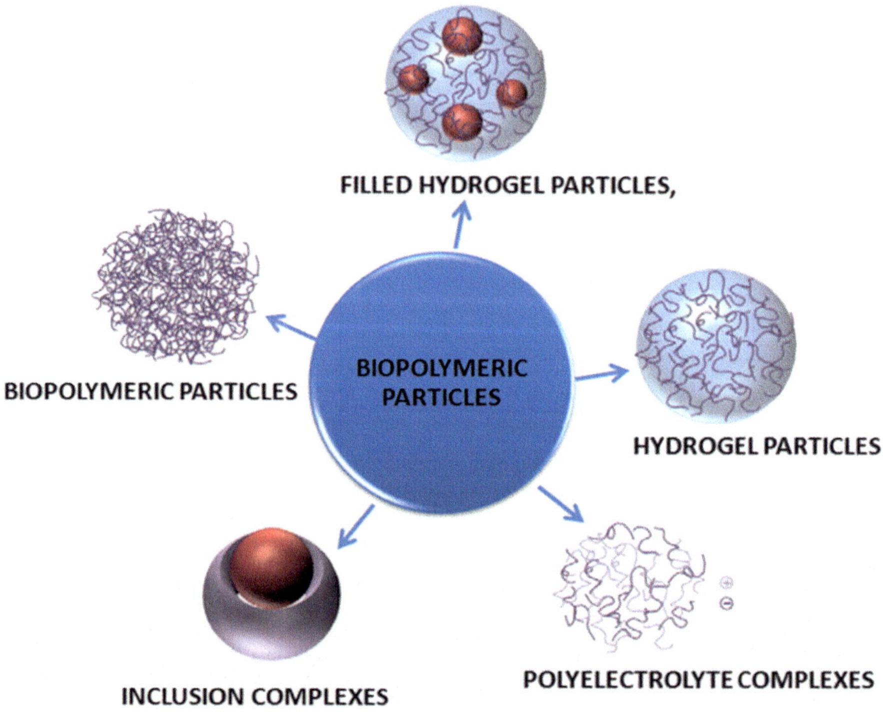

Fig. 7.6 Various forms of biopolymeric particles

Fraceto et al. [57] developed a modified controlled release system for the herbicide ametryn by using poly(hydroxybutyrate) (PHB) or poly(hydroxybutyrate-valerate) (PHBV) to improve the activity of herbicide and reduce the toxicity. They evaluated the release kinetics for encapsulated active substances from poly (hydroxybutyrate) (PHB) or poly (hydroxybutyrate-valerate) (PHBV)-based microparticles and found that the release of ametryn occurred by diffusion process due to anomalous transport and indicated that this system may be used as a safer system to reduce the impacts caused by the herbicide.

7.3.2 *Polymer-Based Nanopesticides*

Polymeric nanoparticles have gained considerable attention because they can be easily synthesized as stable and biodegradable nanospheres or nanocapsules for the encapsulation of pesticides in a uniform manner to prepare release systems. Polymeric nanoparticles do not aggregate easily due to electrostatic repulsion, steric exclusion, or a hydration layer formed by binding of polymeric molecules on the surface of the nanoparticle by some attractive interaction, either chemisorption, electrostatic attraction, or hydrophobic interaction [58, 59].

Hydrogel-based controlled delivery nanoformulations have gained considerable attention for the development of controlled release pesticide formulations because they provide not only increased water holding capacity of the soil, controlled erosion, reduced irrigation frequency but also the slow release of pesticides due to their swelling behavior. The swelling capacity of hydrogel-based nanoformulations can be improved by mixing more than one type of polymers. These systems can be classified as physical combination (laminated, reservoir, and monolithic systems) for controlling the rate of release agents, as well as chemical combination (bind pesticides through labile link) to deliver pesticides either by hydrolysis or by enzymatic degradation. Polar polymeric molecules are soluble in water while apolar polymeric chain molecules are soluble in non-polar solvents, e.g., hexane, toluene, or chloroform [59, 60].

Fernandez-Perez et al. [61] prepared poly (ethylene glycols) based nanomicelles as controlled release (CR) formulations of imidacloprid [1-(6 chloro-3-pyridinyl methyl)-N-nitro imidazolidin-2-ylideneamine] and evaluated its encapsulation efficiency, loading capacity, and stability. They found that prepared nanoformulation not only reduces the release of pesticides but also can be used for efficient pest management in different crops.

The polymeric nanoparticles can be synthesized by conjugated polymers such as polyaniline, polypyrrole, polyacetylene, polythiophenes, polyfluorenes, poly (p-phenylenevinylene)s, and poly(p-phenyleneethynylene)s derivatives as nanopesticide formulations to deliver active ingredient safely, slowly, and continuously for longer duration to at specified targets at a desired rate [62].

Bajpai et al. [63, 64] developed biodegradable delivery systems by using calcium alginate, starch, gelatin for the release of insecticides such as cypermethrin, chlorpyriphos, and evaluated in vitro release kinetics. The soil–pot experiments results clearly suggested that the as-prepared delivery systems released at lower rates in comparison with pesticides applied directly.

Recently, various types of amphiphilic-block-copolymer-(bi-block, tri-block)-based nanomaterials have been synthesized by the polymerization of more than one type of monomers possessing contrast nature, i.e., one hydrophilic and another hydrophobic to encapsulate a wide range of pesticides. Polyethylene (PEG)-based amphiphilic copolymers have more potential as controlled release systems for a series of pesticides such as thiram, thiamethoxam, carbofuran, imidacloprid, and β-cyfluthrin; because, the release rate of such nanoformulations can be controlled by varying the amount of PEG for period from 1 to 20 days and show high effectiveness to target organisms and lower toxicity to non-target organisms.

7.3.3 *Inorganic Nanoparticles as Active Ingredients*

Inorganic nanoparticles containing inorganic metal elements as the core and a protective organic coating on their surfaces that prevent it from degradation and aggregation have been suggested as controlled release systems of high performance in biomedical and environmental applications as these particles possess various size- and shape-dependent physical and chemical properties such as magnetic, optical, and electrical, as well as good stability over broad range of temperatures and pH that make it suitable for agriculture applications to deliver active ingredients with lower toxicity [65, 66] (Fig. 7.7).

Many inorganic nanoparticles such as magnetic iron oxide nanoparticles (IONPs), semiconducting quantum dots (QDs), silica nanoparticles, titanium oxide, metallic gold nanoparticles (AuNPs) exhibit unique antibacterial, light scattering, low toxicity, and good electrical properties. Thus, they are widely studied as magnetic resonance imaging (MRI) contrast agents [67], as optical probes for intracellular organelles and biomolecules [68], for molecular imaging [69], as antibacterial materials to suppress bacterial and fungal pathogens. The physical and chemical properties of inorganic nanoparticles can be improved by using surface coating of suitable polymers to achieve high surface-to-volume ratio and multi-functional behavior for possible applications in agriculture [70, 71].

Scientists and researchers have been paying much attention and showing interest in developing high-performance metallic nanoparticles with well-defined chemical composition, size, and morphology as biosensors to detect pesticides, monitoring

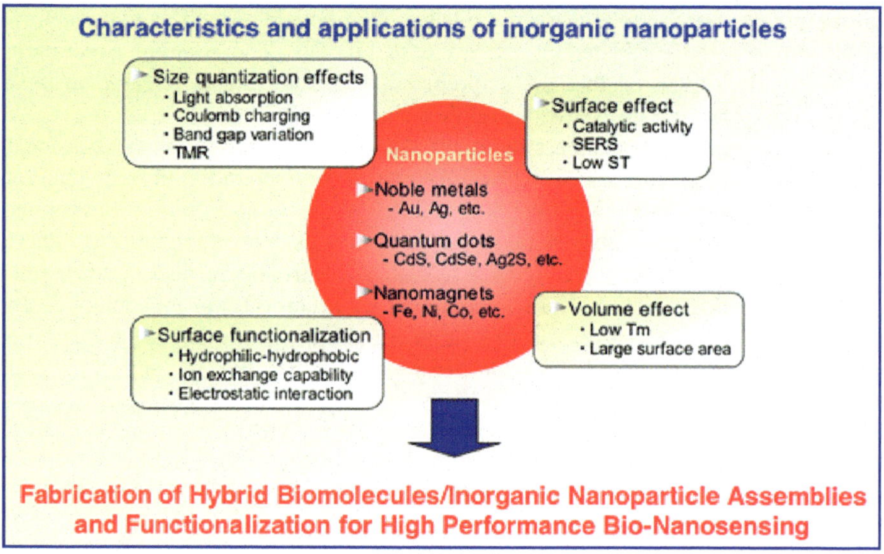

Fig. 7.7 Applications of inorganic nanoparticles

diseases, pathogens, and therapy monitoring [72–74]. Such metallic nanoparticles can be synthesized by different microorganisms and plants as green and efficient materials with unique properties such as biocompatible, nontoxic, self-cleaning, skin-compatibility, antimicrobial, and dermatological behaviors and used for various biomedical and industrial applications.

7.3.4 Hybrid Nanoformulations

The main challenges in the release of pesticides are enzymatic and biological barriers that can lead faster degradation and metabolization, as well as its physiochemical nature that separated the administration site from the site of action which may cause poisoning to healthy tissues and cause pollution. Therefore, ideal delivery systems must possess the properties of releasing the active ingredients to the target site with a slow and controlled rate to avoid its excess use reducing pollutions. Recently, various delivery devices such as carbon nanotubes, liposomes, dendrimers, and micelles have been developed as hybrid formulations (Fig. 7.8).

Liposomes are ideal candidates for molecular vectorization in biological media because they can release both hydrophilic and hydrophobic active ingredients to specific action sites. In liposomes, hydrophilic active ingredient can be encapsulated at the central aqueous core surrounded by one or several phospholipid bilayers and hydrophobic active ingredients anchored into their bilayers. However, the main limitations in vectorization of liposomes are their physical and chemical instability which is due to easy hydrolysis of their ester bond and formation of aggregate particles, requirement of organic solvents that leave toxic residual traces, leakage, releasing the encapsulated drugs before the site of action [75, 76].

Dendrimers are highly branched molecules synthesized from the core by the repetition of a sequence of reactions, containing a low-density small core and outer multiple branching units of a large number of surface groups that create multivalent interactions with hydrophobic active molecules [77].

Carbon nanomaterials such as carbon nanotubes, fullerene consist of single and multilayer of carbon, considered as optimal vehicles for the delivery of genes, peptides, and proteins but possess inert surface, hydrophobic nature, and poor biocompatibility [78, 79].

Micelles are aggregates or spherical amphiphilic molecules that range in size from 1 to 100 nm at critical micelle concentration and contain core of hydrophobic moieties and polar groups ends are in contact with surrounding water. There are two types of micelles—surfactant micelles and polymeric micelles (block or tri-block copolymers). Out of these two, polymeric micelles are more stable, monodisperse and have a diameter of 20–50 nm and show slow degradation rate which makes it ideal candidates for controlled and slow release of pesticides [80].

Solid lipid nanoparticles are lipids (surfactants, triglycerides, glycerides) based partially crystalline solid nanoparticles and can be easily synthesized by various methods such as high-pressure homogenization, microemulsion, and

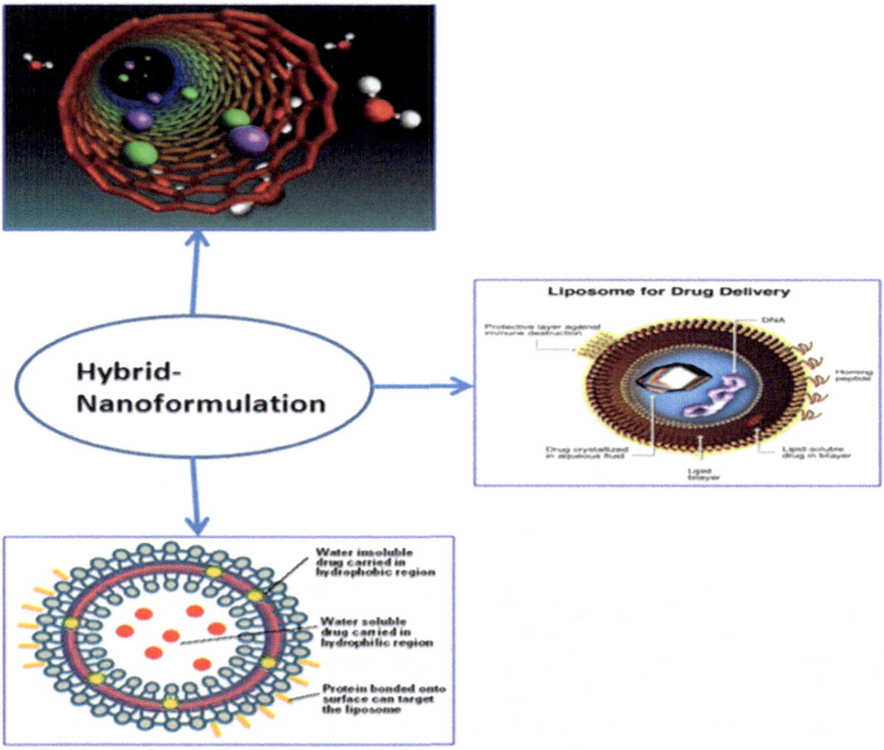

Fig. 7.8 Images showing some hybrid nanoformulation

nanoprecipitation and active ingredient molecules can be entrapped into oily compartments present in amorphous regions that allows an on-command release [81].

7.3.5 *Nanoparticles Used in Biocides Controlled Release Formulations*

"Controlled release (CR) can be defined as a method that uses chemical, mechanical, or electromechanical devices to deliver a predetermined amount of active ingredient molecules from a reservoir to a target site for definite period of times" [82, 83]. It is applied to deliver active substances such as pesticides, herbicides, fertilizers, biomolecules, and drugs in agriculture, biomedical, food, and pharmaceutical industries [84]. Controlled release technology is used in crops so that environmental pollution can be reduced. It offers several advantages over the conventional methods of pesticide applications such as sustainable constant release

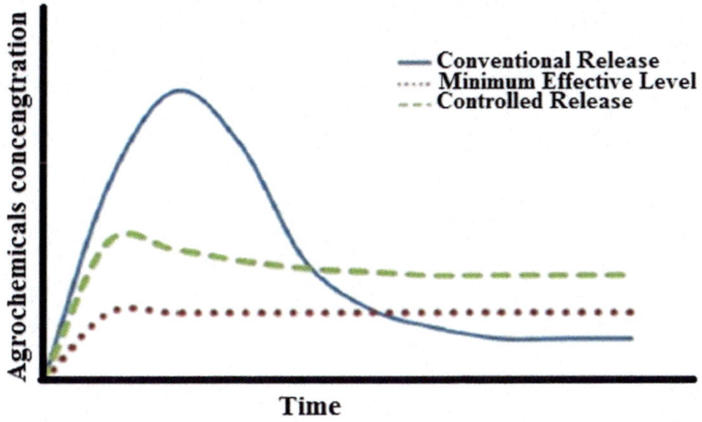

Fig. 7.9 Variation in agrochemical concentrations with time from conventional and controlled release

for long time, ease of handling, reduces excess use of active ingredients, lowers the evaporation rate, phytotoxicity, and pollution. CR technology has many applications in agriculture, where the protection of crops from pests is required for an extended period of time with less side effects, prolonged efficiency, enhanced safety, and reliability of agrochemicals [85–90].

A safe pesticide formulation could be the one which controls the pests and reduces the environmental pollution as well as checks the availability of the active moiety at any point in time. Encapsulation technology provides a handy tool which helps to achieve these properties by using polymer-matrix-based nanoformulations. Polymer-based formulations have been developed to enhance the activity of entrapped active ingredients as they protect the active ingredient molecules from biological and enzymatic degradation, evaporation, leaching and reduce the toxicity, maintain the optimum concentration of AI for a specified period, reduce mobility of pesticide into soil, and minimize excess dosages [91–99] (Fig. 7.9).

7.4 Methods for Preparation of Nanomaterials Controlled Release Formulations for Biocides Application

7.4.1 Ionotropic Gelation

An ionic gelation method involves the complexation between low molecular weight polycation and polyanion molecules. It was first introduced by Bodmeier and coworkers [100]. As compared with covalent crosslinking, the ionotropic gelation has many advantages because of the mild preparation conditions and simple procedure. Nowadays, the most widely used biopolymers for the ionotropic gelation

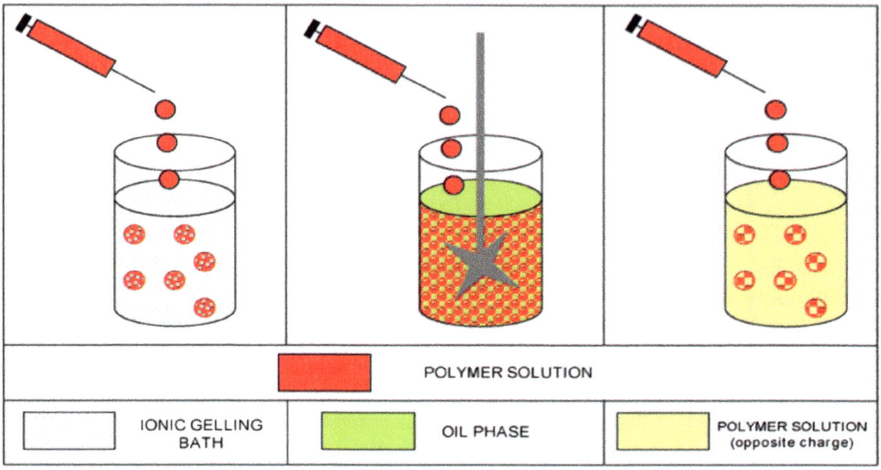

Fig. 7.10 Schematic representation of ionotropic gelation technique

method are alginate nanocarriers which can be prepared by adding $CaCl_2$ or other polycationic crosslinkers. The active ingredient can be entrapped into the prepared calcium alginate crosslinked nanocarriers as a potential AI carrier for the slow release. Due to the crosslinking and gelation of different ions, it is called as ionotropic gelation method. The AI encapsulation efficiency is affected neither by the amount of alginate nor the crosslinking ion used. The neutralization between oppositely charged alginate decreases the solubility of the alginate nanocarriers (Fig. 7.10).

7.4.2 Solvent Evaporation Method

The solvent evaporation techniquc is a very common and popular technique for synthesizing nanocarriers for controlled release as there is no need for washing steps, and there is a minimum possibility of agglomeration. In this method, a mixture of polymer and AI solution is emulsified in different oil/water emulsions with continuous stirring and allow evaporating the solvent by increasing temperature [101, 102]. This results in precipitation of polymers as nanocarriers with dispersed AI in the polymer matrix network as shown in Fig. 7.11. Frequently used polymers are PLGA [poly (D, L-lactidecoglycolide) acid] [103], ethyl cellulose [104] and cellulose acetate phthalate and drugs are albumin [105] testosterone [106] cyclosporine A [107, 108], etc.

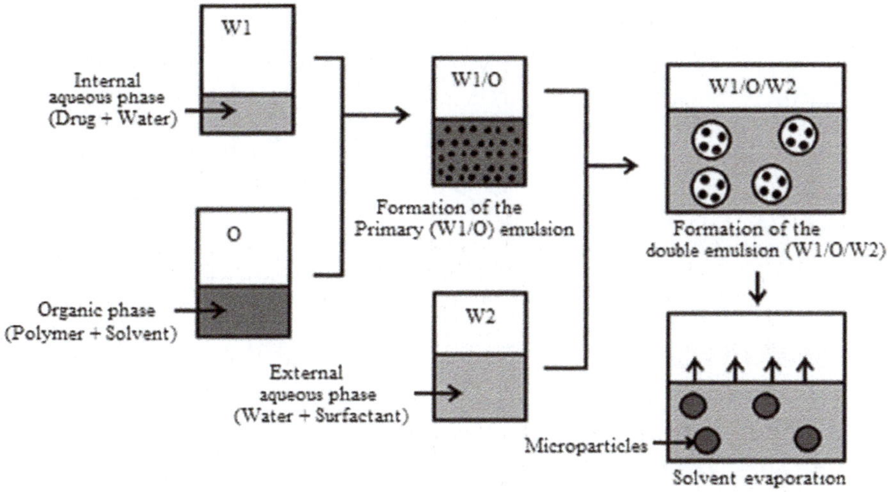

Fig. 7.11 Schematic representation of the solvent evaporation technique

7.4.3 Emulsion Crosslinking Method

Emulsion of polymer and crosslinker is prepared with the help of suitable oil, such as paraffin oil (Fig. 7.12). Both the emulsions are stirred together to form a homogenous solution, after continuous stirring the solution is filtered with toluene, acetone, distilled water, respectively, and then dried in an oven [109].

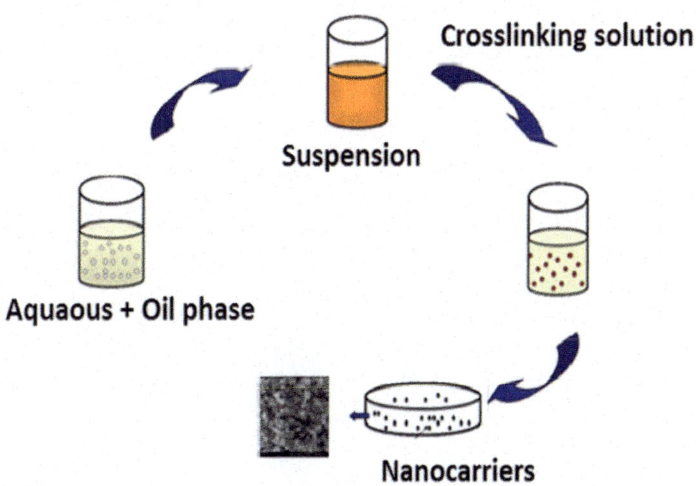

Fig. 7.12 Schematic representation of the preparation of alginate nanocarriers by the emulsification crosslinking technique

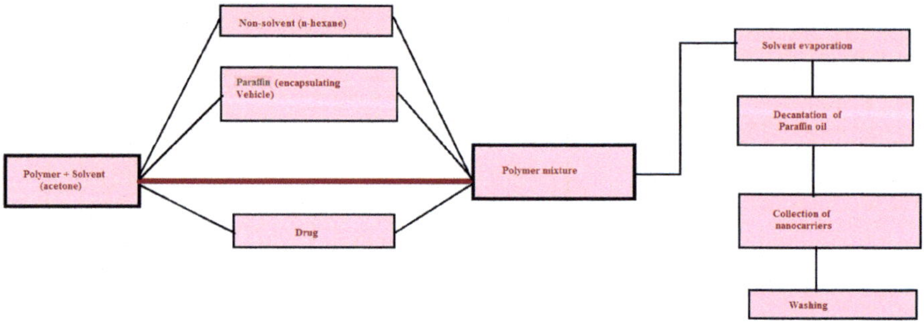

Fig. 7.13 Schematic representation of the non-solvent addition technique

7.4.4 Non-solvent Addition Method

In this method, nanocarriers can be prepared using various solvents and non-solvents. Solvents like acetone, dimethyldigol, and 1, 4-dioxane and non-solvents like n-hexane and chloroform can be used [110, 111]. The polymer is dissolved in a solvent, by stirring the mixture and then drug, liquid Paraffin (encapsulating vehicle), and suitable non-solvents are added slowly and dropwise. The mixture is then agitated at room temperature (25 °C) so that the solvent gets evaporated. Liquid paraffin is decanted (non-solvent helps in washing paraffin oil) and the nanocarriers are collected, washed, and dried at 40 °C in an oven and stored in an airtight container at room temperature. The whole scheme is represented in Fig. 7.13.

7.4.5 Spray Drying Method

Spray drying method is used as a potential method to synthesize nano-controlled release systems. In this method, polymer and AI solutions are suspended in distilled water and converted into very small nanosize droplets using an atomizing device and allow drying in the presence of hot air. During evaporation, the solvent is removed and dry powder is formed which is separated from the moist air by centrifugal action when the mixture of particles and air enters the cyclone system and heavy dense powder particles settle down to the bottom of the cyclone, removed through a discharging device, while the lighter, moist air is directed away through the exhaust pipes [112, 113] (Fig. 7.14).

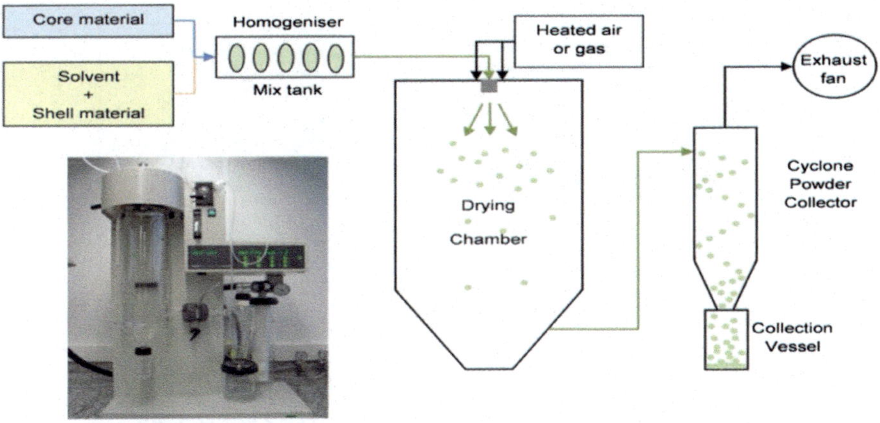

Fig. 7.14 Schematic representation of cyclone dryer's technique

7.5 Mechanism of Biocide Release Profile

7.5.1 Release Mechanisms

For decades, polymeric nanostructures have been developed and investigated as controlled release systems with aim to maintain zero-order release or steady-state release of the pesticide in such a manner that release rate is not affected by the amount of pesticide remaining in the device at any time and thus maintains an effective concentration level for the prolonged period. However, most of the release systems do not follow the zero-order release mechanism kinetically because it is affected by various factors such as the composition of the device, nature of interactions, and methods of preparation [114]. The mechanism of release kinetics can be classified as diffusion, solvent, and erosion-controlled release as summarized below.

7.5.1.1 Diffusion-Controlled Release Systems

Many pesticides encapsulated polymeric nanomaterials (matrix devices or reservoir system) follow the diffusion-controlled release mechanism and the release rate of pesticide molecules depends on the difference of its concentration across the membrane, not on the chemical composition as well as intermolecular interaction between polymers and loaded pesticides and migration of water into preexisting or dynamically formed spaces between the polymer chains that allow the diffusion of loaded pesticides from the matrix (Fig. 7.15a). Hydrophilic polymers show faster release rate than hydrophobic polymers due to their degradation in water.

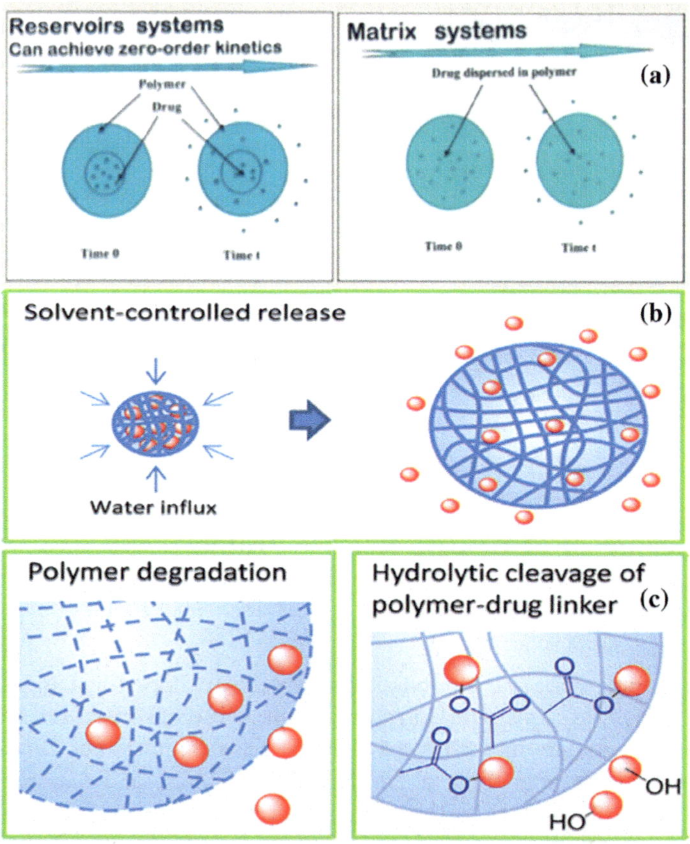

Fig. 7.15 Pictorial representation of various controlled delivery systems

Carboxylic acid group containing polymeric nano-matrices do not show diffusion-controlled release because their composition controlled the rate [115–117].

In case of matrix devices bioactive molecules are dispersed throughout into polymer with no diffusion barrier membrane, and they are the most common systems for designing controlled release nanoformulations for bioactive molecules because of their ease of preparation, release dependence of rate on solubility of the bioactive agent, and pore network of matrix may take place either by diffusion or leaching along with diffusion [118]. However, reservoir systems contain bioactive molecules at the centre surrounded by polymer membrane and their diffusion-controlled release rate is affected by additives, polymer functionality, pH, porosity, film-casting condition, etc., and bioactive molecules are released through the micropores of the capsule walls [119].

7.5.1.2 Solvent-Controlled Release

The transport of solvent molecules into polymeric nanoformulation can affect the release of bioactive molecules either by swelling or osmosis until it is placed in an appropriate biological environment. Semi-permeable polymeric membranes containing polymeric nanoformulations release the entrapped bioactive molecules through osmosis due to the presence of difference of vapor pressure originating from the lower concentration outside and higher concentration inside the matrix. The water molecules penetrate into the central core and apply hydrostatic pressure on the entrapped bioactive molecules and force it to come out and a zero-order release profile is maintained till the concentration gradient is maintained constant across the membrane [120] (Fig. 7.15b).

In swelling-controlled release mechanism, water molecules diffuse into hydrophilic functional groups glassy polymeric hydrogel nanoformulations and relax the polymeric chains and force entrapped bioactive molecules in the matrix to diffuse out through the swollen network into the external environment. The rate of bioactive molecules release from matrix depends on the relaxation rate of polymer chains as well as diffusion rate of water molecules, composition of the polymer, and concentration of entrapped bioactive molecules. The swelling-controlled release can be evaluated by Peppas model (e.g., Fickian or non-Fickian diffusion) [121, 122].

7.5.1.3 Erosion or Chemical Reaction Controlled Release

Biodegradable polymer-based nanoformulations have gained much attention as erosion-controlled release devices for bioactive molecules via hydrolytic and/or enzymatic degradation of the linkage between the drug and the polymer and mostly follow bulk degradation followed by the surface erosion. When bioactive molecules are physically combined with polymer chains, erosion-controlled release takes place while chemical–reaction-controlled release involves chemical bonding between the polymer and the bioactive molecules and bond breaking is required for the separation of molecules from polymeric matrix [123, 124] (Fig. 7.15c).

7.5.2 Models Used for Nanopesticide Release

The mechanism of nanopesticides release can be evaluated by various mathematical models such as Berens and Hofenberg model, Korsmeyer-Peppas or Ritger-Peppas model, zero-order, first-order, and Higuchi Model [125, 126]. The release profile of encapsulated nanopesticides can be evaluated by zero-order release kinetic model (Eq. 7.1) that would be followed by polymeric nanopesticides formulations to release bioactive ingredients with constant rate and not depend on the concentration of bioactive ingredient at any time.

$$Q_t = Q_0 + K_0\, t \tag{7.1}$$

where Q_0 and Q_t represent the initial amount and released amount of bioactive molecules, respectively, and K_0 is the zero-order release constant [127].

The first-order release kinetic model can be represented by Eqs. (7.2) and (7.3) as given below,

$$\frac{\mathrm{d}c}{\mathrm{d}t} = -K_c \tag{7.2}$$

$$\log C = \log C_0 - K_t/2.303 \tag{7.3}$$

where C_0 represents initial concentration and C the released concentration of bioactive molecules at time t, and K_t is the rate constant [128].

The release kinetic of bioactive molecules from semi-solid and solid polymeric devices can be determined by using Higuchi model (Eq. 7.4) [129].

$$Q = K_\mathrm{H} \cdot t^{1/2} \tag{7.4}$$

where Q is the release concentration of bioactive molecules at time t, and K_H is Higuchi dissolution constant.

In order to evaluate the release kinetics of bioactive molecules from swellable or non-swellable polymeric release systems, Ritger-Peppas and Korsmeyer-Peppas models (Eqs. 7.5 and 7.6) can be used as mathematical models to analyze Fickian ar non-Fickian release.

$$W_t/W_\infty = k \cdot t^n \tag{7.5}$$

$$W_t/W_\infty = (D \cdot t)^{0.5}/\pi L^2 \tag{7.6}$$

where W_t and W_∞ represent concentrations of bioactive molecules released at t time and at equilibrium, respectively, and n is release exponent, k is swelling factor, D is diffusion constant, and L being the diameter of the dry nanoparticles (Table 7.1).

Table 7.1 Data showing the types of the mechanism of release of agrochemicals from nanoformulation and their diffusion exponents

Type of transport	Diffusional exponent (n)	Time dependence
Fickian diffusion	0.5	$t^{1/2}$
Anomalous transport	$0.5 < n < 1$	t^{n-1}
Case II transport	1	Time independent
Super case II transport	$n > 1$	t^{n-1}

7.5.3 Factors Affecting Release

The rate of release and mechanism depends on the compositions of polymers, i.e., hydrophilic and hydrophobic nature, functional groups because they controlled the swelling behavior and rate of diffusion of water molecules into polymer matrix due to shrinking or swelling in polymeric chains in response to environmental media, and ionization of functional groups in the presence of other stimuli such as pH, temperature, light, magnetic, and electric fields [130–132]. Crosslinking density plays an important role in the relaxation of polymeric chains and affects the swelling by decreasing the faster relaxation of polymeric chains [133, 134]. Hydrophobic polymers may collapse and hydrophilic polymers expend with change of temperature and may affect the rate of relaxation of polymer chains and also the release rate of bioactive molecules from the central core of polymeric materials [135]. The swelling of polyelectrolyte systems depends on the ionic strength of the swelling medium because that changes the rate of ionization which due to screening effect may enhance the release of loaded bioactive molecules [136].

7.6 Conclusions and Future Prospects

The controlled release nanoformulations of pesticides offer numerous advantages over the conventional methods of pesticide application. Besides other benefits, the major advantage with these release systems is that they warrant minimum harm to the environment including soil fertility, agricultural fields, and underground water. The abundance of a variety of synthetic and naturally occurring polymers provides wide options to design nanoformulations suitable for short-term and long-term applications. Furthermore, depending on the nature of the soil and surrounding environment, these controlled release systems may accordingly be designed for optimum usage [137, 138]. However, there are numerous challenges that must be realized and addressed before a commercially viable nanoformulation is fabricated. Some of the challenges may be exemplified as given below:

1. The cost of controlled release system is normally very high in comparison with conventional formulation used which is obviously due to the technology and materials involved to impart slow releasing property to these formulations. Thus, economically favorable materials and processes are required for large-scale production of these slow releasing nanoformulations.
2. Mostly, inert polymers have been used for developing controlled release formulations for herbicides which are non-biodegradable as well as toxic in nature. In place of such materials, biodegradable polymer-based controlled release formulation should be developed because their degraded products not only stop pollutions but also increase the fertility of soil and crops.

References

1. Sopena F, Maqueda C, Morillo E (2009) Controlled release formulations of herbicides based on micro-encapsulation. Cien Inv Agr 35:27–42
2. Moros J, Armenta S, Garrigues S, Guardia M (2007) Comparison of two vibrational procedures for the direct determination of mancozeb in agrochemicals. Talanta 72:72
3. Pretty JN, Morison JIL, Hine RE (2003) Reducing food poverty by increasing agricultural sustainability in developing countries. Agr Ecosyst Environ 95:217–234
4. Bajpai AK, Giri A (2002) Swelling dynamics of a macromolecular hydrophilic network and evaluation of its potential for controlled release of agrochemicals. React Funct Polym 53:125
5. Ali D, Nagpurea NS, Kumar S, Kumar R, Kushwaha B, Lakra WS (2009) Assessment of genotoxic and mutagenic effects of chlorpyrifos in freshwater fish *Channa punctatus* (Bloch) using micronucleus assay and alkaline single-cell gel electrophoresis. Food Chem Toxicol 47:650
6. Singh B, Sharma DK, Dhiman A (2013) Environment friendly agar and alginate-based thiram delivery system. Toxicol Environ Chem 95:567–578
7. Ahmed RZ, Siddiqui K, Arman M, Ahmed N (2012) Characterization of high molecular weight dextran produced by *Weissella cibaria* CMGDEX3. Carbohydr Polym 90:441–446
8. Campos EVR, de Oliveira JL, Fraceto LF (2014) Applications of controlled release systems for fungicides, herbicides, acaricides, nutrients, and plant growth hormones: a review. Adv Sci Eng Med 6:373–387
9. Grillo R, Pereira AES, Nishisaka CS, de Lima R, Oehlke K, Greiner R, Fraceto LF (2014) Chitosan/tripolyphosphate nanoparticles loaded with paraquat herbicide: an environmentally safer alternative for weed control. J Hazard Mater 278:163–171
10. Gilden RC, Huffling K, Sattler B (2010) Pesticides and health risks. J Obstet Gynecol Neonatal Nurs 39:103–110
11. Carter AD (2000) Herbicide movement in soils: principles, pathways and processes. Weed Res 40:113–122
12. Abhilash PC, Singh N (2009) Pesticide use and application: an Indian scenario. J Hazard Mater 165:1–12
13. Swinnen JFM, de Gorter H, Rausser GC, Banerjee AN (2002) The political economy of public research investment and commodity policies in agriculture: an empirical study. Agric Econ 22:111–122
14. Lin PC, Lin HJ, Liao YY, Guo HR, Chen KT (2013) Acute poisoning with neonicotinoid insecticides: a case report and literature review. Basic Clin Pharmacol Toxicol 112:282–286
15. Akar E, Altinisik A, Seki Y (2012) Preparation of pH- and ionic-strength responsive biodegradable fumaric acid crosslinked carboxymethyl cellulose. Carbohydr Polym 90:1634–1641
16. Chowdhury MA (2014) The controlled release of bioactive compounds from lignin and lignin-based biopolymer matrices. Int J Biol Macromol 65:136–147
17. Jaiswal M, Dudhe R, Sharma PK (2015) Nanoemulsion: an advanced mode of drug delivery system. 3 Biotech 5:123–127
18. Kah M, Hofmann T (2014) Nanopesticide research: current trends and future priorities. Environ Int 63:224–235
19. Solans C, Izquierdo P, Nolla J, Azemar N, Garcia-Celma MJ (2005) Nano-emulsions. Curr Opin Colloid Interface Sci 10:102–110
20. NICNAS Information Sheet (2006) Australian Government Department of Health and Aging, September
21. Whatmore RW (2006) Nanotechnology—what is it? Should we be worried. Occup Med 56:295–299
22. ISO (2004) Occupational ultrafine aerosol exposure characterization, assessment. Draft technical report number 6. ISO/TC146/SC2WG1 particle size selective sampling and analysis (workplace air quality)

23. Wigginton NS, Rosso KM, Lower BH, Shi L, Hochella MF Jr (2007) Electron tunneling properties of outer-membrane decaheme cytochromes from *Shewanella oneidensis*. Geochim Cosmochim Acta 71:543–555
24. Singh P, Kim YJ, Zhang D, Yang DC (2016) Biological synthesis of nanoparticles from plants and microorganisms. Trends Biotechnol 34:588–599
25. Brandl MA, Schumann M, Przyklenk M, Patel A, Vidal S (2016) Wireworm damage reduction in potatoes with an attract-and-kill strategy using *Metarhizium brunneum*. J Pest Sci 1–15
26. Rogge SA, Mayerhofer J, Enkerli J, Bacher S, Grabenweger G (2017) Preventive application of an entomopathogenic fungus in cover crops for wireworm control. Biocontrol 62(5): 613–623
27. European Food Safety Authority (2012) Conclusion on the peer review of the pesticide risk assessment of the active substance *Metarhizium anisopliae* var. anisopliae BIPESCO5/F52. EFSA J 10:2498–2542
28. Saborío HU, Carballo CM, Sandoval J, José R, Baudrit V, Morales AR (2017) Encapsulation of bacterial metabolic infiltrates isolated from different Bacillus strains in chitosan nanoparticles as potential green chemistry-based biocontrol agents against *Radopholus similis*. J Renew Mater 5:290–299
29. Carvalho FP (2017) Pesticides, environment, and food safety. Pest Environ Food Saf 6: 48–60
30. Melo CA, Oliveira LK, Goveia D, Fraceto LF, Rosa AH (2014) Enrichment of tropical peat with micronutrients for agricultural applications: evaluation of adsorption and desorption processes. J Braz Chem Soc 25:36–49
31. Dourado Junior SM, Nunes ES, Marques RP, Rossino LS, Quites FJ, Siqueira JR Jr, Moreto JA (2017) Controlled release behavior of sulfentrazone herbicide encapsulated in Ca-ALG microparticles: preparation, characterization, mathematical modeling and release tests in field trial weed control. J Mater Sci 52(16):9491–9507
32. Nair R, Hanna S, Baiju V, Nair G, Maekawa T, Yoshida Y, Sakthi Kumar D (2010) Nanoparticulate material delivery to plants. Plant Sci 179:154–163
33. Kanhed P et al (2014) In vitro antifungal efficacy of copper nanoparticles against selected crop pathogenic fungi. Mater Lett 115:13–17
34. Tabata J, Ikada Y (1989) Protein precoating of polylactide microspheres containing a lipophilic immunopotentiator for enhancement of macrophage phagocytosis and activation. Pharm Res 6:296–301
35. Bodmeier R, Chen H (1990) Indomethacin polymeric nanosuspensions prepared by microfluidization. J Control Release 12:223–233
36. Landry FB, Bazile DV, Spenlehauer G, Veillard M (1996) Influence of coating agents on the degradation of poly(D, L-lactic acid) nanoparticles in model digestive fluids. STP Pharma Sci 6:195–202
37. Gurny R, Peppas NA, Harrington DD, Banker GS (1981) Development of biodegradable and injectable latices for controlled release of potent drugs. Ind Pharm 7:1–25
38. Jaiswal J, Gupta SK, Kreuter J (2004) Preparation of biodegradable cyclosporine nanoparticles by high-pressure emulsification-solvent evaporation process. J Control Release 96:169–178
39. Saanchez A, Vila Jato JL, Alonso MJ (1993) Development of biodegradable microspheres and nanospheres for the controlled release of cyclosporin A. Int J Pharm 99:263–273
40. Calvo P, Remunan-Lopez C, Vila-Jato JL, Alonso MJ (1997) Chitosan and chitosan/ethylene oxide-propylene oxide block copolymer nanoparticles as novel carriers for proteins and vaccines. Pharm Res 14:143–146
41. Sunitha S, Amareshwar P, Kumar S (2010) Preparation and characterization of ondansetron hydrochloride microspheres using various cellulose polymers. Int J Curr Pharm Res 2:44–49
42. Palmieri GF, Bonacucina G, Di Martino P, Martelli S (2001) Spray-drying as a method for microparticulate controlled release systems preparation: advantages and limits. I. Water-soluble drugs. Drug Dev Ind Pharm 27(3):195–204

43. Chu M, Zhu SQ, Li HM, Huang ZB, Li SQ (2006) Synthesis of poly(acrylic acid)/sodium humate superabsorbent composite for agricultural use. J Appl Polym Sci 102:5137–5143
44. Puoci F, Iemma F, Spizzirri UG, Cirillo G, Curcio M, Picci N (2008) Polymer in agriculture: a review. Am J Agric Biol Sci 3:299–314
45. Kamat M, Malkani R (2003) Disposable diapers: a hygienic alternative. Indian J Pediatr 70:879–881
46. Kosemund K, Schlatter H, Ochsenhirt JL, Krause EL, Marsman DS, Erasala GN (2009) Safety evaluation of superabsorbent baby diapers. Regul Toxicol Pharmacol 53:81–89
47. Kasgoz H, Durmus AZ, Kasgoz A (2008) Enhanced swelling and adsorption properties of AAm-AMPSNa/clay hydrogel nanocomposites for heavy metal ion removal. Polym Adv Technol 19:213–220
48. Wang YQ, Zhang HF, Wu YP, Yang J, Zhang LQ (2005) Preparation and properties of natural rubber/rectorite nanocomposites. Eur Polym J 41:2776–2783
49. Sadeghi M, Hosseinzadeh HJ (2008) Synthesis of starch—poly(sodium acrylate-co-acrylamide) superabsorbent hydrogel with salt and pH-responsiveness properties as a drug delivery system. J Bioact Compat Polym 23:381–404
50. Ray SS, Bousmina M (2005) Biodegradable polymers and their layered silicate nanocomposites: in green the 21st century materials word. Prog Mater Sci 50:962–1079
51. Pourjavadi A, Hosseinzadeh H, Sadeghi M (2007) Synthesis, characterization and swelling behavior of gelatin-g-poly(sodium acrylate)/kaolin superabsorbent hydrogel composites. J Compos Mater 41:2057–2069
52. Yoshimura T, Uchikoshi I, Yoshiura Y, Fujioka R (2005) Synthesis and characterization of novel biodegradable superabsorbent hydrogels based on chitin and succinic anhydride. Carbohydr Polym 61:322–326
53. Zhang JP, Wang Q, Wang AQ (2007) Synthesis and characterization of chitosan-g-poly (acrylic acid)/attapulgite superabsorbent composites. Carbohydr Polym 68:367–374
54. Aries JL (2008) Novel strategies to improve the anticancer action of 5-fluorouracil by using drug delivery systems. Molecules 13:2340–2369
55. Racovita S, Vasiliu S, Popa M, Lucab C (2009) Polysaccharides based on micro- and nanoparticles obtained by ionic gelation and their applications as drug delivery systems. Rev Roum Chim 54:709–718
56. Couvreur P, Gref R, Andrieux K, Malvy C (2006) Nanotechnologies for drug delivery: application to cancer and autoimmune diseases. Prog Solid State Chem 34:231–235
57. Grillo R, Rosa AH, Fraceto LF (2014) Poly(e-caprolactone) nanocapsules carrying the herbicide atrazine: effect of chitosan-coating agent on physico-chemical stability and herbicide release profile. Int J Environ Sci Technol 11:1691–1700
58. Pyun J (2007) Nanocomposite materials from functional polymers and magnetic colloids. Polym Rev 47:231–263
59. Xu C, Ohno K, Ladmiral V, Composto RJ (2008) Dispersion of polymer-grafted magnetic nanoparticles in homopolymers and block copolymers. Polymer 49:3568–3577
60. Allemann E, Leroux JC, Gurny R, Doelker E (1993) In vitro extended-release properties of drug-loaded poly(DL-lactic acid) nanoparticles produced by a salting-out procedure. Pharm Res 10:1732
61. Flores-Cespedes F, Figueredo-Flores CI, Daza-Fernandez I, Vidal-Pena F, Villafranca-Sanchez M, Fernandez-Perez M (2012) Preparation and characterization of imidacloprid lignin-polyethylene glycol matrices coated with ethylcellulose. J Agric Food Chem 60:1042–1051
62. http://www.springer.com/978-1-4471-4212-6
63. Roy A, Bajpai AK, Bajpai J (2009) Dynamics of controlled release of chlorpyrifos from swelling and eroding biopolymeric microspheres of calcium alginate and starch. Carbohydr Polym 76:222–229
64. Roy A, Bajpai AK, Bajpai J (2009) Designing swellable beads of alginate and gelatin for controlled release of pesticide (cypermethrin). J Macromol Sci Res Part A Pure Appl Chem 46:847–859

65. Wang HQ, Nann T (2009) Monodisperse upconverting nanocrystals by microwave-assisted synthesis. ACS Nano 3:3804–3808
66. Sun S, Zeng H (2002) Size-controlled synthesis of magnetite nanoparticles. J Am Chem Soc 124:8204–8205
67. Sperling RA, Parak WJ (2010) Surface modification, functionalization and bioconjugation of colloidal inorganic nanoparticles. Phil Trans R Soc A 368:1333–1383
68. Medintz IL, Uyeda HT, Goldman ER, Mattoussi H (2005) Quantum dot bioconjugates for imaging, labelling and sensing. Nat Mater 4:435–446
69. Barnes WL, Dereux A, Ebbesen TW (2003) Surface plasmon subwavelength optics. Nature 424:824–830
70. Cushing BL, Kolesnichenko VL, O'Connor CJ (2004) Recent advances in the liquid-phase syntheses of inorganic nanoparticles. Chem Rev 104:3893–3946
71. Cai W, Chen X (2007) Nanoplatforms for targeted molecular imaging in living subjects. Small 3:1840–1854
72. White RJ, Luque R, Budarin VL, Clark JH, Macquarrie DJ (2009) Supported metal nanoparticles on porous materials. Methods and applications. Chem Soc Rev 38:481–494
73. Campelo JM, Luna D, Luque R, Marinas JM, Romero AA (2009) Sustainable preparation of supported metal nanoparticles and their applications in catalysis. ChemSusChem 2
74. John AB (2009) Metallic nanoparticles. Elsevier
75. Samad A, Sultana Y, Aqil M (2007) Liposomal drug delivery systems: an update review. Curr Drug Deliv 4(4):297–305
76. Forssen E, Willis M (1998) Ligand-targeted liposomes. Adv Drug Deliv Rev 29(3):249–271
77. Allouche J, Chanéac C, Brayner R (2014) Design of magnetic gelatine/silica nanocomposites by nanoemulsification: encapsulation versus in situ growth of iron oxide colloids. Nanomaterials 4:612–627
78. Taze PT, John S, Dennis Mark B (2015) Iron-filled multiwalled carbon nanotubes surface-functionalized with paramagnetic Gd (III): a candidate dual-functioning MRI contrast agent and magnetic hyperthermia structure. Carbon 87:226–232
79. Nuraje N, Su K (2013) Perovskite ferroelectric nanomaterials. Nanoscale 2013(5):8752–8780
80. Agarwal T, Gupta KAS, Alam S, Zaidi MGH (2012) Fabrication and characterization of iron oxide filled polyvinyl pyrrolidone nanocomposites. Int J Compos Mater 2:17–21
81. Muller RH, Mader K, Gohla S (2000) Solid lipid nanoparticles (SLN) for controlled drug delivery—a review of the state of the art. Eur J Pharm Biopharm 50(1):161–177
82. Tefft J, Friend DR (1993) Controlled release herbicide formulations based on polymeric microspheres. J Control Release 27:27–35
83. Raloff J (2007) Common pesticide clobbers amphibians. Sci News 154:10–150
84. Yang L, Yang Y, Chena Z, Guoc C, Li S (2014) Influence of super absorbent polymer on soil water retention, seed germination and plant survivals for rocky slopes eco-engineering. Ecol Eng 62:27–32
85. Crawford SL, Fiedler ER (1992) Childhood physical and sexual abuse and failure to complete military basic training. Mil Med 157(12):645–648
86. Hodgson E, Levi PE (1996) Pesticides: an important but underused model for the environmental health sciences. Environ Health Perspect 104(1):97–106
87. Daly HV, Doyen JT, Purcell AH (1998) Introduction to insect biology and diversity, 2nd edn. Oxford University Press, pp 279–300
88. Godish T (2000) Indoor environment quality. CRC Press, pp 325–326
89. Sarisozen C, Vural I, Levchenko T, Hincal A, Torchilin VP (2012) PEG-PE-based micelles co-loaded with paclitaxel and cyclosporine A or loaded with paclitaxel and targeted by anticancer antibody overcome drug resistance in cancer cells. Drug Deliv 19(4):169–176
90. Chena Y, Lo CL, Lin YF, Hsiue GH (2013) Rapamycin encapsulated in dual-responsive micelles for cancer therapy. Biomaterials 34:1115–1127

91. Rezaei SJT, Amani V, Nabid MR, Safari N, Niknejad H (2015) Folate-decorated redox/pH dual-responsive degradable prodrug micelles for tumor triggered targeted drug delivery. Polym Chem 6:2844–2853
92. Wu H, Zhua L, Torchilin V (2013) pH-sensitive poly(histidine)-PEG/DSPE-PEG co-polymer micelles for cytosolic drug delivery. Biomaterials 34:1213–1222
93. Kenawy ER, Sherrington DC, Akelah A (1992) Controlled release of agrochemical molecules chemically bound to polymers. Eur Polym J 28:841–862
94. Zhang M, Gao B, Chen J, Li Y, Creamer AE, Chen H (2014) Slow-release fertilizer encapsulated by graphene oxide films. Chem Eng J 255:107–113
95. Zhang W, He S, Liu Y, Geng Q, Ding G, Guo M, Deng Y, Zhu J, Li J, Cao Y (2014) Preparation and characterization of novel functionalized prochloraz microcapsules using silica–alginate–elements as controlled release carrier materials. ACS Appl Mater Interfaces 6:11783–11790
96. Akelah A, Kenawy ER, Sherrington DL (1992) Agricultural polymers with herbicide/fertilizer function—III. Polyureas and poly(Schiff base)s based systems. Eur Polym J 29:1041–1045
97. Kenawy ER, Akelah A, Sherrington DC (1992) Agricultural polymers with herbicide/fertilizer function—II. Glutarate-based systems. Eur Polym J 28:615–621
98. Gao JH, Xu B (2009) Application of nanomaterials inside cells. Nano Today 4:37–51
99. Galaev IY, Mattiasson B (1999) 'Smart' polymers and what they could do in biotechnology and medicine. Trends Biotechnol 17:335–340
100. Caldorera-Moore M, Guimard N, Shi L, Roy K (2010) Designer nanoparticles: incorporating size, shape and triggered release into nanoscale drug carriers. Expert Opin Drug Deliv 7:479–495
101. Cho K, Wang X, Nie S, Chen Z, Shin DM (2008) Therapeutic nanoparticles for drug delivery in cancer. Clin Cancer Res 14:1310–1316
102. Batzill M, Diebold M (2005) The surface and materials science of tin oxide. Prog Surf Sci 79:47–154
103. Kilic C, Zunger A (2002) Origins of coexistence of conductivity and transparency in SnO_2. Phys Rev Lett 88:95–501
104. Fang LM, Zu XT, Li ZJ, Zhu S, Liu CM, Zhou WL, Wang LM (2008) Synthesis and characteristics of Fe^{3+}-doped SnO_2 nanoparticles via sol–gel-calcination or sol–gel-hydrothermal route. J Alloys Compd 454:261–267
105. Janko C, Stephan Dürr S, Luis EM, Stefan Lyer S, Chaurio R, Rainer Tietze R, Löhneysen SV, Schorn C, Herrmann M, Alexiou C (2013) Magnetic drug targeting reduces the chemotherapeutic burden on circulating leukocytes. Int J Mol Sci 14:7341–7355
106. Akbarzadeh A, Samiei M, Davaran S (2012) Magnetic nanoparticles: preparation, physical properties, and applications in biomedicine. Nano Res Lett 7:144
107. Zhang XJ, Jiang W, Li FS, Sun ZD, Ou'yang Z (2010) Controllable preparation of magnetic polymer nanospheres with high saturation magnetization by miniemulsion polymerization. Mater Lett 64:119–121
108. Rahimi M, Yousef M, Cheng Y, Meletis EI, Eberhart RC (2009) Formulation and characterization of covalently coated magnetic nanogel. J Nanosci Nanotechnol 9:4128–4134
109. Thorek DLJ, Chen A, Czupryna J, Tsourkas A (2006) Superparamagnetic iron oxide nanoparticle probes for molecular imaging. Ann Biomed Eng 34:23–38
110. Batlle X, Labarta A (2002) TOPICAL REVIEW: finite-size effects in fine particles: magnetic and transport properties. J Phys D Appl Phys 35:15–42
111. El Bahri Z, Taverdet J-L (2007) Elaboration and characterisation of microparticles by pesticide model. Powder Technol 172:30–40
112. Bodmeier R, Oh KH, Pramar T (1989) Preparation and evaluation of drug-containing chitosan beads. Drug Dev Ind Pharm 15:1475–1494
113. Yurdasiper A, Sevgi F (2010) An overview of modified release chitosan, alginate and eudragit RS microparticles. J Chem Pharm Res 2:704–721

114. Bajpai AK, Shukla SK, Bhanu S, Kankane S (2008) Responsive polymers in controlled drug delivery. Prog Polym Sci 33:1088–1118
115. Huang X, Brazel CS (2001) On the importance and mechanisms of burst release in matrix-controlled drug delivery systems. J Control Release 73:121–136
116. Cappello J, Crissman JW, Crissman M, Ferrari FA, Textor G, Wallis O, Whitledge JR, Xia Z, Burman D, Aukerman L, Stedronsky ER (1998) In-situ self-assembling protein polymer gel systems for administration, delivery, and release of drugs. J Control Release 53:105–117
117. Beasley ML, Collins RL (1970) Water-degradable polymers for controlled release of herbicides and other agents. Science 169:769–770
118. Cauchetier E, Deniau M, Fessi H, Astier A, Paul M (2003) Atovaquone-loaded nanocapsules: influence of the nature of the polymer on their in vitro characteristics. Int J Pharm 250:273–281
119. Colombo P (1993) Swelling-controlled release in hydrogel matrices for oral route. Adv Drug Deliv Rev 11(1–2):37–57
120. Herrlich S, Spieth S, Messner S, Zengerle R (2012) Osmotic micropumps for drug delivery. Adv Drug Deliv Rev 64:1617–1627
121. Lin C-C, Metters AT (2006) Hydrogels in controlled release formulations: network design and mathematical modeling. Adv Drug Deliv Rev 58:1379–1408
122. Peppas NA, Bures P, Leobandung W, Ichikawa H (2000) Hydrogels in pharmaceutical formulations. Eur J Pharm Biopharm 50:27–46
123. Lee SH, Mok H, Lee Y, Park TG (2011) Self-assembled siRNA–PLGA conjugate micelles for gene silencing. J Control Release 152:152–158
124. Burkersroda FV, Schedl L, Göpferich A (2002) Why degradable polymers undergo surface erosion or bulk erosion. Biomaterials 23:4221–4231
125. Razavilar N, Choi P (2013) In-vitro modeling of the release kinetics of micron and nano-sized polymer drug carriers. Int J Drug Deliv 5:362–378
126. Chime Salome AC, Onunkwo Godswill C, Onyishi Ikechukwu I (2013) Kinetics and mechanisms of drug release from swellable and non swellable matrices: a review. Res J Pharm Biol Chem Sci. ISSN: 0975-8585
127. Yadav G, Bansal M, Thakur N, Khare S, Khare P (2013) Multilayer tablets and their drug release kinetic models for oral controlled drug delivery system. Mid-East J Sci Res 16: 782–795
128. Dash S, Murthy PR, Nath L, Chowdhury P (2010) Kinetic modeling on drug release from controlled drug delivery systems. Acta Pol Pharm 67:217–223
129. Higuchi T (2010) Mechanism of sustained-action medication. Theoretical analysis of rate of release of solid drugs dispersed in solid matrices. J Pharm Sci 52:1145–1148
130. Peppas NA, Khare A (1993) Preparation, structure and diffusional behavior of hydrogels in controlled release. Adv Drug Deliv Rev 11:1–35
131. Perera DI, Shanks RA (1996) Swelling and mechanical properties of crosslinked hydrogels containing N-vinylpyrrolidone. Polym Int 39:121–127
132. Wu W, Zhou S (2013) Responsive materials for self-regulated insulin delivery. Macromol Biosci 13:1464–1477
133. Sershen S, West J (2002) Implantable, polymeric systems for modulated drug delivery. Adv Drug Deliv Rev 54:1225
134. Demirci S, Çaykara T (2012) RAFT-mediated synthesis of cationic poly[(ar-vinylbenzyl) trimethylammonium chloride] brushes for quantitative DNA immobilization. Mater Sci Eng C 33(111)
135. Zhong K, Lin Z-T, Zheng X-L, Jiang GB, Fang YS, Mao XY, Liao ZW (2013) Starch derivative-based superabsorbent with integration of water-retaining and controlled-release fertilizers. Carbohydr Polym 92:1367–1376

136. Price R, Poursaid A, Ghandehari H (2013) Controlled release from recombinant polymers. J Control Release 190:304–313
137. Kumar A, Srivastava A, Galaev IY, Mattiasson B (2007) Smart polymers: physical forms and bioengineering applications. Prog Polym Sci 32:1205–1237
138. Roy D, Cambre JN, Sumerlin BS (2010) Future perspectives and recent advances in stimuli-responsive materials. Prog Polym Sci 35:278–301

Chapter 8
Polymer Formulations for Pesticide Release

Abhilasha Mishra, Rajesh K. Saini and A. K. Bajpai

Abstract Pesticides are used to control any form of plant or organisms that can cause damage to human health or property. Agricultural products are attacked by a variety of pests during the production and storage. Despite usefulness and popularity of pesticides in controlling a variety of pests, they can cause many health risks arising from their exposure and residues in food and water. Also, the effective availabilities of traditional pesticide are usually less than 30% due to losses. Polymers, in the form of micro-nanocarriers, beads, granules and gels, are very important materials for the development of controlled release formulations (CRF) of pesticides which provide slow and controlled release of pesticides and also enhance the water-holding capacity of the soil. In this chapter, various types of natural and synthetic polymers used in the preparation of polymeric formulations and their release behavior are discussed. Various ways by which the diffusion of pesticides in polymer takes place and different formulation methods for controlled release pesticides were also discussed in this chapter.

Keywords Pesticides · Controlled release · Human health · Formulation

8.1 Introduction

Pesticides are the substances used to kill, suppress, or control any form of plant or organisms that can cause damage to human health or property. Agricultural products such as cereals, vegetables, and fruits are an essential part of the diet which

A. Mishra
Department of Chemistry, Graphic Era (Deemed to be University), Dehradun, Uttarakhand, India
e-mail: abhi1680@gmail.com

R. K. Saini · A. K. Bajpai (✉)
Department of Chemistry, Bose Memorial Research Laboratory, Government Autonomous Science College, Jabalpur, MP, India
e-mail: akbmrl@yahoo.co.in

Rakhimol K. R. et al. (eds.), *Controlled Release of Pesticides for Sustainable Agriculture*, https://doi.org/10.1007/978-3-030-23396-9_8

provides a variety of nutrients that are required for metabolic reactions of the body. During the production and storage, these products are attacked by a variety of pests. Pests cause damaging of the agricultural products and reduce the quality and quantity as well. Various pesticides along with other pest management techniques are used to reduce the losses of these products and maintain their quality [1]. The uses of pesticides have attracted much attention as they have rapid action and require less labor than other pest control methods. Pesticides can also improve the nutritional value and quality of food [2]. Despite their usefulness and popularity, pesticides can cause many health risks arising from their exposure and from residues in food and drinking water [3]. Pesticide residues are the main source of soil and water pollution. The greater amount of pesticides is used for a longer period of time in conventional agrochemical application methods [4].

After the application of traditional pesticide formulations to the crops, the effective availabilities are usually less than 30% due to losses. More than 70% of pesticide does not reach to the target organisms and spreads in a wide area through water, soil, and air [5, 6]. These losses occur by wash off by rainwater, leaching, precipitation, and volatilization in the environment. Waxy cuticle and root surfaces of the plant also absorb pesticides, and they enter into the transport system of the plant. Degradation due to photolysis in the presence of sunlight, etc., is among the routes that reduce the amount of unused pesticide, and they also leave their residues in the environment and pose health hazards. The presence of pesticide residues in the environment is of great concern for researchers as pesticides have the potential to pose harmful effects and cause diseases to humans and other non-targeted organisms. Pesticides can also interfere with the reproductive systems thus initiating the fetal development. Pesticides also have the potential to cause cancer and asthma [7]. Some of the pesticides are sustained and remain in the body causing long-term exposure.

In the current scenario, the plant protection is based on the proper utilization of pesticide, and it is the most economical way of getting a high yield of good quality food but not compromising with the environment. The use of pesticides indiscriminately would result in harmful effects to the environment, people, and animals [8]. Researchers have been constantly trying to produce new formulations for the controlled release of pesticides for the protection of the environment and human health. Various mixtures of chemical agents to control pests are known as pesticide formulations. Special formulations of pesticides improve effectiveness, storage, safety, and handling. A pesticide formulation has an important role in its effectiveness and safety concerns. An appropriate formulation for a particular application can be obtained by considering type of pesticide, applicators safety, pest biology, available equipment for application, and final product cost. Various classes of polymers such as plastics, elastomers, and fibers are largely used in agriculture for increasing water-holding capacity and achieving controlled release of pesticides and nutrients [9, 10]. Polymers, in the form of micro-nanocarriers, beads, granules and gels, are very important materials for the development of controlled release formulations (CRF). CRF provide slow and controlled release of pesticides and also enhance the water-holding capacity of the soil [11]. After degradation, these formulations convert into compost and enhance soil nutrients [12]. CRF of pesticides are defined as depot systems which continuously release pesticide into the environment for a long time (months to years). According to this definition, such formulations can be successfully employed where a chronic exposure to biologically active compounds is required over a longer period. The use of polymeric formulations to solve the problem of pesticide loss and their accumulation in the environment is shown in Fig. 8.1.

A variety of polymers have been extensively used to control the release rates, molecular motilities, and the period of effectiveness depending on their end application. Polymeric dispersants offer superior stability by strong adsorption to the surface along with multiple anchoring points. The resultant effect is that the dispersant will not be disrupted by the incorporation of an adjuvant leading to a stable and highly efficacious formulation. Polymer-based controlled release systems offer various advantages over conventional formulations to avoid excessive use of active agrochemicals thus providing the most suitable technical solution of the pesticide residues [13]. The major benefit of the CRF is that if fewer amounts of agrochemicals are used for the protection of plants for a predetermined period, then it will reduce volatilization, leaching, and degradation of pesticide.

In 2016, the global market for controlled release pesticide size was estimated to be 1.7 billion USD, and up to 2025, it is expected to grow at a CAGR of 7.3%. The increasing demand of food for growing population is expected to drive more quanta of research in the field of CRF in the future [14].

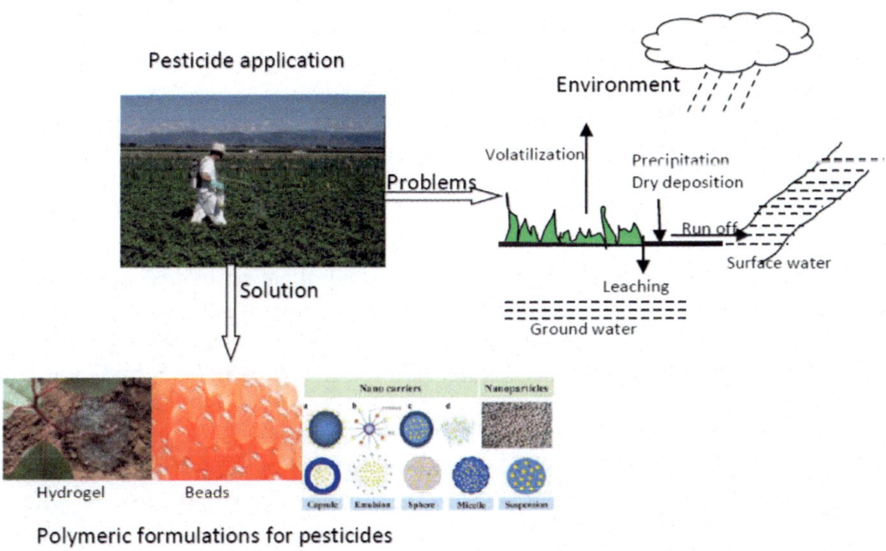

Fig. 8.1 Use of polymeric formulations to solve the problems of pesticide loss and their accumulation in environment

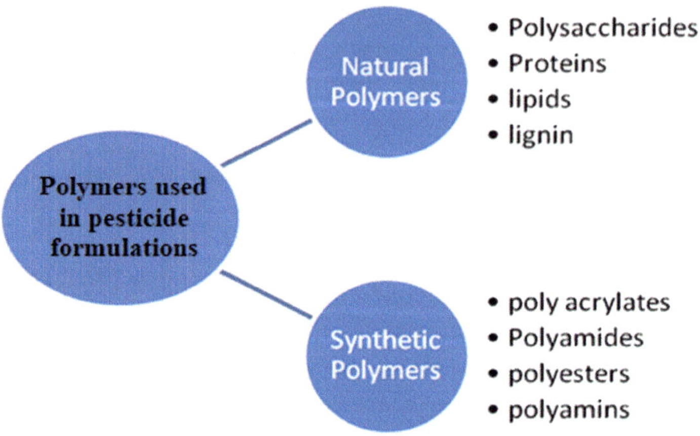

Fig. 8.2 Major classes of some natural and synthetic polymers used in making pesticide formulations

8.2 Types of Polymers Used in Formulation

For the release of pesticides, many natural and synthetic polymers have been largely used in designing their formulations. Natural polymers are environment-friendly and easily degradable whereas synthetic polymers provide better stability to pesticide release carriers. Major classes of some natural and synthetic polymers used in making pesticide formulations are shown in Fig. 8.2.

List of some natural and synthetic polymers which are used in formulation for pesticides are given in Table 8.1.

8.2.1 Natural Polymers

Nowadays, natural polymers are gaining increasing attention over synthetic polymers in pesticide release formulations because of their easy availability, eco-friendly nature, cost-effectiveness, and biodegradability [35]. The following are major classes of natural polymers which are used in polymeric formulations of pesticides.

8.2.1.1 Polysaccharides

Polysaccharides are most widely used in controlled release pesticide formulations. Due to their ready availability, fast degradation, low cost and wide variety polysaccharides facilitate large-scale production of these formulations. Some of the

Table 8.1 Some natural and synthetic polymers used in pesticide formulations

Polymers	Pesticides	Formulation type	References
Sodium alginate	Carbaryl	Hydrogel beads	[15]
Polyacrylamide-*g*-guar gum/sodium alginate	Chlorpyrifos and fenvalerate	IPN beads	[16]
Alginate, chitosan	Paraquat	Nanoparticle	[17]
Carboxymethylchitosan	4-Azidobenzaldehyde	Nanocapsules	[18]
Lignin PAA	Paraquat, cyfluthrin, cyhalofop-butyl	Hydrogel	[19]
Polylactic acid PAA	Emamectin benzoate	Microemulsion	[20]
Poly(hydroxybutyrate) (PHB) or poly (hydroxybutyrate-valerate) (PHBV)	Ametryn	Microparticle	[17]
Polylactic acid PAA	Lambda-cyhalothrin	Microcapsules	[21]
Polydopamine	Triazolone	Mesoporous particle	[22]
Poly-3-hydroxybutyrate	Metribuzin	Degradable matrix	[23]
Poly(oxy- ethylene) lauryl ether and methyl decanoate	β-cypermethrin	Nanoemulsion	[13]
Lignin-polyethylene glycol-ethylcellulose	Imidacloprid	Nanocapsule	[24]
Polyethylene	Piperonyl butoxide and deltamethrin	Nanocapsule	[25]
Polyethyleneglycol-dimethyl esters	Carbofuran	Micelle	[26]
Poly(methyl methacrylate)-poly (ethylene glycol) Polyvinylpyrrolidone	Carbofuran	Nanosuspension	[27]
Polyvinylpyrrolidone	Triclosan	Nanoparticle	[28]
Chitosan/alginate/gelatin	Malathion and spinosad	Capsules	[29]
Montmorillonite-chitosan	Clopyralid	Nanocomposite	[30]
Ethylcellulose	Norflurazon	Microparticles	[31]
Alginate	Azadirachtin	Nanoemulsion	[32]
Poly(ε-caprolactone)	Carbendazim, tebuconazole	Nanocapsule	[33]
Sodium alginate	E-β-caryophyllene	Beads	[34]

known examples are amylose, cellulose, pectin, alginate, etc. On the basis of degradation behavior, polysaccharides are classified into (a) starch and systems based on amylase which are readily degradable, and (b) other polysaccharides such as cellulose and derivatives, dextran, chitin, chitosan, alginate, and guar gum which degrade slowly as compared to starches [36]. These macromolecules can be cyclic (cyclodextrin) or linear (chitosan) with positive or negative charges or even neutral [37]. Many polysaccharides can be used for the controlled release of agrochemicals by ionotropic gelation process by using multivalent metal ions [38–40]. Polysaccharides are suitable for all type of formulations including hydrogel, micro/nanoparticles, micro/nanocapsules, beads, and emulsions. Polysaccharides are hydrophilic in nature and are most suitable for hydrophilic pesticides as a matrix system. Hydrophobic pesticides can also be loaded into cross-linked micro/nanocapsules or used in the form of nano-emulsions.

8.2.1.2 Proteins

Proteins are another major class of natural polymers which are used in controlled release formulations of pesticides. Gelatin [29], casein, and albumin [31] are some of the common proteins which are largely used in these formulations. A herbicide [(4-chloro-2-methylphenoxy) acetic acid] formulation with gelatin clay composite was prepared and release of herbicide was investigated [41]. Proteins are also used in combination with polysaccharide to give structural stability.

8.2.1.3 Phospholipids

Many natural phospholipids from plants and animal origin such as lecithin are used in the preparation of liposomal delivery systems for pesticides. Natural phospholipids are preferred over synthetic phospholipids for the preparation of liposomes as they are cheap and largely available with reproducible results. Natural phospholipids obtained from vegetable sources such as soybeans, canola seed, wheat germ, sunflower, and flaxseed, and from animal sources, such as milk and egg yolk, can be used to produce low-cost liposomes for controlled release of pesticides. Natural phospholipids are also well accepted by regulatory authorities and are produced using less chemicals and solvents at higher yields [42].

8.2.1.4 Other Natural Polymers

Lignin which is a complex organic molecule found in support tissues of vesicular plants, and some algae are also widely used in agriculture for controlled release formulations [43]. Polylactic acid (PLA) which is a FDA-approved material is used as carriers for active agents [44, 45]. Liu et al. used polylactic acid for preparing carriers for controlled release of Lambda-Cyhalothrin via premix membrane emulsification. They prepared three types of carriers including microspheres, microcapsules, and porous microcapsules which are shown in Fig. 8.3.

Fig. 8.3 SEM images of the **a** microspheres, **b** microcapsules, and **c** porous microspheres before and after cut by a super thin blade [21]

8.2.2 Synthetic Polymers

Many biodegradable synthetic polymers such as polyvinyl alcohol, polyacrylamide [46] are frequently used in the fabrication of slow-release formulations for pesticides. Although polyacrylamide gels are frequently used for encapsulation of pesticides, other polymers such as polyethylene, divynilbenzene, copolymers of acrylic acid and copolymers of cyclopentadiene with a glyceryl ester of an unsaturated fatty acid have also been used by researchers for designing pesticide release applications [47]. Some pesticides containing monomers are copolymerized with acrylamide and other hydrophilic co-monomers such as 4 vinyl pyridine [48]. Many pesticides such as pentachlorophenol (PCP), 2, 4-dichlorophenoxyacetic acid (2,4-D) and 4-chloro-2-methylphenoxyacetic acid are used as pendant groups in polymers [49]. Vinyl monomers containing pentachlorophenol pesticide via an ester linkage was prepared and homo- and copolymerized with 4-vinylpridine and styrene to induce hydrophilic and hydrophobic nature to the polymers [50]. Researchers combined chloropyriphos with polyethylene which can control mosquito larvae for about 18 months by one application itself [51]. Poly (vinyl alcohol) was widely used in agriculture for studying controlled release of agrochemicals [52]. Furthermore, the pesticides Azadirachtin A was physically bounded to both the PVA and poly (vinyl acetate) for its controlled release [53].

8.3 Diffusion of Pesticide into Polymer

Pesticides can diffuse into a polymer forming monolithic or reservoir-type formulations. In monolithic type, the pesticide is uniformly dispersed in a polymer matrix whereas in reservoir type the pesticide is covered by polymeric membrane as in case of capsules, laminates, liposome, coatings, etc.

8.3.1 Reservoir System

Reservoir-type formulations essentially consist of an enclosure system which exploits the diffusion of pesticides through a non-porous membrane as wall material for adjustment of the migration rates. Encapsulation ranges from macro to micro and nanoscales. In these systems, pesticides in the form of solid particles, liquids, or dispersions of solid particles in liquids, form a nucleus which is enclosed in a polymeric film. In reservoir-type formulation, the migration is controlled by membranes in which the transport of pesticides is controlled by Fick's law

$$J = \mathrm{d}Mt/A\mathrm{d}t = (-D\mathrm{dcm})/\mathrm{d}x$$

where J = Flux in g/cm

cm= Concentration of pesticide in g/cm^2 of the polymeric membrane

dcm/dx = Concentration gradient

D = Diffusion coefficient of the pesticide in cm^2/s in polymeric membrane

A = Surface area in cm^2, through which diffusion takes place

M = Mass of agent releasedd$M$/

dt = steady-state release rate at time t.

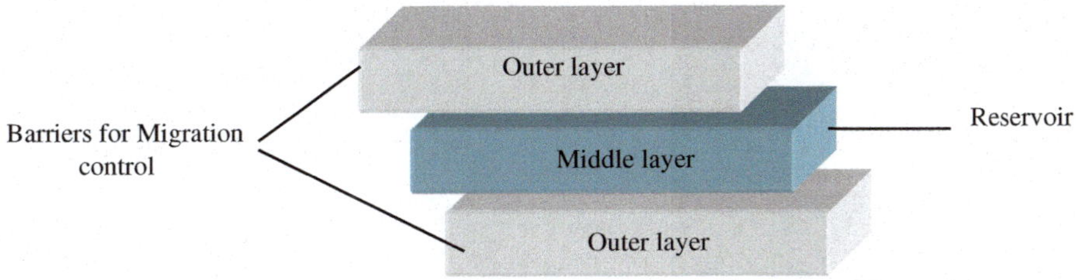

Fig. 8.4 Illustration of laminates-type controlled release system

8.3.1.1 Laminates

In laminate systems, a special type of membrane-controlled pesticide formulations can be produced. Laminates are obtained by the reciprocal bonding of different layers of sheet-like materials. For example in a three-layer formulation, the middle layer act as a reservoir containing pesticide in it and the other two outer layers act as barriers for controlling migration of pesticide. At the surfaces, the pesticides are released by leaching, evaporation, degradation, moisture wind dust, or mechanical contact by human or animals [54]. Laminates are used for the release of insect pheromones and insect attractants for insect control. A typical laminate system is shown in Fig. 8.4.

8.3.1.2 Capsules

Capsules are core–shell type structure where core is made up of pesticide which is surrounded by polymeric shell. They are usually in the range of nano to microsize. In the past decade, various pesticides were encapsulated into micro and nanocapsules [18, 33, 55]. Capsules can be mononuclear or polynuclear [56].

In the mononuclear capsules, only one depot of pesticide is present in core, whereas in multinuclear one, many small depots of pesticide are present. Chuxiang Sun et al. prepared nanocapsules of cross-linked carboxymethyl chitosan for the encapsulation of hydrophilic pesticide methomyl [18]. SEM and tem images of prepared nanocapsules are shown in Fig. 8.5 which clearly reveal core–shell nature of the capsules.

Latheef et al. prepared microcapsules of poly (methyl methacrylate) (PMMA), ethyl cellulose, poly (α-methylstyrene) and cellulose acetate with butyrate insecticide sulprofos contained in the core. The authors found the best results with ethyl cellulose formulations against eggs and larvae of the tobacco budworm Heliothis virescens in cotton plants [57].

Controlled release capsules of chitosan, alginate, and gelatin mixture were prepared by cross-linking biopolymers with glutaraldehyde. The capsules were loaded with the insecticides temephos [29], malathion and spinosad [54] and their pesticticidal activities were studied against Culex pipiens larvae.

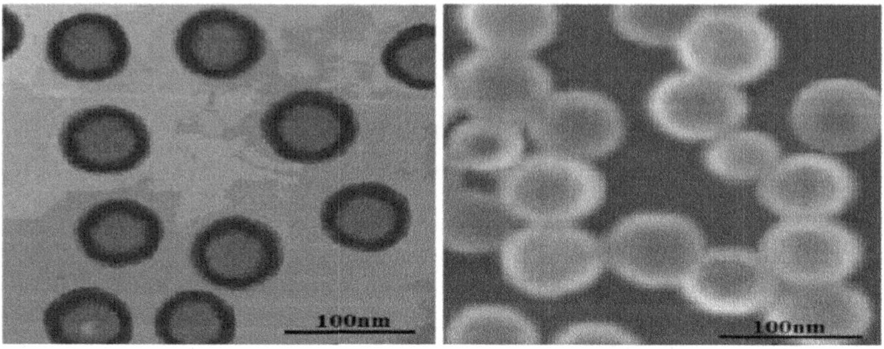

Fig. 8.5 TEM (left) and SEM (right) images of cross-linked nanocapsules

8.3.1.3 Liposomes

Liposomes are lipid-based nonmaterial with hydrophic core encapsulated in a lipid bilayer [58]. Liposomes are also known as fatty acid vesicles [59], and it has been reported that the pesticides encapsulated in liposomes demonstrate superior action due to the prolonged persistence at reduced damage to other non-targeted organisms [60]. Liposomes can carry both hydrophilic and hydrophobic pesticide [61]. Hydrophobic pesticide can be encapsulated between lipid bilayer and hydrophilic in the core. Both types of liposomes can be made sticky so that they remain on leaves for longer times and do not wash off to the ground. Inexpensive liposomes can also be produced from synthetic lipids and natural lipids for the controlled release of pesticides. Hwang et al. prepared liposome of chitosan-coated lecithin for the controlled release of etofenprox [62].

8.3.1.4 Coatings

Mesoporous silica nanoparticles with polymer coatings have been developed to achieve controlled release of pesticides. The tunable pore size, low cost, large surface area, good-loading capacity, and low cost make them ideal carriers for controlled release of pesticides [63, 64]. Lidong et al. successfully coated a water-soluble chitosan (CS) derivative (N-(2-hydroxyl) propyl-3-trimethyl ammonium CS chloride on the surfaces of the mesoporous silica nanoparticle loaded with pyraclostrobin. The loading efficiency was greatly improved by the coating and the material showed excellent fungicidal activity against Phomopsis asparagi (Sacc.) in half dose [65]. Sato et al. developed a slow-release system with a certain lag time for Imidacloprid. They mixed Imidacloprid with bentonite and clay for the preparation of core granules. The core granules were coated with a mixture of high-density polythene and talc by using spouted bead coating system [66].

8.3.2 *Matrix System*

In matrix-type formulations, the pesticide is heterogeneously dispersed or dissolved in a solid polymeric matrix, and this can be either biodegradable or non-biodegradable. The release of pesticide is generally controlled by diffusion through the matrix, chemical, or biological erosion, or through a combination of diffusion and erosion. The release by erosion is surface area-dependent, and the general expression that describes the rate of release *Rr* by an erosion mechanism is given as,

$$\mathrm{Rr} = \mathrm{d}M/\mathrm{d}t = \mathrm{KECo}\,A$$

where KE= erosion rate constant*A* = exposed surface area

Co=concentration of loaded pesticide in the matrix.

Matrix systems are easy to produce than the reservoir systems, but the zero-order release cannot be obtained by these systems. These systems are easy to produce than the reservoir devices [67]. A major factor affecting the release process is the porosity or free volumes within the matrix of the plastic material the magnitude of which is controlled by the processing conditions and incorporated additives [68]. Volova and coworkers used poly(3-hydroxybutyrate) as a matrix for slow-release formulations of the herbicide metribuzin. Physical mixtures of polymer and pesticide in the form of solutions, powders, and emulsions were used to construct different metribuzin formulations in the form of granules, pellets, and microparticles [23].

8.3.2.1 Particles

Micro and nanoparticle prepared from a variety of natural and synthetic polymers have attracted attention of researchers in the area of controlled release of pesticide. The size of particles has great influence on the release behavior. Grillo et al. developed biodegradable polymeric microparticles of poly (hydroxybutyrate) (PHB) or poly(hydroxybutyrate-valerate) (PHBV), and the authors noticed that the herbicidal activity in microparticles formulation was found better as compared to the conventional formulations [69]. Faria et al. prepared calcium alginate microparticles for controlled release of tebuthiuron [70]. Liu et al. incorporated tebuconazole and chlorothalonil fungicides in polymeric nanoparticles prepared from polyvinyl pyridine and polyvinyl(pyridine-co-styrene) with 10 and 30% styrene, respectively. The mean diameter of particles was found in the range of 100–250 nm, and it increases with increase in the styrene content [71].

8.3.2.2 Pellets

PVA and 1,8-cineole mixture were prepared by dry mixing method. The aim of the present study was the evaluation of controlled release time and efficiency of 1 g insecticide pellets prepared using the eucalyptol, botanic constituent, and poly (vinyl alcohol), a biodegradable polymer, under laboratory condition [72].

8.3.2.3 Beads

Beads are the simplest formulations for the controlled release of pesticides. They are normally prepared by cross-linking of preformed polymers. The beads of polyacrylamide-*g*-guar gum and sodium alginate were prepared by cross-linking the IPN of grafted polymer and alginate with gluteraldehyde. The as-prepared beads were loaded with chlorpyrifos and liquid fenvelarate for their controlled release [16]. Hydrogel beads of calcium alginate and nickel alginate were prepared and assessed for the controlled release of carbaryl insecticide [15].

8.4 Formulation Methods

The release rate of pesticide from the polymer matrix is highly influenced by the following factors:

Chemical nature of the pesticide—polymer bond such as esters, amides, and acetals.

- The distance of the pesticide molecule from the polymer backbone
- Rate of breakdown of the bond between pesticide and the polymer by chemical, biological, or environmental agents.
- Biodegradation of polymer.
- Structure and dimension such as degree of polymerization, solubility, degree of cross-linking, co-monomers, and the stereochemistry of the polymer.

In order to control these parameters, the pesticides are either chemically bounded or physically incorporated into a polymer matrix by different techniques. The migration of the substances is, therefore, preceded by chemical reactions for bond cleavage or physical transport processes in and through polymers.

8.4.1 Physical Combination of Polymer and Pesticide

The active agent is dissolved, dispersed, or encapsulated within the polymeric matrix or coating. The release takes place through diffusion or biological/chemical degradation of the polymer. The polymer acts as a rate-controlling device in physical combination [73]. The physical combination broadly comprises laminated, reservoir, and monolithic systems [54].

8.4.1.1 Interfacial Polymerization

Morgan and co-researchers first described the principles of interfacial polymerization [74]. In interfacial polymerization, the polymerization reaction occurs at the interface of the two immiscible liquids [75]. For interfacial polymerization, two reactive monomers are dissolved in two immiscible solvents to facilitate their contact only at interface. In this way, the polymerization occurs only at the interface forming a polymeric film. Interfacial polymerization is largely employed for the formation of microcapsules consisting of liquid pesticide inside the polymeric membrane. In the formulation, generally one phase is an aqueous and the other phase is organic. If the aqueous phase is dispersed in organic phase, the core will be hydrophilic and can be used for encapsulation of hydrophilic pesticide whereas inverting the phases would result in a hydrophobic core which can be used for hydrophobic pesticides. Microcapsules of polyamide, polyurathanes, polyurease, and polyesters can be prepared by interfacial polymerization [76]. Moghbeli et al. prepared polyurea microcapsule containing ethion pesticide by interfacial polycondensation [77].

8.4.1.2 Coacervation

Encapsulation of pesticide by the coacervation phase-separation technique works in the liquid phase and generally consists of three steps.

- Formation of three immiscible chemical phases
- Deposition of coating
- Hardening of the coating.

The pesticide and the polymer materials are initially present as different phases in the form of an emulsion or suspension. During mixing, the polymer encloses pesticide, through adsorption, around a nucleus of the pesticide. The polymer phase is subsequently precipitated (solidified) by means of thermal treatment, cross-linking reaction or desolvation [78]. The solvent used in this process must not dissolve the pesticide to obtain discrete capsules. Particle sizes of capsules are significantly affected by the mixing rate. The best example of simple coacervation is the encapsulation of core material in gelation [79].

The coacervation-phase separation can be divided into two categories, simple coacervation and complex coacervation. In simple coacervation, a strongly hydrophilic substance is added to a colloidal solution which results in the formation of two phases. The complex coacervation is a pH-dependent process where the microcapsules are produced by manipulating acidity or basicity of the system. The microcapsule formation takes place only above a critical pH. Complex coacervation deals with the system containing more than one colloid [80].

8.4.1.3 Ionic Gelation

Ionic gelation involves the interaction of an ionic polymer with oppositely charged ion to initiate cross-linking [81]. Pesticide-loaded matrix of various size ranges can be prepared by ionic gelation method in which the pesticide containing polymer solution is poured dropwise into the solution of metal ions. The best example of ionic gelation is the solution of sodium alginate and pesticide added drop-wise into calcium chloride solution. This method is very useful for the preparation of hydrogel-type release system where release of pesticide depends on swelling of polymer matrix. Isıklan encapsulated carbaryl in the alginate beads by the ionotropic gelation of sodium alginate (NaAlg) with calcium and nickel ions [15]. Kumbar prepared interpenetrating polymer network beads by ionic gelation technique. Solution of polyacrylamide-grafted-guar gum with sodium alginate containing chlorpyrophos/fenvelarate pesticides was dropped into the aqueous solution of gluteraldehyde and hydrochloric acid [82]. Micro and nanoparticles of polysaccharides were also prepared by ionic gelation process, but some defects such as improper surface morphology, high dispersibility index, and fragile particulate system can be encountered [83, 84].

8.4.1.4 Spray Drying and Spray Congealing

Spray drying is a single-step microencapsulation technique for different ingredients in various applications such as cosmetics, food materials, and agrochemicals [85–87]. In this technique, a liquid is rapidly transformed into a dried powder [88]. The four important steps of spray drying are atomization, contact of droplet with hot gas, evaporation, and powder separation. The main advantage of this technique is that one can use different forms such as emulsion, suspension, solution, and slurries via this technique [89, 90].

Spray drying has been extensively used for many years for microencapsulation of active agents. Both synthetic and natural polymers such as ethyl cellulose, gums, maltodextrin, polylactic acid (PLA), poly(lactic-co-glycolic acid) (PLGA), chitosan, and poly(ε-caprolactone) (PCL), are widely used materials to encapsulate different pesticides [91, 92]. For microencapsulation of pesticide, the pesticide is dispersed in a liquid-coating material forming an emulsion. The emulsion is then sprayed into the heated chamber of a spray drier where rapid solidification of the coating takes place [93]. The size of the encapsulated particles depends upon the droplet size of emulsion. Spray drying process is beneficial for microencapsulation of biopesticides as spray-dried encapsulated biopesticides can be preserved for long time [94, 95].

Basic principal and equipment used for both spray drying and spray congealing are same and the only difference between the two methods is coating solidification. Coating solidification in the case of spray drying the polymeric material used for coating is dissolved in a suitable solvent whereas in spray congealing molten coating material is used. Waxes, fatty acids, and alcohols, polymers which are solids at room temperature but meltable at reasonable temperature are applicable to spray congealing [96].

8.4.2 Chemical Combination of Polymer and Pesticide

In chemical combination of polymer and pesticide, the pesticide is chemically attached to polymeric chain through covalent or ionic bonds. Here the pesticide either constitutes a part of the macromolecular backbone or is attached to it as a pendent group as shown in Fig. 8.6.

The polymer acts as a carrier for the pesticide in chemical combination [97] and the release of pesticide is dependent on the biological or chemical degradation by cleavage of the bonds between the polymer and the pesticide molecules [98]. The release of the pesticide is dependent on environmental conditions that break the linkages via chemical attack (hydrolytic by moisture; thermal/photo by sunlight) or biological degradation (enzymatic by microorganisms). pH of the medium, electrolyte concentration, temperature and ionic strength of the dissolution medium have great influence on degradation rate and hence on the release of pesticide also.

Only the pesticides containing a structural moiety with at least one reactive functional group can be used in this technique. For chemical combination, two approaches are utilized; in the first approach, a preformed polymer is used to form chemical linkage with pesticide; while in the second approach, a pesticide containing monomer unit or a pesticide and monomer mixture is polymerized. Amide, ester, and anhydride linkages are most common linkages formed between polymer and pesticide in their chemical combination [99].

8.4.2.1 Chemical Attachment of Pesticide with Preformed Polymer

It involves chemical modification of a preformed polymer with the pesticide via a chemical bond, leading to a polymer having the pesticide linked to the main chain as a pendant group. Pesticides containing carboxylic acid groups can be easily attached to a polymer containing labile hydrogen as found in polymers with alcohol or amine side chain. Here the pesticide is bonded to polymeric main chain as pendent group. This type of reaction is given in equation,

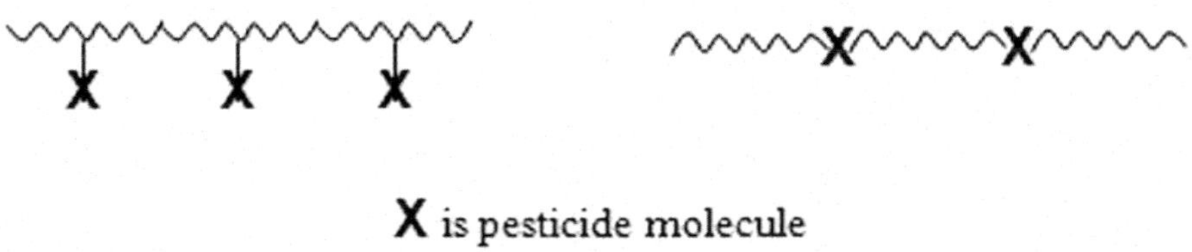

Fig. 8.6 Illustration of chemical attachments of pesticide molecule with polymer backbone as pendent group (left) and as a part of backbone (Right)

$$\text{Polymer}-\text{X}-\text{H} + \text{HCOOR} \rightarrow \text{Polymer}-\text{X}-\text{COR} + \text{H}_2\text{O}$$

X can be sulfur, oxygen, or nitrogen.

8.4.2.2 Chemical Attachment of Pesticide with Polymerization of Monomer

This type of reaction leads to the formation of both type of polymers, either having pesticide group in main chain or pesticide group attached with main chain as pendent group. The main advantages of this method are the ability to control the molecular weight of the polymer, the weight ratio of polymer and pesticide, the hydrophobic-hydrophilic balance via appropriate co-monomers, and the distribution of groups along the backbone.

A. **Pesticide attached to polymer chain as pendent group**

Copolymers containing pendant pentachlorophenol have been prepared by free radical copolymerization of the corresponding monomer. A series of vinyl monomers containing PCP via an ester linkage have been prepared which are shown in Scheme 8.1. These monomers have been homo-and copolymerized with styrene and 4-vinylpridine [100].

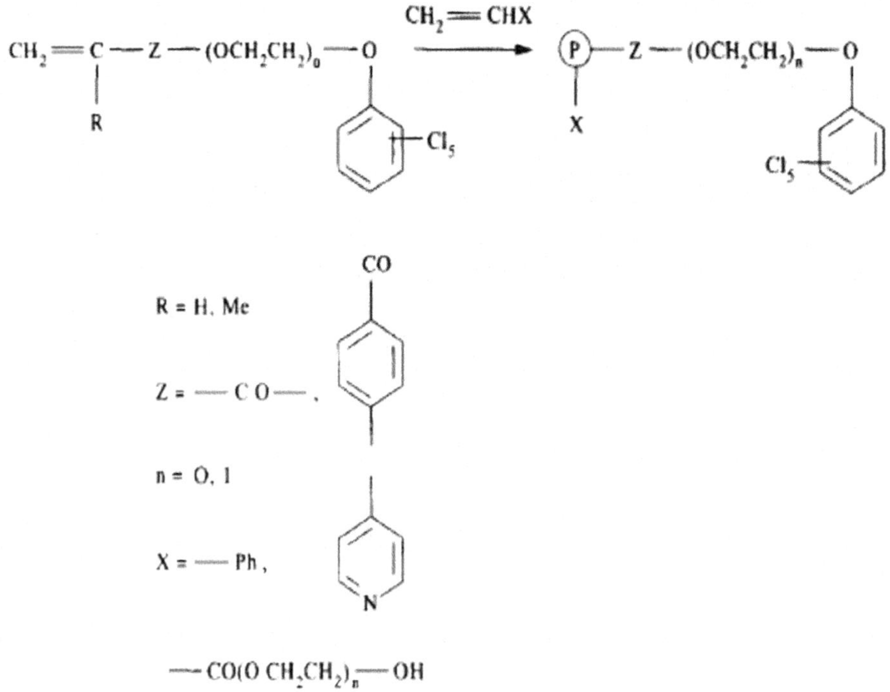

Scheme 8.1 Examples of polymerization of biologically active monomers. [97]

B. **Polymers having the pesticide in the main polymeric chain**

Heptachlor is a chlorinated dicyclopentadiene insecticide which is used to control soil insects and termites. Three copolymers of heptachlor with methacrylic acid (MAA), ethylene glycol dimethacrylate (EDMA), 4-vinyl pyridine (4-VP), divinyl-benzene (DVB) and styrene were prepared by Singh et al. The prepared copolymers are heptachlor-co-MAA-co-EDMA, heptachlor-co-4-VP-co-DVB, and heptachlor-co-Styrene-co-DVB [101].

References

1. Cooper J, Dobson H (2007) The benefits of pesticides to mankind and the environment. Crop Prot 26(9):1337–1348
2. Narayanasamy P (2006) Postharvest pathogens and disease management. John Wiley & Sons, New York
3. Maroni M, Fanetti AC, Metruccio F (2006) Risk assessment and management of occupational exposure to pesticides in agriculture. Med Lav 97:430–437
4. Bajpai AK, Giri A (2003) Water sorption behaviour of highly swelling (carboxy methylcellulose-g-polyacrylamide) hydrogels and release of potassium nitrate as agrochemical. Carbohyd Polym 53(3):271–279
5. Gamon M, Saez E, Gil J, Boluda R (2003) Direct and indirect exogenous contamination. Arch Environ Contam Toxicol 44:141–151
6. Shalaby SEM, Abdou GY (2010) The influence of soil microorganisms and bio-or-organic fertilizers on dissipation of some pesticides in soil and potato tube. J Plant Prot Res 50 (1):86–92
7. Gilden RC, Huffling K, Sattler B (2010) Pesticides and health risks. J Obstet Gynecol Neonatal Nurs 39(1):103–110
8. CICOPLAFEST (2004). Catalogo Oficial de Plaguicidas. Comisión Intersecretarial para el Control de Proceso y Uso de plaguicidas, Fertilizantes y Sustancias Tóxicas
9. Maghchiche A, Haouam A, Immirzi B (2010) Use of polymers and biopolymers for water retaining and soil stabilization in arid and semiarid regions. J Taibah Univ Sci 4(1):9–16
10. Huang B, Chen F, Shen Y, Qian K, Wang Y, Sun C, Zhao X, Cui B, Gao F, Zeng Z, Cui H (2018) Advances in targeted pesticides with environmentally responsive controlled release. Nanomaterials 8(2):1–18
11. Roy A, Singh SK, Bajpai J, Bajpai AK (2014) Controlled pesticide release from biodegradable polymers. Cent Eur J Chem 12:453–469
12. Kumar S, Bhanjana G, Sharma A (2014) Synthesis, characterization and on field evaluation of pesticide loaded sodium alginate nanoparticles. Carbohydr Polym 101:1061–1067
13. Wang L, Li X, Zhang G, Dong J, Eastoe J (2007) Oil-in-water nanoemulsions for pesticide formulations. J Colloid Interface Sci 314:230–235
14. https://www.grandviewresearch.com/industry-analysis/slow-controlled-release-pesticides-market
15. Isıklan N (2007) Controlled release study of carbaryl insecticide from calcium alginate and nickel alginate hydrogel beads. J Appl Polym Sci 105:718–725
16. Kumbar SG, Ashok Dave AM, Aminabhavi TM (2003) Release kinetics and diffusion coefficients of solid and liquid pesticides through interpenetrating polymer network beads of polyacrylamide-*g*-guar gum with sodium alginate. J Appl Polymer Sci 90:451–457
17. Grillo R, Pereira ADES, Melo NFS, Porto RM, Feitosa LO, Tonello PS, Filho NLD, Rosa AH, Lima R, Fraceto LF (2011) Controlled release system for ametryn using polymer

microspheres: Preparation, characterization and release kinetics in water. J Hazard Mater 186 (2–3):1645–1651
18. Sun C, Shu K, Wang W, Ye Z, Liu T, Gao Y, Zheng H, He G, Yin Y (2014) Encapsulation and controlled release of hydrophilic pesticide in shell cross-linked nanocapsules containing aqueous core. Int J Pharm 463:108–114
19. Sun Y, Ma Y, Fang G, Fu Y (2016) Controlled pesticide release from porous composite hydrogels based on lignin and polyacrylic acid. Bio Res 11(1):2361–2371
20. Wang Y, Wang A, Wang C, Cui B, Sun C, Zhao X, Zeng Z, Shen Y, Gao F, Liu G, Cui H (2017) Synthesis and characterization of emamectin-benzoate slow-release microspheres with different surfactants. Sci Rep 7(1):1–9
21. Liu B, Wang Y, Yang F, Wang X, Shen H, Cui H, Wu D (2016) Construction of a controlled-release delivery system for pesticides using biodegradable PLA-based microcapsules. Colloid Surf B 144:38–45
22. Chen H, Huang G, Zhou H, Zhou X, Xu H (2018) Highly efficient triazolone/metal ion/polydopamine/MCM-41 sustained release system with pH sensitivity for pesticide delivery. R Soc Open Sci 5(7):1–10
23. Volova TG, Zhila NO, Vinogradova ON, Nikolaeva ED, Kiselev EG, Shumilova AA, Shershneva AM, Shishatskaya EI (2015) Constructing herbicide metribuzin sustained-release formulations based on the natural polymer poly-3-hydroxybutyrate as a degradable matrix. J Environ Sci Health B 51(2):113–115
24. Frandsen MV, Pedersen MS, Zellweger M, Gouin S, Roorda SD, Phan TQC (2010) Piperonyl butoxide and deltamethrin containing insecticidal polymer matrix comprising HDPE and LDPE. Patent Number WO 2010015256(A2):20100211
25. Cespedes FF, Flores CIF, Fernandez ID, Pena FV, Sanchez MV, Perez MF (2012) Preparation and characterization of imidacloprid lignin-polyethylene glycol matrices coated with ethylcellulose. J Agric Food Chem 60:1042–1051
26. Shakil NA, Singh MK, Pandey A, Kumar J, Parmar VS, Singh MK, Pandey RP, Watterson AC (2010) Development of poly(Ethylene Glycol) based amphiphilic copolymers for controlled release delivery of carbofuran. J Macromol Sci A Pure App Chem 47:241–247
27. Chin CP, Wu HS, Wang SS (2011) New approach to pesticide delivery using nanosuspensions: research and applications. Ind Eng Chem Res 50:7637–7643
28. Paula HCB, Sombra FM, Abreu FOMS, de Paula RCM (2010) Lippia sidoides essential oil encapsulation by angico gum/chitosan nanoparticles. J Brazilian Chem Soc 21:2359–2366
29. Badawy MEI, Taktak NEM, Awad OM, Elfiki SA, El-Ela NEA (2015) Larvicidal activity of temephos released from new chitosan/alginate/gelatin capsules against Culex pipiens. Int J Mosq Res 2(3):45–55
30. Celis R, Adelino MA, Hermosin MC, Cornejo J (2012) Montmorillonite—chitosan bionanocomposites as adsorbents of the herbicide clopyralid in aqueous solution and soil/water suspensions. J Hazard Mater 209–210:67–76
31. Sopena F, Villaverde J, Maqueda C, Morillo E (2011) Photostabilization of the herbicide norflurazon microencapsulated with ethylcellulose in the soil-water system. J Hazard Mater 195:298–305
32. Jerobin J, Sureshkumar RS, Anjali CH et al (2012) Biodegradable polymer based encapsulation of neem oil nanoemulsion for controlled release of Aza-A. Carbohydr Polym 90:1750–1756
33. Campos EVR, De Oliveira JL, Da Silva CMG, Pascoli M, Pasquoto T, Lima R, Abhilash PC, Fraceto LF (2015) Polymeric and Solid Lipid Nanoparticles for Sustained Release of Carbendazim and Tebuconazole in Agricultural Applications, Scientific Reports vol 5, Article number: 13809
34. Daems F, Béra F, Lorge S, Fischer C, Brostaux Y, Francis F, Lognay G, Heuskin S (2016) Impact of climatic factors on the release of E-β-caryophyllene from alginate beads. Biotechnol Agron Soc Environ 20(2):130–142
35. Azwa ZN, Yousif BF, Manalo AC, Karunasena W (2013) A review on the degradability of polymeric composites based on natural fibres. Mater Des 47:424–442

36. Wilkins RM (ed) (1990) Controlled delivery of crop-protection agents. Taylor and Francis Ltd, London
37. Hassani LN, Hendra F, Bouchemal K (2012) Auto-associative amphiphilic polysaccharides as drug delivery systems. DrugDiscov Today 7:608–614. https://doi.org/10.1016/j.drudis.2012.01.016
38. Fernandez-Perez M (2007) Controlled release systems to prevent the agroenvironmental pollution derived from pesticide use. J Environ Sci Health B 42(7):857–862
39. Singh B, Sharma DK, Dhiman A (2013) Environment friendly agar and alginate-based thiram delivery system. Toxicol Environ Chem 95(4):567–578
40. Wlodarczyk M, Siwek H (2013) Clomazone release kinetics from alginate matrix to the water environment. Przemysl Chem 92(8):1513–1516
41. Alromeed AA, Scrano L, Bufo S, Undabeytia T (2015) Slow-release formulations of the herbicide MCPA by using clay-protein composites. Pest Manage Sci 71(9):1303–1310
42. Hoogevest PV, Wendel A (2014) The use of natural and synthetic phospholipids as pharmaceutical excipients. Eur J Lipid Sci Technol 116(9):1088–1107
43. Chowdhury MA (2014) The controlled release of bioactive compounds from lignin and lignin-based biopolymer matrices. Int J Biol Macromol 65:136–147
44. Huang D, Li W, Wang T, Shen H, Zhao P, Liu BX, You Y, Ma Y, Yang F, Wu D, Wang S (2015) Isoniazid conjugated poly (lactide-co-glycolide): long-term controlled drug release and tissue regeneration for bone tuberculosis therapy. Biomaterials 52:417–425
45. Duan Y, Zhang B, Chu L, Tong HH, Liu W, Zhai G (2016) Colloids Surf B Biointerfaces 141:345–354
46. Rudzinski WE, Chipuk T, Dave AM, Kumbar SG, Aminabhavi TM (2003) pH-sensitive acrylic-based copolymeric hydrogels for the controlled release of a pesticide and a micronutrient. J Appl Polym Sci 87:394–403
47. Puoci F, Iemma F, Spizzirri UG, Cirillo G, Curcio M, Picci N (2008) Polymer in agriculture: a review. Am J Agric Biol Sci 3(1):299–314
48. Raheb A, Akelah A, Issa R, Solaro R, Chiellini E (1991) Herbicide containing methacrylate, synthesis, polymerization and release investigation. J Contr Release 17:113–122
49. Kashyap PL, Xiang X, Heiden P (2015) Chitosan nanoparticle based delivery systems for sustainable agriculture. Int J Bio Macromolecules 77:36–51
50. Sun G, Wheatley WB, Worley SD (1994) A new cyclic N-Halamine. Biocidal polymer. Indian Eng Chem Res 33:68–170
51. Miller T, Nelson LL, Young WW, Roberts LW, Roberts DR, Wilkinson RN (1973) Polymer formulations of mosquito larvicides. I. Effectiveness of polyethylene and polyvinyl chloride formulations of chlorpyrifos applied to artificial field pools. Mosq News 33(2):148–155
52. Chandra R, Rustgi R (1998) Biodegradable polymers. Prog Polym Sci 23(7):1273–1335
53. Riyajan SA, Sakdapipanich JT (2009) Encapsulated neem extract containing Azadirachtin-A within hydrolyzed Poly(vinyl acetate) for controlling its release and photodegradation stability. Chem Eng J 152(2–3):591–597
54. Scher HB (1999) Controlled-release delivery systems for pesticides. CRC Press, Boca Raton
55. Badawy ME, Taktak NE, Awad OM, Elfiki SA, Ela-Ela NEA (2016) Evaluation of released malathion and spinosad from chitosan/alginate/gelatin capsules against culex pipiens larvae. Res Rep Trop Med 7:23–38
56. Jyothi SS, Seethadevi A, Prabha KS, Muthuprasanna P, Pavitra P (2012) Microencapsulation: a review. Int J Pharma Bio Sci 3(1):509–531
57. Latheef MA, Dailey OD, Franz E (1993) Efficacy of polymeric controlled release formulations of sulprofos against tobacco budworm, heliothis virescens (Lepidoptera: Noctuidae) on cotton. In: Berger PD, Devisetty BN, Hall FR. (eds) Pesticide formulations and applications systems, vol 13, ASTM STP 1183. American Society for Testing and Materials, Philadelphia, pp 300–311
58. Gregoriadis G, Bacon A, Wanderley WC, Mccormack B (2002) A role for liposomes in genetic vaccination. Vaccine 20:B1–B9

59. WALDE 2006. Formation and properties of fatty acid vesicles (liposomes) In: Liposomes technology. Inform Healthcare, New-York
60. Tahibi A, Sakurai JD, Mathur R, Wallach DFH (1991) Novasome vesicles in extended pesticide formulation. Proc Symp Contr Rel Bioact Mat 18(1):231–232
61. Chrai SS, Murari R, Ahmed I (2002) Liposomes (a review). part two: drug delivery systems. Bipharm 15:40–43
62. Hwang C, Kim TH, Bang SH, Kim KS, Kwon HR, Seo MJ, Youn YN, Park HJ, Yasunaga-Aokt C, Yu YM (2011) Insecticidal effect of controlled release formulations of etofenprox based on nano-bio technique. J Fac Agric Kyushu Univ 56(1):33–40
63. Lehman SE, Larsen SC (2014) Zeolite and mesoporous silica nanomaterials: greener syntheses, environmental applications and biological toxicity. Environ Sci Nano 1(3):200–213
64. Sun R, Wang W, Wen Y, Zhang X (2015) Recent advance on mesoporous silica nanoparticles-based controlled release system: intelligent switches open up new horizon. Nanomaterials 5(4):2019–2053
65. Cao L, Zhang H, Cao C, Zhang J, Li F, Huang Q (2016) Quaternized chitosan-capped mesoporous silica nanoparticles as nanocarriers for controlled pesticide release. Nanomaterials 6(7):1–13
66. Sato C, Watanabe M, Nakamura Y, Okuda S (1982) Jpn. Kokai Tokkyo Koho JP 57-126402 (in Japanese)
67. Anka Rao A, Narasimha Rao V, Seetha Devi A, Anil K, Vasu Naik V, Rajesh A (2015) Oral controlled release drug delivery system: an overview. Int J Pharma Chem Res 1(1):6–15
68. Sfirakis A, Rogers CE (1981) Sorption and diffusion of alcohols in amorphous polymers. Polym Eng Sci 21(9):542–547
69. Grillo R, Clemente Z, Oliveira JLD, Campos EVR, Chalupe VC, Jonsson CM, Lima RD, Sanches G, Nishisaka CS, Rosa AH, Oehlke K, Greiner R, Fraceto LF (2015) Chitosan nanoparticles loaded the herbicide paraquat: the influence of the aquatic humic substances on the colloidal stability and toxicity. J Hazard Mater 286:562–572
70. Faria DM, Junior SMD, Nascimento JPLD, Nunes EDS, Marques RP, Rossinoc LS, Moretoa JA (2016) Development and evaluation of a controlled release system of TBH herbicide using alginate microparticles. Mate Res 20(1):225–235
71. Liu Y, Yan L, Heiden P, Laks P (2001) Use of nanoparticles for controlled release of biocides in solid wood. J Appl Polym Sci 79(3):458–465
72. Sharifian I, Farahani S, Shafiei SE, Sanchooli M (2015) Evaluation of release time and efficiency of a botanical insecticide pellet under laboratory conditions. J Entomol Zool Stud 3(2):355–358
73. Kenawy ER (1998) Recent advances in controlled release of agrochemicals. J Macromol Sci C Polym Rev 38:365–390
74. Morgan PW, Kwolek SL (1959) Interfacial polycondensation. II. Fundamentals of polymer formation at liquid interfaces. J Polym Sci 11:299–327
75. Sinha VR, Goyal V, Bhinge JR, Mittal BR, Trehan A (2003) Diagnostic microspheres: an overview. Crit Rev Ther Drug Carrier Syst 20(6):431–460
76. Perignon C, Ongmayeb G, Neufeld R, Frere Y, Poncelet D (2014) Microencapsulation by interfacial polymerisation: membrane formation and structure. J Microencapsul 32(1):1–15
77. Moghbeli MR, Abedi V, Dekamin MG (2011) Microencapsulation of Ethion by interfacial polymerization utilizing potassium phthalimide-N-oxyl (PPINO) as a promoter. Iran J Chem Eng 8(4):34–42
78. Bakan JA (1975) Microencapsulation of pesticides and other agricultural materials. In: Harris
79. Mohanty B, Aswal VK, Kohlbrecher J, Bohidar HB(2005) Synthesis of Gelatin nanoparticles via simple coacervation. 21(3):149–160
80. Kiyoyama S, Shiomori K, Kawano Y, Hatate Y (2003) Preparation of microcapsules and control of their morphology. J Microencapsul 20(4):497–508

81. Ahirrao SP, Gide PS, Shrivastav B, Sharma P (2014) Ionotropic gelation: a promising cross linking technique for hydrogels. Res Rev J Pharma Nanotechnol 2(1):1–6
82. Kumbar SG, Dave AM, Aminabhavi TM (2003) Release kinetics and diffusion coefficients of solid and liquid pesticides through interpenetrating polymer network beads of polyacrylamide-*g*- Guar Gum with sodium alginate. J Appl Polymer Sci 90:451–457
83. Badawi AA, El-Laithy HM, El Qidra RK, El Mofty H, El Dally M (2008) Chitosan based nanocarriers for indomethacin ocular delivery. Arch Pharm Res 31(8):1040–1049
84. Wu P, He X, Wang K, Tan W, He C, Zheng M (2009) A novel methotrexate delivery system based on chitosan-methotrexate covalently conjugated nanoparticles. J Biomed Nanotechnol 5(5):557–564
85. Garza CB, Fernández JY, Huerta BEB (2013) Thermal and pH stability of spray-dried encapsulated astaxanthin oleoresin from Haematococcus pluvialis using several encapsulation wall materials. Food Res Int 54(1):641–649
86. Sen D, Khan A, Bahadur J, Mazumder S, Sapra BK (2010) Use of small-angle neutron scattering to investigate modifications of internal structure in self-assembled grains of nanoparticles synthesized by spray drying. J Colloid Interface Sci 347(1):25–30
87. Gong P, Zhang L, Han X, Shigwedha N, Song W, Yi H, Du Cao C M (2014) Injury mechanisms of lactic acid bacteria starter cultures during spray drying: a review. Drying Technol 32(7):93–800
88. Ribeiro RF, Motta MH, Flores FC, Beck RCR, Harter APG, Schaffazick SR, Silva CDBD (2016) Spray-dried powders improve the controlled release of antifungal tioconazole-loaded polymeric nanocapsules compared to with lyophilized products. Mater Sci Eng C 59:875–884
89. Abbaspourrad A, Datta SS, Weitz DA (2013) Controlling release from pH-responsive microcapsules. Langmuir 29(41):12697–12702
90. Li X, Anton N, Arpagaus C, Belleteix F, Vandamme T (2010) Nanoparticles by spray drying using innovative new technology: the Buchi Nano Spray Dryer B-90. J Control Release 147 (2):304–310
91. Schafroth N, Arpagaus C, Jadhav UY, Makne S, Douroumis D (2012) Colloids Surf B Biointerfaces 90:8–15
92. Ungaro F, D'Angelo I, Coletta C, Di Villa Bianca RD, Sorrentino R, Perfetto B, Tufano MA, Miro A, Rotonda MIL, Quaglia F (2011) Dry powders based on PLGA nanoparticles for pulmonary delivery of antibiotics: modulation of encapsulation efficiency, release rate and lung deposition pattern by hydrophilic polymers. J Controlled Release 157(1):149–159
93. Desai KGH, Park HJ (2005) Preparation of cross-linked chitosan microspheres by spray drying: Effect of cross-linking agent on the properties of spray dried microspheres. J Microencapsul 22(4):377–395
94. Horaczek H, Viernstein H (2004) Beauveria brongniartii subjected to spray-drying in a composite carrier matrix system. J Microencapsul 21(3):317–330
95. Kurmen JEC, Alvarez MIG, Rivero LFV (2015) microencapsulation of a colombian spodoptera frugiperda nucleopolyhedrovirus with Eudragit® S100 by spray drying. Braz Arch Biol Technol 58(3):468–476
96. Albertini B, Passerini N, Sabatino MD, Vitali B, Brigidi P, Rodriguez L (2008) Polymerlipid based mucoadhesive microspheres prepared by spray-congealing for the vaginal delivery of econazole nitrate. Eur J Pharm Sci 36:591–601
97. Kenawy ER, Sherringtont DC (1992) Controlled release of agrochemicalmolecules chemically bound to polymers. Eur Polym J 28(8):841–862

98. Mitrus M, Wojtowicz A, Moscicki L (2009) Biodegradable polymers and their practical utility. In: Janssen LPBM, Leszek Moscicki L (eds) Thermoplastic starch: a green material for various industries, 1st edn. Wiley, Germany SAGARPA, SEDESOL, México, D. F, p 481
99. Allan GG, Chopra CS, Neogi AN, Wilkins R (1972) Design and synthesis of controlled release pesticide-polymer combinations. Nature 234(5328):349–351
100. Akelah A (1990) Applications of functionalized polymers in agriculture. J Islamic Acad Sci 3(1):49–61
101. Singh K, Pasha A, Rani BEA (2013) Preparation of molecularly imprinted polymers for heptachlor: an organochlorine pesticide. Chron Young sci 4(1):46–50

Chapter 9
Molecularly Imprinted Sensors for Detecting Controlled Release of Pesticides

Fatma Yılmaz, Nilay Bereli, Ali Derazshamshir, Duygu Çimen, Semra Akgönüllü, Yeşeren Saylan, Aykut Arif Topçu and Adil Denizli

Abstract Increasing of the world population and food demand has led to the development of new agricultural practices to improve food production using more effective pesticides. The controlled release formulations of pesticides are used to delay the release of the biologically active components into their environment for a defined period of time. In these formulations, compounds are incorporated in a polymer matrix by different chemical or physical incorporation techniques. Reducing the toxicity of the compounds and unwanted side effects on non-target organisms are prominent characteristics of these formulations. The development of an ideal controlled release pesticide formulation is just as difficult as developing the ideal pesticide. For these reasons, further research works are needed. In this chapter, the recent applications of sensor systems to detect pesticides in environmental studies with particular attention to the fate of pesticides once introduced in water and soil, and also the advantages of their use are discussed. Current trends in the development of sensor technology for rapid assessment of the environment, as well as challenges for practical implementation and future research directions, are discussed. Findings show that the use of sensors to detect pesticides can improve the quality of the environment and help detect polluted sites.

Keywords Pesticides · Controlled release · Molecularly imprinted · Sensors

F. Yılmaz
Chemistry Technology Division, Abant Izzet Baysal University, Bolu, Turkey

N. Bereli · A. Derazshamshir · D. Çimen · S. Akgönüllü · Y. Saylan · A. Denizli (✉)
Department of Chemistry, Hacettepe University, Beytepe, Ankara, Turkey
e-mail: denizli@hacettepe.edu.tr

A. A. Topçu
Department of Chemistry, Aksaray University, Aksaray, Turkey

Rakhimol K. R. et al. (eds.), *Controlled Release of Pesticides for Sustainable Agriculture*, https://doi.org/10.1007/978-3-030-23396-9_9

9.1 Introduction

Today, approximately two million tons of pesticides are applied to control pathogens and pests in the world every year. However, 90% of pesticides are estimated to be actually lost due to degradation, photolysis, evaporation and surface run-off. Only 0.1% of pesticides are finally deposited to harmful biological targets. Due to their loss and degradation, pesticides are used more than recommended doses, which results in a large amount of waste and serious environmental pollution. These long-term widespread and inefficient use of pesticides has led to serious social concerns on food safety and ecological environment [1]. Researchers are motivated to develop effective, safe and green pesticide formulations. Controlling of pests is an unquestionably crucial task if goals for the future concerning food and energy are to be achieved.

Recent estimates suggest that inefficient pest control techniques lead to large financial losses. Although the realization about the danger of pesticides is constantly increasing, the frightening fact is: in the case of pesticide bans, the starvation will become more common in the world population. Scientists have tried to solve pest control problems by designing new, more potent agents, but desired biological response is often inefficient. These agents were not accomplished in delivering to their targets on precise time and in optimum quantities. Recognizing the cost and limitations in the design of new pesticides, scientists began to improve the delivery of the agents in the 1960s. The main advantage of controlled release formulations is that they allow less use of pesticides for the same period of activity [2]. Furthermore, release formulations are particularly advantageous when the normal half-life of a potential pesticide is short. In a controlled delivery system, the minimum effective level of drugs for a specified period of time is maintained by the continuous and stable release of active ingredients [3]. Controlled-release system improves the utilization of pesticide via prolonging the effective duration with reduced spraying times and pesticide dosages and achieves economic, safe and effective insect control with reducing environmental pollution [4].

9.2 Pesticides

Pesticides are substances or mixtures which play a significant role in the high efficiency achieved in the agriculture field, and they used to destroy, reduce and control of pests such as plants and insects. Pesticides are used to secure the requests of the growing human population and to improve food production. Significant impacts on pest control are useful in preventing hazardous diseases in agricultural products [5, 6]. Pesticides have been broadly applied to protect agricultural products since the 1940s, and the use of pesticides has increased significantly during the last decades [7, 8]. The use of these compounds may benefit agriculture, but many can arrive at a different destination than targets species and contaminate food, soil and water. Some pesticides are carcinogenic and therefore can be very harmful to human health even at low concentrations. The continued use of these compounds is a source of concern worldwide due to the risks to food safety [9]. Pesticides are firstly used to control, prevent or eliminate weeds and pets to increase crop productivity in modern agricultural fields. In the literature, the use of pesticides has been reported to help secure approximate one-third of product producing worldwide. The residue of pesticide even at trace levels can critically reason food contamination and seriously disrupt the ecosystem and pose a great danger to people's daily life. Consequently, pesticide pollution has become one of the most alarming challenges. In order to ensure proper use and management of pesticides, governments regulated maximum residue levels on food and agricultural products for pesticide use. Although most pesticides are found to be within detection limits, the bioaccumulation effect and continuous exposure may increase the security risks for human health. In addition, some new generation of pesticides with highly active, which have not been clearly elucidated by the toxic mechanism, are constantly on the market. For this reason, analysis of pesticide residues is essential to protect the ecosystem, to ensure food quality and safety and to protect human health from potential hazards [10].

The application of pesticides can be seen in non-agricultural fields such as insect control in buildings, industrial vegetation control (roads, railways), pet care or grass management. Pesticide application is constantly increasing; it is very difficult to prevent contact with the environment. With the increase in the human population, the increase in food demand will continue, and the increase in pesticide application is expected to continue in the coming years. Moreover, rapid urbanization means that less land will be allocated to agricultural activities, so society is expected to produce highly productive crops [11]. The use of harsh pesticides has been criticized and accused of human health problems and environmental damage. The use of pesticides for a long time, pests will slowly show resistance to pesticides. For this reason, numerous researches are carried out for more effective chemicals to fight against large insect groups. People are exposed to pesticides in three ways: first, by diet or by ingestion; this is the most serious way of exposure to these chemicals. Second, dermal contact; the use of pesticides in the home though. Finally, breathe; it can be carried out by inhaling contaminated air, especially for those living near agricultural areas [6]. Pesticide excessive application may leave harmful residues, including metabolites and degradation products into the environment, such as soil, water, plants and food. Significant amounts of residues in these complex matrices have become an important issue and are a serious concern for health and the environment. Until today, various techniques have been used for the detection of pesticides. Pesticides are traditionally detected by conventional chromatographic techniques such as gas chromatography (GC) [12], high-performance liquid chromatography (HPLC) [13], mass spectrometry (MS) [14], enzyme-linked immune-absorbent assays (ELISA) [15] and capillary electrophoresis (CEC) [16]. As depicted in Fig. 9.1, these techniques have a strong trace analysis with excellent high reproducibility and sensitivity and low detection limits but have limited many disadvantages, such as time-consuming, sophisticated equipment, costly

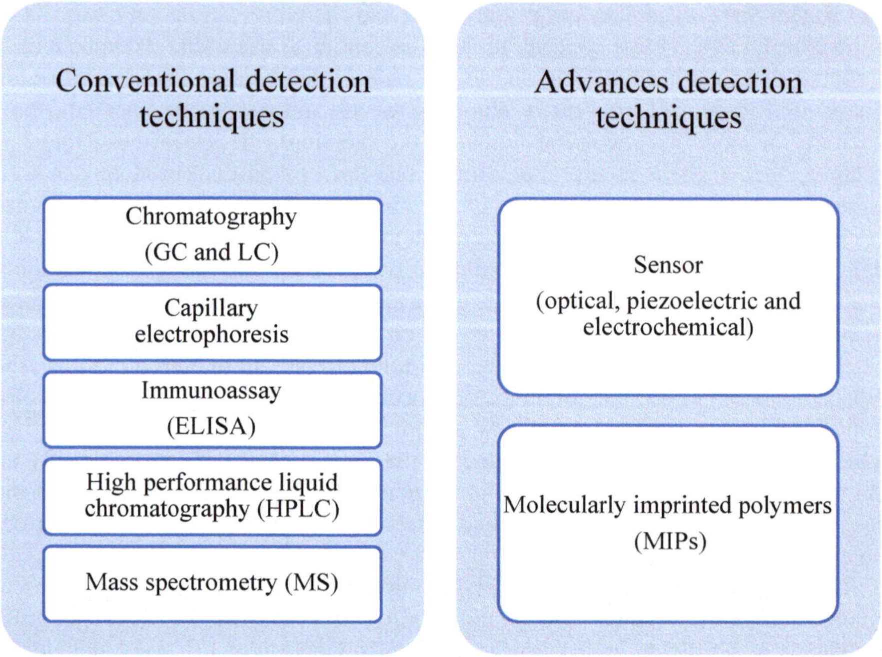

Fig. 9.1 Available techniques for pesticides detection

instruments, laborious, boring sample preparation, high-skilled people power, on-site and real-time applications, especially in emergency situations [6]. The other approach is to detection pesticides via an advanced technique based on sensors principles. These methods offer various advantages including high sensitivity and selectivity, simple, rapid and low-cost operation, user-friendly and on-site detection. For pesticide detection, numerous strategies have been established using biorecognition elements, such as an aptamer, antibody, enzyme, host–guest recognizer and molecularly imprinted polymers, which utilized to directly identify and capture the target pesticide. Besides, combination of biorecognition elements with nanomaterials enables highly sensitive and selective real-time detections which is in high demand for pesticides detection [17].

9.2.1 Classification and Toxicity

Pesticides are classified according to chemical structure, target organism and origin. It can form in biological (biopesticides), synthetic (synthetically that produced) or inorganic samples. There are two main groups of pesticides: chemical and biological pesticides. Chemical pesticides are originally synthetic materials that

Table 9.1 Classification of pesticides

Typical action	Subtype
Insecticide	Organochlorines Organophosphates Carbamates Pyrethroids Neem products Others: thiocyanates, phenols, formamides
Herbicide	Carboxylic acid Derivatives Substituted ureas Triazines
Fungicide	Dithiocarbamates Organometallics

inactivate or directly kill the pests. They are mostly classified as herbicides, insecticides, fungicides, nematicides and rodenticide [18]. Biological pesticides are derived from natural sources like bacteria, plants, animals and certain minerals. For example, baking soda and canola oil are used as pesticidal biopesticides. Pesticides separated into four important types; organophosphorus, organochlorines, pyrethroids and carbamates. Pesticides are classified in Table 9.1 according to their general chemical properties.

Pliny suggested using arsenic molecule to kill insects as early as about AD 70. At the end of the sixth century, the Chinese also used arsenic sulphide. In the early nineteenth century, some inorganic compounds were used as insecticides such as sodium fluorosilicate, zinc phosphide, paris green (copper acetoarsenite) and lead arsenate. At that time, it was not realized that arsenical pesticides would remain in the soil for up to 40 years. It was noted that the period of synthetic organic pesticides started around 1940. The dichloro-diphenyl-trichloroethane (DDT), commonly known as an insecticide, was first synthesized in 1874, and it is rediscovered as an insecticide in 1939. Benzene hexachloride (BHC) was discovered after DDT . Thousands of chemical compounds were synthesized and used for many years [19].

Organochlorines (OCs) pesticides, the most residual chemicals, have high toxicity, bioaccumulation and lipophilicity. OCs are carcinogenic, estrogenic and also found to be resistive to environmental degradation process [20, 21]. OCs has estimated that the ranges of half-life is about 10–30 years and have long half-lives in tissues. OCs has relatively persistent, and it can also be transported at unlimited distances with water and air. Organophosphates (OPs) pesticides are broadly used in agricultural fields globally. The first invention of OPs was hexaethyl tetraphosphate (HETP) which applied as an insecticide in agriculture industries in 1942 [22]. OPs compounds can be applied as insecticides besides as herbicides and fungicides in these days. OPs usually exist in the form of esters that simply can degrade. They are low soluble in water; however, it is well dissolved in fats and inorganic solvents. Carbamates, another pesticide type, largely applied to protect crops against agricultural pests and household pests in agriculture [23, 24]. OPs and carbamates are preferred as they are low persistence in the environment (shorter half-life), easy to obtain, best price and a wide spectrum of applications (wider utilizations), they can also destroy large pest groups (Table 9.2).

Pesticides, which have contributed greatly to the protection of agricultural productivity, have been applied worldwide in the control of weeds, pests and plant diseases [25]. Pesticide loss and release cause serious adverse effects on the ecosystem and serious hazards to human health. As a result, it is of great importance to develop new technologies to control pesticide release, to increase utilization efficiency and to reduce related pollution. Among the traditional and analytical

Table 9.2 Classification of pesticides according to pesticidal actions [19]

Pesticide	Pesticidal action
Algicides	Algae control in swimming pools, lakes, canals
Antifouling agents	Organisms kill or repel that attach to underwater surfaces such as boat bottoms
Antimicrobials	Organisms kill such as viruses and bacteria
Attractants	Pests attract tolure an insect or rodent into a trap, and also food is considered a pesticide when used as the attractant
Biopesticides	Pesticides derived from natural materials such as plants, bacteria, animals and some minerals
Biocides	Microorganisms kill
Disinfectants and sanitizers	Kill or inactivate microorganisms that produce disease-producing on inanimate objects
Fungicides	Kill fungi including mildews, moulds, rusts and blights
Fumigants	Produce gas or vapour to destroy pests in soil or buildings
Herbicides	Kill plants and weeds growing in unwanted places
Insecticides	Kill insects and arthropods
Miticides/acaricides	Kill mites which feed on animals and plants
Microbial pesticides	Microorganisms that kill, inhibit or defeat pests
Nematicides	Kill nematodes
Ovicides	Kill eggs of mites and insect
Pheromones	Biochemicals used to disrupt the mating behaviour of insects
Repellents	Repel pests, including insects such as mosquitoes and birds
Rodenticides	Mice and other rodents control
Defoliants	Cause leaves or other foliage to drop from a plant
Desiccants	Promote drying of living tissues, such as unwanted plant tops
Insect growth regulators	Disrupt the moulting, maturity from pupal stage to adult or other life processes of insects
Plant growth regulators	Substances excluding fertilizers or other plant nutrients that alter the expected growth, flowering or reproduction rate of plants

methods, molecularly imprinted may be an advantageous method for the detection of pesticides, as it is highly selective and sensitive compared to the other methods [26].

9.2.2 *Molecularly Imprinted Polymers (MIPs)*

Molecularly imprinted polymers (MIPs) are based on the polymerization of a cross-linker and a functional monomer around a target molecule. Firstly, target-monomer precomplexes are formed between a target and a complementary functional monomer. After, the polymerization process is carried out around the precomplexes after addition of initiator and cross-linker. Finally, the target molecule is removed for creating three-dimensional specific cavities for functional groups of the target (Fig. 9.2).

The target molecule can interact several times with the imprinted sites [27, 28]. In a synthetic polymer, the specific sites can be created for various targets such as amino acids [29], proteins [30], antibodies [31, 32], bacteria [33–35] and organic molecules [17, 26, 36]. Moreover, molecularly imprinted polymers are easy to synthesize, stable, cost-effective and can be produced in large quantities with perfect reusability. These perfect biorecognition surfaces are interesting materials that are useful in many various areas, such as analytical, biomedical, enantiomeric separations and sensors applications [28, 37]. The molecularly imprinted methodology is currently a skilled and cost-effective approach to the design of synthetic polymer. Because of their high selectivity and sensitivity, interest to MIPs has increased. When a target molecule is connected to a MIP-based sensor, the generation of an optical response (e.g. a refractive index using surface plasmon resonance), mass-sensitive frequency change (e.g. mass change in quartz crystal sensors) or electrochemical signal change (e.g. an impedance in electrochemical sensors is obtained [38]). For this reason, the ease of adaptation of MIPs to practical applications in sensors has led increased interest in applications of MIP-based sensors in various fields. Recent developments in biotechnology for the detection of specific target molecules in a short time can provide the design of more effective and highly selective sensors.

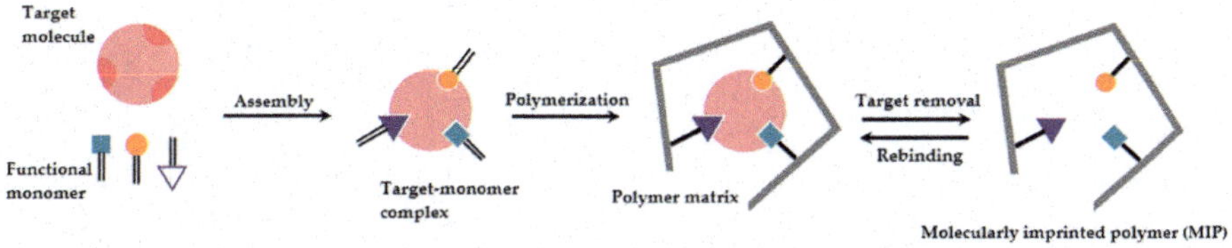

Fig. 9.2 Schematic illustration of molecularly imprinted polymer's polymerization steps

9.3 Sensing Techniques for Detecting Pesticides

A sensor is a self-contained integrated device which is able to convert a chemical signal depending on analyte concentration into a measurable analytical signal (Fig. 9.3) [39, 40]. Sensors consist of two main components; the receptor recognizes the analyte and transducer is responsible for converting the binding event into a measurable signal [39, 41]. Sensors possess many advantages such as specificity, sensitivity, portability and quick response time, thereby, which are fabricated for various purposes including pesticide detection [17, 39, 42, 43].

9.3.1 *Classification of Sensors*

Sensors are basically categorized according to bioreceptor and transducer units [44]. Selectivity of a sensor plays a crucial role during the analysis; so, enzymes (enzymatic sensors), antibodies (immunosensors), DNA (DNA-based sensors), cells

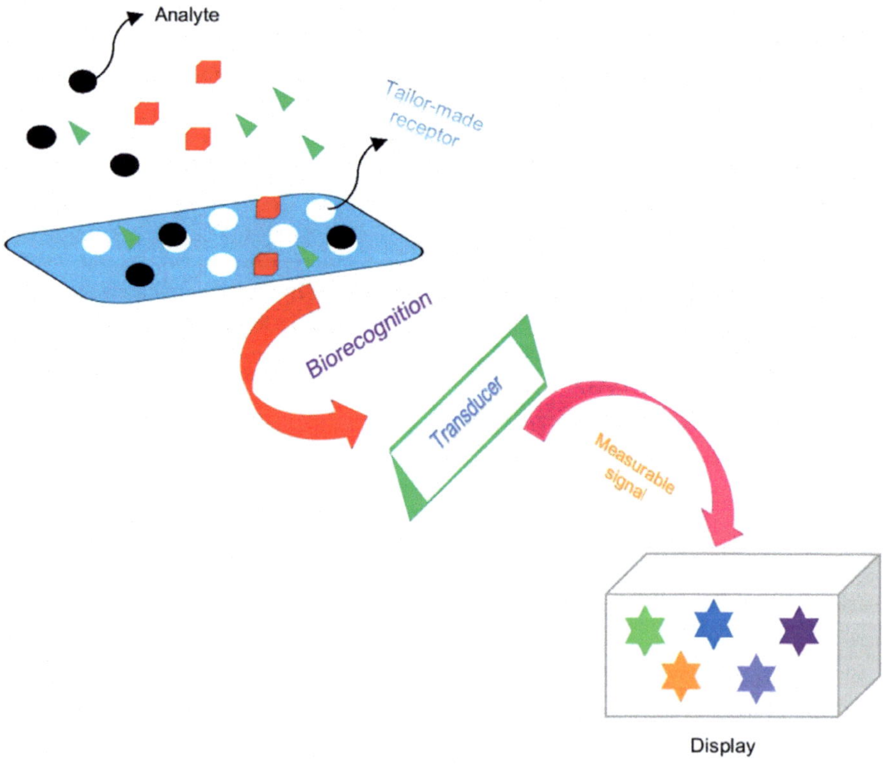

Fig. 9.3 Schematic representation of a sensor system

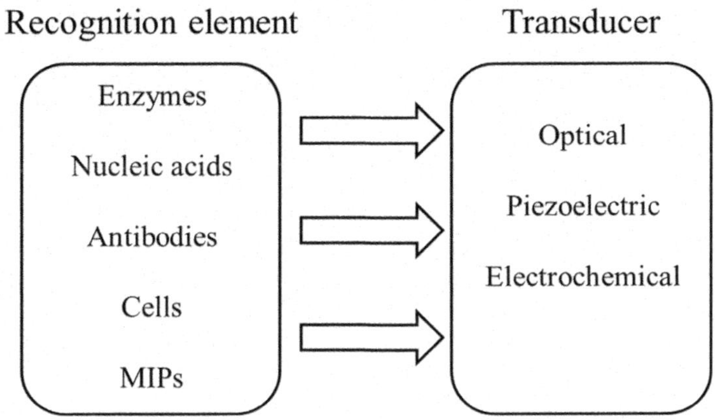

Fig. 9.4 Basic concept of a sensor

(microbial sensors) and MIPs are used as biorecognition units, but several parameters such as immobilization procedure, chemical or physical conditions and stability of receptors can directly affect the performance of developed sensor platforms (Fig. 9.4) [45–47].

9.3.1.1 Categorization of Sensors According to the Receptor

For the fabrication of enzymatic sensors, oxidoreductase and hydrolases enzymes are extensively used and catalyzing products for instance ions, CO_2 and O_2 are measured with a transducer [44, 45]. Entrapment, microencapsulation and covalent binding methods are used to immobilize an enzyme on a solid support [39]. Immunosensors are basically depended on antigen–antibody interactions, and antibodies are immobilized on a support material where the specific interactions occur at the surface of a solid and liquid phase [44]. But immobilization of antibody plays a vital role to avoid the functional loss of the antibodies [39]. Microbes and yeasts can be used as the receptor for the fabrication of whole-cell-based sensors, and these receptors are immobilized on a transducer, and in addition, whole cells (containing periplasmic enzymes) can be alternatively used as enzymes [39, 45]. DNA-based sensors are classified into two groups: affinity-based and catalytic-based sensors [39]. In affinity-based sensors, probe selection and immobilization can directly affect the analytical performance of the developed sensor, and when compared with monoclonal antibodies, aptamers (single-stranded DNA or RNA oligonucleotides) are able to bind their targets with high specificity and are more resistant towards denaturation and degradation [39, 45]. MIPs are tailor-made receptors that allow creating specific binding cavities towards the target molecule in a polymer matrix. Owing to their selectivity, low-cost and stability of MIPs, which are utilized as a receptor for sensor studies [48, 49].

9.3.1.2 Categorization of Sensors According to the Transducers

Sensors are classified according to the transducer unit which are designed as electrochemical, optical, and piezoelectric. Electrochemical sensors are utilized three electrodes: A reference electrode, a counter electrode and working electrode, and these sensors basically measure the electrochemical changes on electrode interfaces [39, 50]. Optical sensors are mainly depended on fluorescence, colorimetric, surface-enhanced Raman scattering (SERS), surface plasmon resonance (SPR) and enzymes, antibodies, aptamers, and MIPs can be acted as a receptor unit for designing of optical sensors [10]. Piezoelectric sensors are mass-sensitive devices that are based on the piezoelectric effect, and these platforms allow to analysis in a real time and label-free [51, 52]. Quartz crystal microbalance (QCM) is a well-known piezoelectric-based sensor, in which receptor is immobilized on a quartz crystal electrode [39, 53]. For designing of thermal- or calorimetric-based sensors, molecules are immobilized on temperature based sensor which measures the reaction heat according to analyte concentration [45].

9.4 Electrochemical Sensors for the Determination of Pesticides

The most rapidly developing type of chemical sensors is the electrochemical sensors. Electrochemical sensors have advantages such as remarkable detectability, low cost and experimental simplicity which compared to thermal, mass and optical sensors. Electrochemical sensors have a leading position among other sensors and which have arrived at the commercial stage [54]. The three classes of electrochemical sensors are conductometric, potentiometric and voltammetric. They have important applications in the agricultural, industrial, environmental and clinical fields. Conductometric sensors are based on changes in electric conductivity of a film or a bulk material. Conductometric methods are not essentially selective. These sensors have become more available methods only with the advent of modified surfaces for improved instrumentation and selectivity. Improved instrumentation is used only on the measurement of conductivity. No reference electrode is needed in conductometric sensors. Conductometric methods have significant advantages such as low cost and simplicity [55].

Pardieu et al. designed an electrochemical sensor based on molecularly imprinted conducting polymer for the recognition of a small pesticide target molecule, atrazine. In this study, poly(3,4-ethylenedioxythiophene-co-thiophene-acetic acid) MICP was synthesized in two steps by electrochemically on a platinum electrode. These two steps are the polymerization of comonomers in the presence of atrazine and the removal of atrazine from the prepared polymer, respectively. The prepared MICP was highly specific for added atrazine. The linear range of atrazine detection was 1.0×10^{-9}–1.5×10^{-2}mol/L, and the low detection limit was 10^{-7}mol/L [56].

Potentiometric sensors have been used frequently since the early 1930s due to their practical applicability, simplicity, familiarity and low costs. In a basic principle of the potentiometric sensor, local equilibrium is established at the sensor interface. The potential difference between the two electrodes in the potentiometric sensors in which the composition of a sample, electrode or membrane potential is measured to get information about the composition of a sample [57, 58].

Anirudhan et al. developed a potentiometric sensor with high selectivity for the determination of lindane, γ-hexachlorocyclohexane (γ-HCCH). In this study, the modification of γ-HCCH imprinted polymer film onto the surface of the Cu electrode was performed for the potentiometric sensor. A multiwalled carbon nanotube imprinted polymer (MWCNT-MIP) was synthesized with methacrylic acid as a monomer, an organochlorine pesticide molecule, as the template, ethylene glycol dimethacrylate as the cross-linker and α,α′-azobisisobutyronitrile as the initiator. The optimizations of operational parameters for the designed potentiometric sensors were done and were characterized by means of Raman spectra, TEM, FTIR and XRD analyses. The limit of detection for organochlorine pesticide was as 1.0×10^{-10} M, and the linear range was from 1×10^{-10} to 10^{-3} M [59].

Anirudhan et al. prepared a new potentiometric sensor on ion-imprinted polymer containment membrane. In this study, the multiwalled carbon nanotube-based molecularly imprinted polymer was investigated for the detection of 2,4-dichlorophenoxyaceticacid pesticide in natural water samples. For MWCNT-imprinted polymer was synthesized by 2,4-D organochlorine pesticide used as the template, α,α'-azobisisobutyronitrile as the initiator, trimethylolpropane trimethacrylate as the cross-linker and methacrylic acid as the monomer. The characterization of MWCNT-imprinted polymer was done with SEM, FTIR and XRD analyses. The optimization of amount and nature plasticizer sensing material, pH and response time parameters was investigated for MWCNT-imprinted polymer. The imprinted polymer inclusion membrane and non-imprinted sensors were compared under optimum conditions. The limit of detection was found to 1.2×10^{-9} M in the range of 1×10^{-9} –1×10^{-5} M. The prepared potentiometric sensor was performed with the ground and tap water samples [60].

Prasad et al. designed a biomimetic potentiometric sensor for atrazine detection. Firstly, the atrazine imprinted polymer particles were dispersed in di-n-octyl phthalate plasticizer and then embedded into polyvinyl chloride matrix. This prepared potentiometric sensor was used to atrazine detection in pH 2.5–3.0, in atrazine concentration of 0.0001–10 mM. The limit of detection was found to be 0.5 μM. The response time was ~2 min. The important characteristic of the potentiometric sensor has high selectivity towards herbicides and other pesticides. The stability of sensor was investigated for the detection of atrazine in real samples [61].

Abdel-Ghany et al. investigated five novel potentiometric membrane sensors for the detection of the dinotefuran levels in real samples. A prepared four potentiometric sensors were used a molecularly imprinted polymer material with methacrylic acid or acrylamide as the functional monomer. Other potentiometric sensors were prepared with a carboxylated PVC. The application of the prepared sensors was demonstrated for the detection of the dinotefuran in spiked cucumber and soil samples. The important characteristic of prepared potentiometric sensors includes short response time, good precision and selectivity, low limit of detection, simple construction, and high accuracy [62].

Voltammetric techniques involve the application of a potential to an electrode and recording the resulting current flowing through the electrochemical cell as a function of the concentration of the analyte. In most cases, the applied potential is varied or the current is monitored over a period of time [63, 64]. Voltammetric sensors continue to be the most popular ones among electrochemical sensors due to their simplicity, ease of production and the low cost. The application of voltammetric molecularly imprinted polymer-based sensors has attracted much attention as reflected by many references in the literature.

Yang et al. prepared a sensor for the determination of parathion based on coupled grafting of the functional macromolecule polyethyleneimine (PEI) on surfaces of silica gel particles using molecularly imprinted technology. The electrochemical behaviour of parathion at the imprinted film sensor was characterized by cyclic voltammetry (CV) and linear sweeping voltammetry (SWV). A linear range over parathion concentration from 0.015 to 15 mg/kg found to be the detection limit of 3×10^{-3} mg/kg. They applied the imprinted electrochemical sensor to the determination of parathion in spiked cucumber and cabbage samples, and validation studies were performed by HPLC analyses [65].

Mazzotta et al. performed the electrochemical synthesis of a metal complex-based molecularly imprinted polymer (MIP). They used this system for the development of an electrochemical sensor for a chlorophenoxy carboxylic acid (4-(2,4-dichlorophenoxy)butyric acid (2,4-DB)) commonly used as a pesticide. The developed MIP was successfully applied to the electrocatalytic detection of the template by cyclic voltammetry (CV) and at a fixed potential. Linear response over 2,4-dichlorophenoxybutyric acid concentration in the range of 200 μM–2 mM was exhibited with a detection limit of 40 μM [66].

Motaharian et al. developed an electrochemical sensor based on molecularly imprinted polymer nanoparticles for selective and sensitive determination of diazinon (DZN) pesticides. They synthesized diazinon imprinted nanoparticles using suspension polymerization and then used for modification of carbon paste electrode (CPE) composition in order to prepare the sensor. Cyclic voltammetry (CV) and square wave voltammetry (SWV) methods were applied for electrochemical measurements. Under the optimized experimental conditions, the calibration curve of the molecularly imprinted carbon paste electrode sensor had two linear concentration ranges from 2.5×10^{-9} to 1.0×10^{-7}mol/L ($R^2 = 0.9971$) and 1.0×10^{-7} to 2.0×10^{-6}mol/L ($R^2 = 0.9832$) with a limit of detection of 7.9×10^{-10} mol/L. The diazinon imprinted carbon paste electrode sensor was applied for the determination of diazinon in well water and apple fruit samples with recovery values in the range of 92.53–100.86% [67].

Imidacloprid was also used as a target molecule pesticide in molecular imprinted electrochemical sensor studies in the literature. Ghodsi and Rafati have published a study on the detection of imidacloprid (IMD). In this study, an electrochemical sensor having high selectivity and sensitivity was developed for IMD determination based on an imprinted poly(levodopa) electropolymerized on electrodeposited TiO_2 nanoparticles (TiO_2NPs) modified glassy carbon electrode (GCE). They found high selectivity and sensitivity for imidacloprid in 2–400 μM concentration of IMD. The limit of detection (LOD) and the limit of quantification (LOQ) of the imprinted TiO_2 nanoparticles modified glassy carbon electrode sensor were found as 0.3 and 1 μM by SWV measurements, respectively [68].

Carbofuran is widely used in agriculture because of its high effectiveness and low cost. Due to environmental pollution and the frequent use of carbofuran, an effective detection method for carbofuran needs to be developed. One recent study on Au nanoparticles modified molecularly imprinted electrochemical sensor was reported by Qi et al. for the detection of carbofuran. As shown in Fig. 9.5, they developed an electrochemical sensor for the detection of carbofuran using MIP as the recognition element and AuNPs on a glassy carbon electrode (GCE). Cyclic voltammetry (CV) and differential pulse voltammetry (DPV) methods were applied for electrochemical measurements. In this study, the detection limit of carbofuran was reported as 2.4×10^{-8} mol/L [69].

In a study reported by Hassan et al. molecularly imprinted polymer produced on the magnetic-MIP electrode by square wave voltammetry was used for the direct electrochemical detection of methyl parathion in fish samples (Fig. 9.6). The detection limit was found 1.22×10^{-6} mg/L for methyl parathion, and the recovery for tuna and catfish samples was in the range of 89.4–94.7% [70].

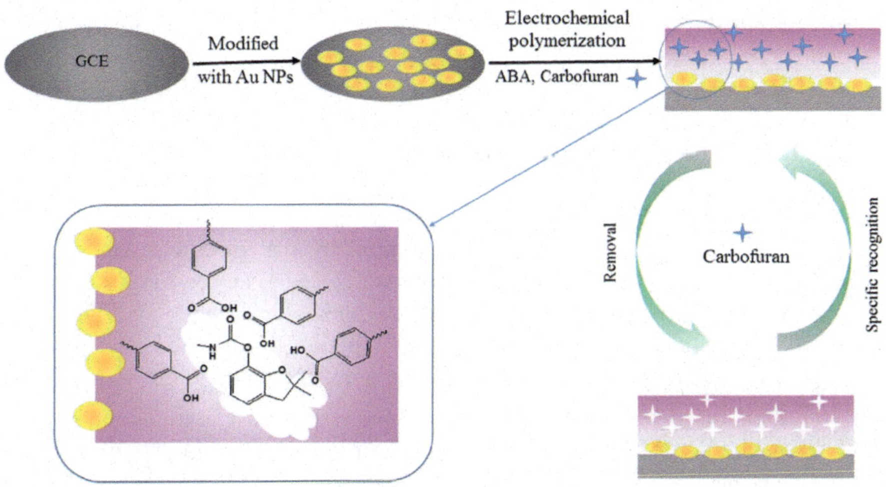

Fig. 9.5 Schematic diagram of carbofuran imprinted polymer membrane. Modified electrode preparation procedure [69]

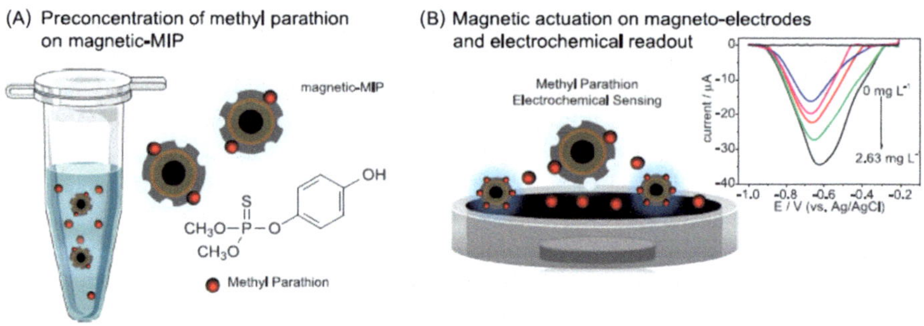

Fig. 9.6 Scheme of the electrochemical sensing of methyl parathion on magnetic-MIP [70]

9.5 Optical Sensors for the Determination of Pesticides

Optical sensors focus on the measurement of a change in the optical properties of the transducer when the target and recognition element form a complex. These types of sensors can be divided into two groups: (i) direct and (ii) indirect. In the direct optical sensors, signal generation is based on the complex formation on the surface of the transducer. The indirect optical sensors are mostly designed with various labels such as fluorophores or chromophores to detect the bindings and amplify the signals. Although indirect sensing methods can produce higher signal levels, they suffer from non-specific binding and high reagent cost of labelling step [71]. In the literature and the market, there are multiple optical sensors, including optrode-based fibre, evanescent wave fibre, time-resolved fluorescence, the resonant mirror, interferometric and surface plasmon resonance sensors. Their detection window is so versatile, and they sense multiple types of molecules [47, 72, 73].

Zor et al. reported a study about the preparation and characterization of a multifunctional composite material (magnetic silica beads/graphene quantum dots/molecularly imprinted polypyrrole). This composite material was engineered to detect tributyltin. Their concept was based on turn-off in photoluminescence of composite material and tributyltin was captured by the specific cavities of the imprinted polymer, photoluminescent quantum dots lead to an optic transducer, that is, quantum dots are quenched upon analyte recognition (Fig. 9.7). They also discussed this composite material capable of a simple, fast and sensitive platform for small molecule detection without any sample treatment due to the magnetic and selective properties [74].

Wang et al. prepared molecularly imprinted fluorescent hollow nanoparticles for λ-cyhalothrin detection in water samples. They reported that this imprinted fluorescent sensor exhibited a significant sensitivity and had a low detection limit (10.26 nM), fast detection (8 min), easy regeneration ability and appreciable selectivity over several structural competitors. They also tested the fluorescent sensor to detect λ-cyhalothrin in real sample solutions from the Beijing–Hangzhou Grand Canal Water. According to the results, the fluorescent sensor demonstrated a

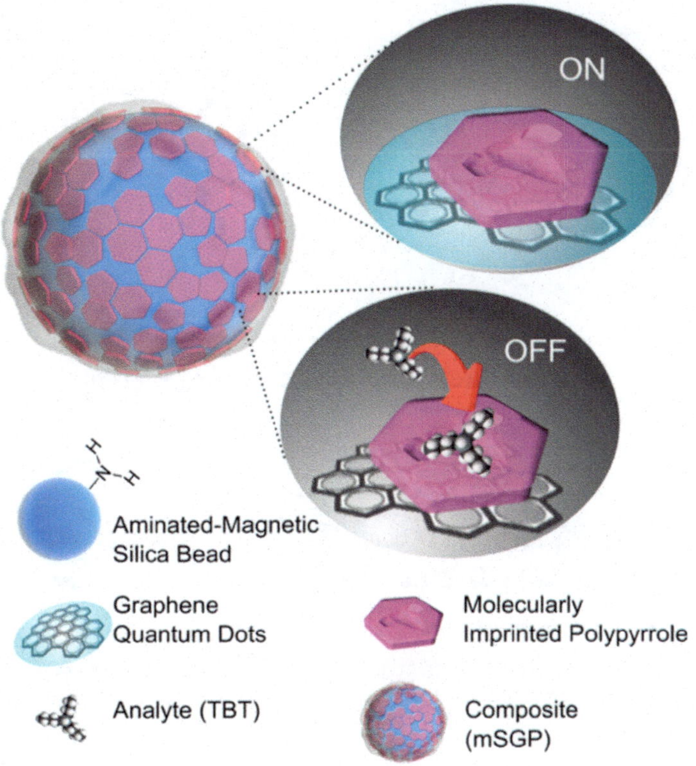

Fig. 9.7 Scheme of the composite material and its sensing mechanism [74]

high recovery that shows the potential application of this sensor in environmental studies [75].

A colorimetric sensor was developed for 3-phenoxybenzaldehyde detection by Ye et al. They prepared surface molecularly imprinted nanoparticles, and the 3-phenoxybenzaldehyde resulted in a characteristic colour fading by potassium permanganate reduction reaction. The 3-phenoxybenzaldehyde could be detected by absorbing potassium permanganate solution with a high correlation coefficient. The colorimetric sensor exhibited a low limit of detection value (0.052 μg/mL) with the range from 0.1 to 1 μg/mL of 3-phenoxybenzaldehyde. They also performed real sample experiments to compare detection performance of 3-phenoxybenzaldehyde [76].

Shrivastav et al. fabricated an optic sensor based on surface plasmon resonance system and molecularly imprinted method to detect profenofos (Fig. 9.8). They applied the concentration range of profenofos towards the sensor from 10^{-4} to 10^{-1} μg/L. They found that the detection limit of the sensor as 2.5×10^{-6} μg/L. They reported that the combination of surface plasmon resonance system and molecularly imprinted method provided high sensitivity and selectivity and had numerous other advantages including fast response, immunity to electromagnetic

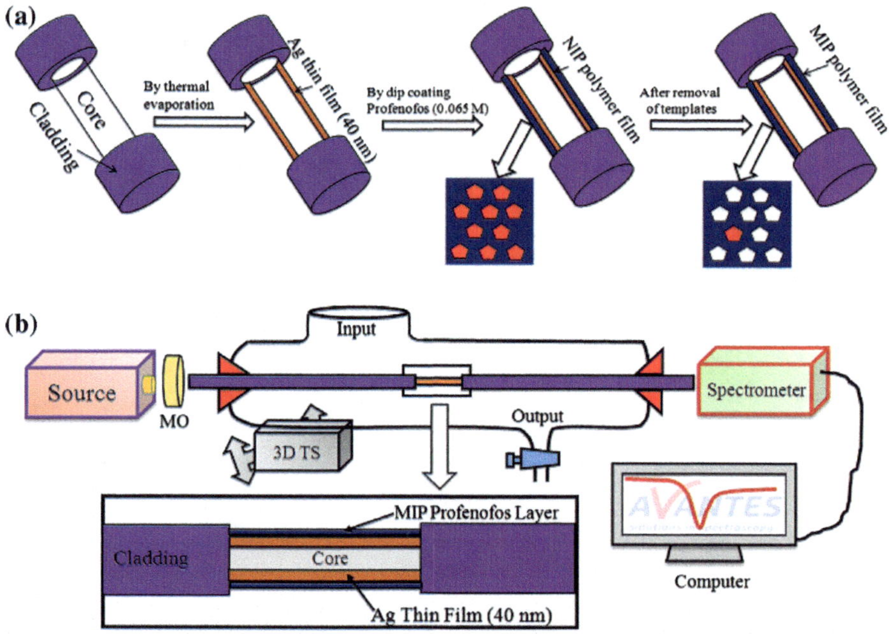

Fig. 9.8 **a** Scheme of sensing probe fabrication and **b** experimental set-up of surface plasmon resonance and molecularly imprinted based sensor system [77]

interference, low cost and capable of real-time monitoring and remote sensing of analyte [77].

Saylan et al. fabricated the molecularly imprinted nanofilms and integrated them with surface plasmon resonance sensors for detection of multiple pesticides, including cyanazine (SNZ), simazine (SMZ) and atrazine (ATZ). The molecularly imprinted nanofilms onto the surface plasmon resonance sensor surfaces were prepared by using functional monomer, co-monomer, and cross-linker. As shown in Fig. 9.9, they performed the real-time measurements on surface plasmon resonance sensor with a detection range from 0.10 to 6.64 nM and found the detection limit values of 0.095, 0.031 and 0.091 nM for cyanazine, simazine and atrazine, respectively [26].

Yılmaz et al. prepared the surface plasmon resonance sensor by using molecularly imprinted nanoparticles to detect pesticides. The schematic representation of the surface plasmon resonance sensor preparation process is displayed in Fig. 9.10. They chose atrazine as a recognition element and applied a wide concentration range (0.5–15 ng/mL) of atrazine solutions to surface plasmon resonance sensor system for the adsorption kinetics determination. They claimed that the imprinted nanoparticles displayed more sensitivity to atrazine than the non-imprinted ones. In addition, their results showed that the surface plasmon resonance sensor was more selective atrazine compared to other pesticides. They also calculated the limit of detection value as 0.7134 ng/mL [78].

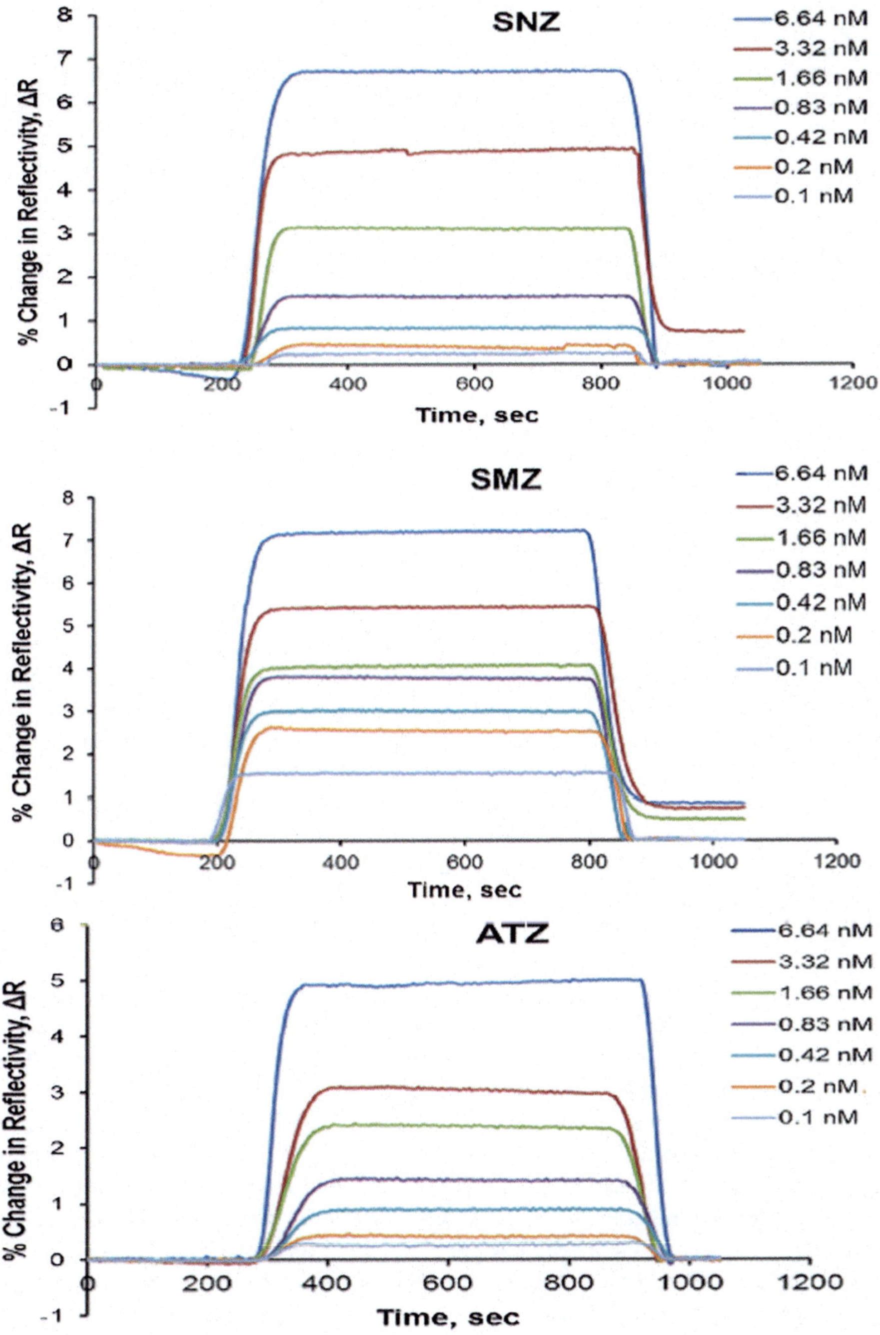

Fig. 9.9 SNZ, SMZ and ATZ detection on a surface plasmon resonance sensor system [26]

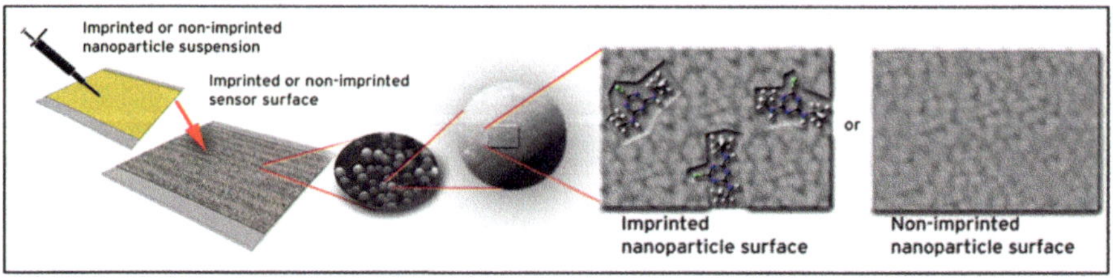

Fig. 9.10 Preparation and schematic representation of the surface plasmon resonance sensor [78]

9.6 Piezoelectric Sensor for Determination of Pesticides

Environmental pollution from industrial and agricultural activities can have serious negative effects on nature and people. Surface waters like rivers and lakes are contaminated by pesticides, which are major pollutants [79, 80]. Pesticide pollution in natural waters due to intensive agriculture and horticulture activities has a negative impact on human and nature. In order to ensure in future human well-being, there is a rapidly growing awareness of the world for environmental protection and sustainable use of natural resources. Governments create legal frameworks and action plans to protect natural water resources and regulate pesticide use [81]. Analytical methods have been developed for the mostly used pollutants such as organophosphorus and organonitrogen pesticides. However, in line with future needs, online monitoring methods or fast sample screening are highly desirable. Therefore, new, cost-effective analytical methods are intensively developed for continuous monitoring of natural waters and on-field sample screening in order to meet future analysis needs [82]. There is a growing interest to develop sensors for the detection of pesticides which are used currently in large quantities. Sensors technology using sensors or chemical sensors are considered the perfect basis for such field-usable devices for pesticide detection [9]. In the domain of chemical sensors, quartz crystal microbalance (QCM) or surface acoustic wave sensors functionalized with organic polymers for pesticides detection have been reported [83–86]. Piezoelectric systems are the most attractive of all physical transducers due to their real-time and label-free detection with high sensitivity, low instrumentation costs and simplicity. QCM sensor, which is a practical and convenient monitoring tool, measures small changes with high specificity, accuracy, stability and reproducibility on the sensor surface. Therefore, QCM as a highly stable and portable transducer element is suitable for chemical sensors. The increased mass on the gold surface associated with the binding reactions causes the frequency to decrease. QCM devices are well suited for converting the recognition process using MIP-based memories into a sensor signal. QCM-based sensors have been used to detect several analytes, even in very different matrix environments. Thus, the combination of a QCM and MIP as synthetic receptors increases the sensitivity by means of MIP process-based multiplexing binding sites having molecular memories prepared towards the target compound. This section aims to highlight and summarize the recent progress and studies in the field of sensor systems based on QCMs combined with molecularly imprinted technology [87]. MIP particles are attached to the surface of the QCM sensor modified with self-assembled monolayers comprising thiol-containing compounds such as 11-mercaptoundecanoic acid, using a chemical, thermal or photochemical initiator [88]. Although photochemical initiation has some advantages, such as easy control and polymerization at room temperature, the MIP layers can also be formed on the QCM sensor surface by electropolymerization [89]. In this strategy, by changing the applied voltage the thickness of the MIP layer can be easily controlled by changing the applied voltage [90]. However, the non-adhesion of the MIP layer prepared on the surface may cause problems during the washing step. Therefore, some pretreatments must be applied to increase the adhesion of the MIP layers to the sensor surface. In this section, it was provided with an overview of the recent progress and applications of MIP-based QCM sensors about pesticide detections.

In the last decade, one of the most promising technical applications based on the use of MIPs is QCM. QCM sensors have been developed for the detection of various pesticide targets. Such as Gurgu and co-workers were able to detect different pesticide types in aqueous solution in order to prevent their illegal use in agriculture. QCM is widely used due to its sensitivity and ability to determine in real time the variation of the adsorbed mass on the sensor surface, in the field of ng/cm^2. They used chrome–gold electrodes (CrAu) covered with self-assembled monolayers (SAM) in order to increase the limit of detection up to 18 ng/cm^2 for a frequency shift of 1 Hz. All pesticide samples were investigated using CrAu sensors covered with ethanethiol SAM, in static mode: the SAM was obtained using ethanethiol (1%). The obtained data were used to calculate the sensor response (frequency shift in time [Hz/s]) and to draw the linear or polynomial regression. For Dithane M-45 was observed linear regression for all samples (i.e. D5, D7, D9, D11) with R^2 determination coefficient values in the range of 0.8945–0.9693. For Fastac and Optimol, the fitting curves show polynomial regression for all samples (i.e. F5, F7, F9, F11 and O5, O7, O9, O11) with R^2 determination coefficient values in the range of 0.9882–0.9946 for insecticide samples, respectively, 0.9826–0.9963 for molluscicide samples. Considering the fact that the adsorbed mass on QCM sensor surface was inversely proportional to sensor response, it could be observed that for fungicide *Dithane M-45* with active substance concentration of 0.01 μg/L, the recorded signal was very poor and probably was under the limit of detection. For the other two pesticides (*Fastac* and *Optimol*), the recorded signal was good even for the lowest concentration of active substances. The pesticides selected for this study (i.e. *Dithane M-45*, *Fastac* and *Optimol*) were banned, as they were classified as persistent organic pollutants (POPs). The proposed solution for the determination of pesticides using QCM techniques was advantageous in terms of the time required for sample preparation, and the time devoted to actual analysis. The fact is that microbalance provided real-time results with its use in various fields. The used sensor showed high sensitivity at low concentrations (10 ng/L) and was considered a promising method for the direct analysis of aqueous samples. The obtained data established the potential of the QCM sensor for the pesticide analysis at low levels [91].

MIP technique is an attractive method for the generation of polymer-based molecular "memory" for a present target molecule or group of target molecules. Cheng et al. prepared MIP-QCM sensor for detection of parathion pesticide. OP compounds, as important components of insecticides, pesticides and chemical threat agents, have become more and more serious threats to the environment. Real-time monitoring of the threats is an urgent demand for environmental safety. They integrated a novel MIP with QCM to build a simple sensor to selectively monitor OP in water samples in real time. The MIP system was built up by the polymerization of methacrylic acid with divinylbenzene, initiated with 2,2'-azobis (2-methylpropionitrile) in the presence of a target molecule. Using parathion as a representative template, the MIP was coated on QCM by casting a semi-polymerized solution. The sensor could detect parathion in the wide range from 290 ppb to 29 ppm. This sensor was stable, easy to prepare which can be reused and has potential to be applied to other analytes, which are threats to the environment [92].

Fast evaluation of the QCM sensor is based on the piezoelectric properties of quartz crystals, which could reflect mass and viscosity changes on the crystal surface. By the aid of the specific nature of MIP, the sensor of MIP-based QCM could detect the mass of the target molecule captured in samples by reading the corresponding frequency changes. The sample solution was pumped to pass through the surface of the MIP-based QCM sensor, and the frequency shift was monitored in real time. To rapidly evaluate the response to the target, they performed the detection of water samples in sequence in the interval of 30 min, including various concentrations of the target (0.1–100 μM or 0.029–29 ppm). After 50 min of water rinsing, the sensor was brought to the resonant frequency (f_0). Then, the frequency of the sensor decreased once it was treated with aqueous parathion solution. The frequency remained steady until reaching 10 μM (2.9 ppm) of parathion. By increasing the concentration of parathion, the frequency shifts dramatically. Compared to the MIP-based QCM sensor, NIP-based QCM sensor showed much smaller frequency shifts to the samples. MIP has a larger capacity for the target as it contained many cavities for binding target molecules, while NIP holds much less molecule by the non-specific binding on the surface. The binding experiment illustrated that the MIP-based QCM showed the superior specific and sensitive response to the target compared with NIP-based QCM. Performance of the sensor was evaluated by using aqueous parathion solution; the sensor was further studied to establish the correlation between the frequency shift and sample concentrations. After 20 min of water flow, the sensor was treated with various concentrations of parathion for 60 min. Compared to the blank sample (0 μM of parathion in water), aqueous parathion solution showed obvious frequency shifts. The limit of detection (LOD) was calculated as 0.83 μM. Subsequently, they evaluated the selectivity of MIP (parathion) against two analogues—paraoxon and methyl parathion. The frequency shift of parathion (−30.28 Hz) is much larger than that of methyl parathion (−16.29 Hz) and paraoxon (−11.02 Hz), which indicated MIP (parathion) has higher selectivity to the template (parathion) than to the two analogues. Comparing parathion with paraoxon, MIP (parathion) showed distinct affinities. Comparing parathion with methyl parathion, the dominant interaction was the differences in three-dimensional matching because of the sizes of ethyl and methyl group. The results indicated that the designed sensor had good selectivity for parathion. In brief, with parathion as the preset template, the selectivity experiment showed that the MIP-based QCM sensor could selectively detect parathion in water samples as the MIP offered specific cavity for the template molecule. The sensor has a wide detection range (290 ppb ~ 29 ppm) and fast response (60 min) to parathion. The sensor was easy to prepare and could be potentially applied to other templates including organophosphorus compounds. The sensor could potentially be useful for monitoring targets in real water samples. Carbamate pesticides are widely used in the agricultural industry due to their high efficiency and low toxicity. Metolcarb, an N-methylcarbamate pesticide, has been extensively used to control the agricultural production of rice leafhoppers, planthoppers and fruit flies in [93]. If metolcarb enters the human body, it inhibits acetylcholinesterase activity. It results from the accumulation of acetylcholine, causing headache, nausea or quadriplegic. Unfortunately, metolcarb residues have been found in fruit and many other food products [94]. Nowadays, high-performance liquid chromatography (HPLC) [95], capillary electrophoresis (CEC) [96], enzyme-linked immunosorbent analysis (ELISA) [97] and electrochemical analyses are the methods used for the determination of metolcarb in food samples [98]. However, these methods have some difficulties in limiting the widespread use of large-scale instrumentation, such as time-consuming pretreatments, large reagent consumption, the need for skilled operators and the need for expensive biomaterials that are difficult to maintain. Low cost, fast, sensitive and selective methods are needed for metolcarb detection.

Wang and co-workers fabricated a novel MIP-based QCM sensor, successfully applied to detect trace metolcarb in food samples with high specificity, high sensitivity and simplicity. They modified on the quartz crystal Au electrode surface for metolcarb detection. The synthesis process of MIP was performed via one-step polymerization by using template molecule metolcarb and functional monomer methacrylic acid. As a control, non-imprinted polymer (NIP) was prepared using the same procedure in the absence of the template [99]. The response of the QCM sensor towards metolcarb was evaluated by submerging the quartz crystal resonator assembled in the Teflon holder. The frequency shift Δf ($f_i - f_0$, Hz) could be obtained, which was attributed to mass change caused by the specific absorption of MIP towards metolcarb on the quartz crystal surface. The average value of Δf was used to calculate the mass of metolcarb adsorbed onto the quartz crystal surface using the Sauerbrey equation (Eq. 9.1) [100].

$$\Delta f = 2.26 \times 10^{-6} \Delta m f_0^2 / A \tag{9.1}$$

where f_o is the original resonant frequency of the quartz crystal (Hz), Δm is the mass change (g) and A is the surface area of the electrode (cm^2). The adsorption capacity of MIP in methanol was almost two times larger than that of NIP, revealing non-specific adsorption could be effectively suppressed in methanol (Fig. 9.11).

They evaluated the binding performance of MIP-based and NIP-based QCM sensors against metolcarb concentrations ranging from 5 to 90 μg/L by measuring the frequency shifts at room temperature, and the results are presented in Fig. 9.12. According to the results, the frequency shifts of the MIP-based and NIP-based QCM were increased with the increasing initial concentration of metolcarb. At any concentration, the MIP-based QCM sensor showed a greater frequency response than the MIP-based QCM sensor showed better binding capability for metolcarb than the NIP-based QCM sensor. At any particular concentration, the frequency change of the MIP sensor was −175.7 Hz, while the NIP sensor was −63.2 Hz. These results are attributed to the MIP imprinting cavities that can specifically recognize the metolcarb in terms of the spatial structure and the binding sites formed during the polymerization process. The results showed that the MIP-based QCM sensor had a linear relationship between the frequency shifts and the concentrations of metolcarb ranging from 5 to 70 μg/L, with a low detection limit of 2.309 μg/L ($S/N = 3$).

The selective recognition property of the MIP-based QCM sensor is important for the practical applications in the complicated food matrix. Some interference with similar structures of metolcarb, such as carbaryl, isoprocarb and propoxur, were used to evaluate the frequency response of the MIP-based QCM sensor at the same concentrations. The frequency response of the proposed MIP-based QCM sensor for metolcarb was significantly higher than the other three structural analogues, and SC values of metolcarb towards carbaryl, isoprocarb and propoxur were calculated to be 2.45, 2.97 and 4.41, respectively, suggesting the good selectivity of the MIP-based QCM sensor.

During the last two decades, pesticide usage has increased dramatically, and their presence in vegetable, fruits and the environment has increased. Nowadays, the more sensitive detection of pesticides has arisen due to the problem of them faster [26]. Herbicides, which are known as one of the pesticides group, are commonly used against crop diseases in agricultural areas. When these chemicals are applied to crops, its droplets are fallen on soil, plant and water. Hence, they may

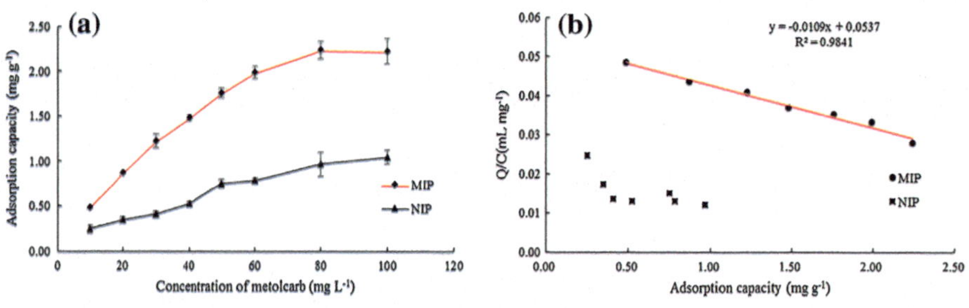

Fig. 9.11 **a** Binding isotherm of MIP and NIP for metolcarb in methanol, **b** Scatchard analysis of MIP and NIP [99]

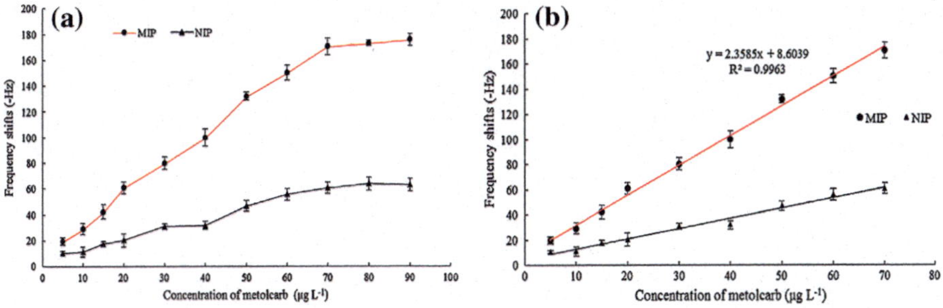

Fig. 9.12 **a** Frequency shifts of MIP- and NIP-based QCM sensors towards different concentrations of metolcarb; **b** calibration curves of the prepared sensors [99]

be harmful, especially when these pesticides are taken directly or indirectly to the body [101]. Denizli and co-workers prepared QCM sensor by using molecularly imprinted method for selective recognition of toxic herbicide molecule, amitrole which is used to control weeds in agriculture [102]. N-methacryloyl-(L)-tryptophan methyl ester (MATrp) was selected as a proper functional monomer and polymerized with ethylene glycol dimethacrylate (EGDMA). Amitrole imprinted poly (ethylene glycol dimethacrylate-N-methacryloyl-(L)-tryptophan methyl ester) [poly (EGDMA-MATrp)] nanofilms were attached to the gold surface of the QCM sensor. The amitrole imprinted QCM sensor was characterized by the contact angle, AFM and ellipsometer measurements. According to AFM images, the surface depth of amitrole imprinted QCM sensor was higher than non-imprinted one, meaning that the imprinting procedure was achieved successfully. On the other hand, the roughness of the surface is well distributed through the surface of the sensor. It indicated that amitrole imprinting on the QCM sensor has been homogeneously achieved, which is an important parameter controlling the specificity, selectivity and recognition rate of the sensor. Besides, kinetic and affinity binding of amitrole was investigated by binding the pesticide imprinted and non-imprinted sensor to QCM sensor. The contact angle values of unmodified, non-imprinted and amitrole imprinted QCM sensor were found to be $60.2° \pm 0.24$, $68.3° \pm 0.47$ and $72.4° \pm 0.12$, respectively. The values of contact angle were increased due to the hydrophobic character of MATrp, which is a hydrophobic functional monomer. The QCM sensor was prepared for real-time detection of amitrole in aqueous solution in the concentration range from 0.06 to 11.9 nM. When the amitrole solution reached to the sensor surface, amitrole imprinted QCM sensor responded quickly. The total measurement time was read as 25 min. The increasing of amitrole concentration caused an increased response, as expected. The selectivity of amitrole imprinted QCM sensor was investigated for the better understanding of the specificity of the interactions between the binding sites of the MIP-based QCM sensor and template molecule. For this aim, benzotriazole and dinoseb were used as a competitor agent. No defined cavities exist to provide selective binding of amitrole with NIP-based QCM sensor. The relative selectivity coefficient for amitrole showed that the cavities created in the pesticide imprinted QCM sensor recognized imprinted template molecule and had structural memory and remarkable molecular size matching for amitrole. The reusability of amitrole imprinted QCM sensor was investigated by six cycles. It was observed that there was a slight difference in the frequency shift. Thus, it was concluded that the amitrole imprinted QCM sensor could be reused with reproducible results. Amitrole imprinted QCM sensor was also used to detect amitrole in a natural source (green pepper). Freshly prepared green pepper samples caused an increase in sensor response as expected by remaining between calibration points. Amitrole concentration was approximately 0.342 nM (28.69 ng/L) in green pepper. It was concluded that amitrole imprinted sensor detects amitrole even in a natural complex mixture.

9.7 Conclusions and Perspectives

The use of pesticide dated back in 1940 in order to improve the food quality; however, bioaccumulation of pesticides is highly toxic for the environment and living organisms. Chromatographic methods are used for pesticide detection but which are time-consuming, high-cost equipment and need skilled person during the analysis. Therefore, rapid, selective, sensitive detection methods can be improved for detection of pesticides which is a global issue around worldwide. Sensors are rapid, selective and sensitive devices for pesticide detection, and enzymes, antibodies, whole cells and MIPs are integrated with sensor platforms in order to push the sensitivity and selectivity of these platforms. Among the receptors, MIPs known as tailor-made receptors have many advantages such as cost-effective easy-prepared, robust, sensitive and selective against the target molecules. Today, different types of biomimetic sensors are fabricated for pesticide detection but the need for better sensitive, selective and reliable sensing platforms are a highly important issue to protect the environment and living organisms.

References

1. Liu B, Wang Y, Yang F, Wang X, Shen H, Cui H, Decheng W (2016) Construction of a controlled-release delivery system for pesticides using biodegradable PLA-based microcapsules. Colloid Surface B 144:38–45
2. Liang J, Tang S, Cheke RA, Wu J (2015) Models for determining how many natural enemies to release inoculatively in combinations of biological and chemical control with pesticide resistance. J Math Anal Appl 422:1479–1503
3. Pastora MA, Garnesa MP, Pradasa MM, Lluch AV (2016) Topologically controlled hyaluronan-based gel coatings of hydrophobic grid-like scaffolds to modulate drug delivery. Colloid Surface B 140:412–420
4. Qian K, Shi TY, He S, Luo LX, Liu XL, Cao SY (2013) Release kinetics of tebuconazole from porous hollow silica nanospheres prepared by miniemulsion method. Micropor Mesopor Mater 169:1–6

5. Nsibande SA, Forbes PBC (2016) Fluorescence detection of pesticides using quantum dot materials-a review. Anal Chim Acta 945:9–22
6. Samsidar A, Siddiquee S, Shaarani SM (2018) A review of extraction, analytical and advanced methods for determination of pesticides in environment and foodstuffs. Trend Food Sci Technol 71:188–201
7. Grung M, Lin Y, Zhang H, Steen AO, Huang J, Zhang G, Larssen T (2015) Pesticide levels and environmental risk in aquatic environments in China-a review. Environ Int 81:87–97
8. Souza Tette PA, Guidi LR, Gloria A, Fernandes MB (2016) Pesticides in honey: a review on chromatographic analytical methods. Talanta 149:124–141
9. Llorent-Martinez EJ, Ortega-Barrales P, Fernandez-de Cordova ML, Ruiz-Medina A (2011) Trends in flow-based analytical methods applied to pesticide detection: a review. Anal Chim Acta 684:30–39
10. Yan X, Li H, Su X (2018) Review of optical sensors for pesticides. Trend Anal Chem 10: 1–20
11. Tankiewicz M, Fenik J, Biziuk M (2011) Solventless and solvent-minimized sample preparation techniques for determining currently used pesticides in water samples: a review. Talanta 86:8–22
12. Grimalt S, Dehouck P (2016) Review of analytical methods for the determination of pesticideresidues in grapes. J Chromatogr A 1433:1–23
13. Wang Q, Zhang X, Xu Z, Gao H (2015) Simultaneous determination of three trace organophosphorus pesticide residues in vegetables using molecularly imprinted solid-phase extraction coupled with high-performance liquid chromatography. Food Anal Method 8:2044–2051
14. Kasiotis KM, Anagnostopoulos C, Anastasiadou P, Machera K (2014) Pesticide residues in honeybees, honey and bee pollen by LC–MS/MS screening: reported death incidents in honeybees. Sci Total Environ 485–486:633–642
15. Wang L, Zhang Q, Chen D, Liu Y, Li C, Hu B, Du D, Liu F (2011) Development of a specific enzyme-linked immunosorbent (ELISA) for the analysis of the organophosphorous pesticide fenthion in real samples based on monoclonal antibody. Anal Lett 44:1591–1601
16. Tang T, Deng J, Zhang M, Shi G, Zhou T (2016) Quantum dot-DNA aptamer conjugates coupled with capillary electrophoresis: a universal strategy for ratiometric detection of organophosphorus pesticides. Talanta 146:55–61
17. Yılmaz F, Saylan Y, Akgönüllü S, Çimen D, Derazshamshir A, Bereli N, Denizli A (2017) Surface plasmon resonance based nanosensors for detection of triazinic pesticides in agricultural foods. New pesticides and soil sensors. Academic Press, New York, pp 679–718
18. Sharma BK (2006) Environmental chemistry. Goel Publication House, New Delhi, India, pp 83–140
19. Nollet ML, Rathore HS (eds) (2010) Handbook of pesticides: methods of pesticide residues analysis. Taylor & Francis Group, London, pp 8–44
20. Taylor KW, Novak RF, Anderson HA, Birnbaum LS, Blystone C, De Vito M (2013) Evaluation of the association between persistent organic pollutants and diabetes in epidemiological studies: a national toxicology program workshop review. Environ Health Persp 121:774–783
21. Shaikh H, Memon N, Bhanger MI, Nizamani SM, Denizli A (2014) Core–shell molecularly imprinted polymer-based solid-phase microextraction fiber for ultratrace analysis of endosulfan I and II in real aqueous matrix through gas chromatography–micro electron capture detector. J Chromatogr A 1337:179–187
22. Tahara MK, Nakazawa RH (2005) Use of cholinesterase activity as an indicator for the effects of combinations of organophosphorus pesticides in water from environmental sources. Water Res 39:5112–5118
23. Zhang YY, Arugula MA, Wales M, Wild J, Simonian AL (2015) A novel layer by layer assembled multi-enzyme/CNT biosensor for discriminative detection between organophosphorus and non-organophosphorus pesticides. Biosens Bioelectron 67:287–295

24. Shaikh H, Andaç M, Memon N, Bhanger MI, Nizamania SM, Denizli A (2015) Synthesis and characterization of molecularly imprinted polymer embedded composite cryogel discs: application for the selective extraction of cypermethrins from aqueous samples prior to GC-MS analysis. RSC Adv 5:26604–26615
25. Chen C, Zhang G, Dai Z, Xiang Y, Liu B, Bian P, Zheng K, Wu Z, Cai D (2018) Fabrication of light-responsively controlled-release herbicide using a nanocomposite. Chem Eng J 349:101–110
26. Saylan Y, Akgönüllü S, Cimen D, Derazshamshira A, Bereli N, Yılmaz F, Denizli A (2017) Development of surface plasmon resonance sensors based on molecularly imprinted nanofilms for sensitive and selective detection of pesticides. Sens Actuator B Chem 241:446–454
27. Say R, Keçili R, Denizli A, Ersöz A (2016) Biomimetic imprinted polymers: theory, design methods, and catalytic applications. Molecularly imprinted catalysts: principles, syntheses, and applications. Elsevier, Amsterdam, pp 103–120
28. Saylan Y, Yılmaz F, Derazshamshir A, Yılmaz E, Denizli A (2017) Synthesis of hydrophobic nanoparticles for real-time lysozyme detection using surface plasmon resonance sensor. J Mol Recognit 2631:1–7
29. Akgönüllü S, Yavuz H, Denizli A (2016) Preparation of imprinted cryogel cartridge for chiral separation of l-phenylalanine. Artif Cell Nanomed B 45:800–807
30. Saylan Y, Denizli A (2018) Molecular fingerprints of hemoglobin on a nanofilm chip. Sensors 18:3016
31. Uzun L, Say R, Ünal S, Denizli A (2009) Hepatitis B surface antibody purification with hepatitis B surface antibody imprinted poly(hydroxyethyl methacrylate-N-methacryloyl-l-tyrosine methyl ester) particles. J Chromatogr B 877:181–188
32. Dibekkaya H, Saylan Y, Yılmaz F, Derazshamshir A, Denizli A (2016) Surface plasmon resonance sensors for real-time detection of cyclic citrullinated peptide antibody. J Macromol Sci Part A Pure Appl Chem 53:585–594
33. Perçin I, Idil N, Bakhshpour M, Yılmaz E, Mattiasson B, Denizli A (2017) Microcontact imprinted plasmonic nanosensors: powerful tools in the detection of salmonella paratyphi. Sensor 17:137
34. Idil N, Hedström M, Denizli A, Mattiasson B (2017) Whole cell based microcontact imprinted capacitive biosensor for the detection of Escherichia coli. Biosens Bioelectron 87:807–815
35. Erdem Ö, Saylan S, Cihangir N, Denizli A (2019) Molecularly imprinted nanoparticles based plasmonic sensors for real-time *Enterococcus faecalis* detection. Biosens Bioelectron 126:608–614
36. Shaikh H, Sener G, Memon N, Bhanger MI, Nizamani SM, Üzek R, Denizli A (2015) Molecularly imprinted surface plasmon resonance (SPR) based sensing of bisphenol a for its selective detection in aqueous systems. Anal Methods 7:4661–4670
37. Erdem Ö, Saylan Y, Andaç M, Denizli A (2018) Molecularly imprinted polymers for removal of metal ions: an alternative treatment method. Biomimetic 3:38
38. Omar SA, Thomas B, Esen C, Garcia-Cruz A, Sergey PA (2019) Molecularly imprinted polymers in electrochemical and optical sensors. Trend Biotechnol 37:294–309
39. Goradel NH, Mirzaei H, Sahebkar A, Poursadeghiyan M, Masoudifar A, Malekshai VZ, Negahdari B (2018) Biosensors for the detection of environmental and urban pollutions. J Cell Biochem 119:207–212
40. Bhalla N, Jolly P, Formisano N, Estrela P (2016) Introductions to biosensors. Essays Biochem 60:1–8
41. Koedrith P, Thasiphu T, Weon JI, Boonprasert R, Tuitemwong K, Tuitemwong P (2015) Recent trends in rapid environmental monitoring of pathogens and toxicants: potential of nanoparticle-based biosensor and applications. Sci World J. 1–12
42. Rodriguez-Mozaz S, de Alda MJL, Barcelo D (2006) Biosensors as useful tools for environmental analysis and monitoring. Anal Bioanal Chem 386:1025–1041

43. Palchetti I, Mascini M (2008) Nucleic acid biosensors for environmental pollution monitoring. Analyst. 133:846–854
44. Alhamri HA (2018) Biosensors: classifications medical applications and future prospective. Biotechnol App Biochem 65:497–508
45. Monosik R, Stredansky M, Stundik E (2012) Biosensors-classification, characterization and new trends. Acta Chim Slov 5:109–120
46. Kissenger PT (2005) Biosensor-a perspective. Biosens Bioelectron 20:2512–2516
47. Saylan Y, Yilmaz F, Özgür E, Derazshamshir A, Yavuz H, Denizli A (2017) Molecular imprinting of macromolecules for sensor applications. Sensor 17:898
48. Yılmaz E, Majidi D, Özgür E, Denizli A (2015) Whole cell imprinting based Escherichia coli sensors: a study for SPR and QCM. Sens Actuator B Chem 209:714–721
49. Bakhshpour M, Özgür E, Bereli N, Denizli A (2017) Microcontact imprinted quartz crystal microbalance nanosensor for protein C recognition. Colloid Surface B 15:264–270
50. Zhu C, Yang G, Li H, Du D, Lin Y (2014) Electrochemical sensors and biosensors based on nanomaterials and nanostructures. Anal Chem 87:230–249
51. Çiçek Ç, Yılmaz F, Özgür E, Yavuz H, Denizli A (2016) Molecularly imprinted quartz crystal microbalance sensor (QCM) for bilirubin detection. Chemosensor 4:1–13
52. Battal D, Akgönüllü S, Yalcin MS, Yavuz H, Denizli A (2018) Molecularly imprinted polymer based quartz crystal microbalance sensor system for sensitive and label-free detection of synthetic cannabinoids in urine. Biosens Bioelectron 111:10–17
53. Mazzara G (2014) Piezoelectric biosensors for organophosphate and carbamate pesticides: a review. Biosensors 4:301–317
54. Stradiotto NR, Yamanaka H, Zanoni MVB (2003) Electrochemical sensors: a powerful tool in analytical chemistry. J Braz Chem Soc 14:159–173
55. Gerard M, Chaubey A, Malhotra BD (2002) Application of conducting polymers to biosensors. Biosens Bioelectron 17:345–359
56. Pardieu E, Cheap H, Vedrine C, Lazerges M, LattachY GF, Remita S, Pernelle C (2009) Molecularly imprinted conducting polymer based electrochemical sensor for detection of atrazine. Anal Chim Acta 649:236–245
57. Privett BJ, Shin JH, Schoenfisch MH (2008) Electrochemical sensors. Anal Chem 80:4499–4517
58. Prathish KP, Prasad K, Prasad Rao T, Suryanarayana MVS (2007) Molecularly imprinted polymer-based potentiometric sensor for degradation product of chemical warfare agents. Part I Methylphosphonic Acid Talanta 71:1976–1980
59. Anirudhan TS, Alexander S (2015) Design and fabrication of molecularly imprinted polymer-based potentiometric sensor from the surface modified multiwalled carbon nanotube for the determination of lindane (γ-hexachlorocyclohexane), an organochlorine pesticide. Biosen Bioelectron 64:586–593
60. Anirudhan TS, Alexander S (2014) Multiwalled carbon nanotube based molecular imprinted polymer for trace determination of 2,4-dichlorophenoxyaceticacid in natural water samples using a potentiometric method. Appl Surf Sci 303:180–186
61. Prasad K, Prathish KP, Mary Gladis J, Naidu GRK, Prasada Rao T (2007) Molecularly imprinted polymer (biomimetic) based potentiometric sensor for atrazine. Sens Actuator B Chem 123:65–70
62. Abdel-Ghany MF, Hussein LA, El Azab NF (2017) Novel potentiometric sensors for the determination of the dinotefuran insecticide residue levels in cucumber and soil samples. Talanta 164:518–528
63. Janata J (2009) Principles of chemical sensors. Springer, New York, pp 99–118
64. Kounaves SP (1997) Handbook of instrumental techniques for analytical chemistry. Prentice Hall PTR, Upper Saddle River, NJ, pp 709–725
65. Yang Q, Sun Q, Zhou T, Shi G, Jin L (2009) Determination of parathion in vegetables by electrochemical sensor based on molecularly imprinted polyethyleneimine/silica gel films. J Agric Food Chem 57:6558–6563

66. Mazzotta E, Malitesta C (2010) Electrochemical detection of the toxic organohalide 2,4-DB using a Co-porphyrin based electrosynthesized molecularly imprinted polymer. Sens Actuator B Chem 148:186–194
67. Motaharian A, Motaharian F, Abnous K, Hosseini MRM, Hassanzadeh-Khayyat M (2016) Molecularly imprinted polymer nanoparticles-based electrochemical sensor for determination of diazinon pesticide in well water and apple fruit samples. Anal Bioanal Chem 408:6769–6779
68. Ghodsi J, Rafati AA (2018) A novel molecularly imprinted sensor for imidacloprid pesticide based on poly(levodopa) electro-polymerized/TiO_2 nanoparticles composite. Anal Bioanal Chem 410:7621–7633
69. Qi P, Wang J, Wang Z, Xu H, Di S, Wang O, Wang X (2018) Sensitive and selective detection of the highly toxic pesticide carbofuran in vegetable samples by a molecularly imprinted electrochemical sensor with signal enhancement by AuNPs. RSC Adv 8:25334–25341
70. Hassan AHA, Moura SL, Ali FHM, Moselhy WA, Sotomayor MPT, Pividori MI (2018) Electrochemical sensing of methyl parathion on magnetic molecularly imprinted polymer. Biosens Bioelectron 118:181–187
71. Suri CR (2003) Immunosensors for pesticide monitoring. In: Advances in biosensors. Perpectives in biosensors, vol 5. Elsevier Science, Amsterdam, pp 161–176
72. Sciacca B, François A, Hoffmann P, Monro TM (2013) Multiplexing of radiative-surface plasmon resonance for the detection of gastric cancer biomarkers in a single optical fiber. Sens Actuator B Chem 183:454–458
73. Saylan Y, Yılmaz F, Özgür E, Derazshamshir A, Bereli N, Yavuz H, Denizli A (2018) Surface plasmon resonance sensors for medical diagnosis. Nanotechnology characterization tools for biosensing and medical diagnosis. Springer-Verlag, Berlin Heidelberg, pp 425–458
74. Zor E, Morales-Narváez E, Zamora-Gálvez A, Bingol H, Ersoz M, Merkoçi A (2015) Graphene quantum dots-based photoluminescent sensor: a multifunctional composite for pesticide detection. ACS Appl Mater Inter 7:20272–20279
75. Wang J, Qiu H, Shen H, Pan J, Dai X, Yan Y, Pan G, Sellergren B (2016) Molecularly imprinted fluorescent hollow nanoparticles as sensors for rapid and efficient detection λ-cyhalothrin in environmental water. Biosens Bioelectron 85:387–394
76. Ye T, Yin W, Zhu N, Yuan M, Cao H, Yu J, Gou Z, Wang X, Zhu H, Reyihanguli A, Xu F (2018) Colorimetric detection of pyrethroid metabolite by using surface molecularly imprinted polymer. Sens Actuator B Chem 254:417–423
77. Shrivastav AM, Usha SP, Gupta BD (2016) Fiber optic profenofos sensor based on surface plasmon resonance technique and molecular imprinting. Biosens Bioelectron 79:150–157
78. Yılmaz E, Özgür E, Bereli N, Türkmen D, Denizli A (2017) Plastic antibody based surface plasmon resonance nanosensors for selective atrazine detection. Mater Sci Eng C 73:603–610
79. Erbahar DD, Gürol I, Gümüs G, Musluoglu E, Öztürk ZZ, Ahsen V, Mika Harbeck M (2012) Pesticide sensing in water with phthalocyanine based QCM sensors. Sens Actuator B Chem 173:562–568
80. Boxall ABA (2011) Hazardous substances in Europe's fresh and marine waters-an overview. EEA technical report No 8/2011, Technical report european environment agency
81. Villaverde JJ, Sevilla-Morán B, Sandín-España P, López-Goti C, Alonso-Prados JL (2014) Biopesticides in the framework of the European Pesticide Regulation (EC) No. 1107/2009. Pest Manage Sci 70:2–5
82. Tankiewicz M, Fenik J, Biziuk M (2010) Determination of organophosphorus and organonitrogen pesticides in water samples. Trend Anal Chem 29:1050–1063
83. Yaqub S, Latif U, Dickert FL (2011) Plastic antibodies as chemical sensor material for atrazine detection. Sens Actuator B Chem 160:227–233
84. Mensah-Brown AK, Wenzel MJ, Josse FJ, Yaz EE (2009) Near real-time monitoring of organophosphate pesticides in the aqueous-phase using SH-SAW sensors including estimation-based signal analysis. IEEE Sens J 9:1817–1824

85. Bi X, Yang KL (2009) On-line monitoring imidacloprid and thiacloprid in celery juice using quartz crystal microbalance. Anal Chem 81:527–532
86. Pinkham W, Wark M, Winters S, French L, Frankel DJ, Vetelino, J (2005) A lateral field excited acoustic wave pesticide sensor. In: Proceedings of the 2005 IEEE Ultrasonics Symposium, pp 2279–2283
87. Sibel ED, Keçili R, Arzu E, Say R (2017) Molecular imprinting technology in quartz crystal microbalance (QCM) sensors. Sensor 17:454
88. Uludağ Y, Piletsky SA, Turner APF, Cooper MA (2007) Piezoelectric sensors based on molecular imprinted polymers for detection of low molecular mass analytes. FEBS J 274:5471–5480
89. Liao H, Zhang Z, Nie L, Yao S (2004) Electrosynthesis of imprinted polyacrylamide membranes for the stereospecific L-histidine sensor and its characterization by AC impedance spectroscopy and piezoelectric quartz crystal technique. J Biochem Biophys Method 59:75–87
90. Dickert F, Tortschano M, Bulst WE, Fischerauer G (1999) Molecularly imprinted sensor layers for the detection of polycyclic aromatic hydrocarbons in water. Anal Chem 71:4559–4563
91. Dulama ID, Radulescu C, Bucurica IA, Teodorescu S, Stirbescu RM, Cimpoca GV, Let DD, Gurgu IV (2018) Quartz crystal microbalance used as sensor for pesticides detection. J Sci Arts 18:445–452
92. Cheng X, Zhang Q, Fong S, Li Y, Liu B (2017) Molecularly imprinted polymer-coated quartz crystal microbalance for detection of parathion. MJAS 21:1327–1334
93. Yang Y, Cao Y, Wang X, Fang G, Wang S (2015) Prussian blue mediated amplification combined with signal enhancement of ordered mesoporous carbon for ultrasensitive and specific quantification of metolcarb by a three-dimensional molecularly imprinted electrochemical sensor. Biosens Bioelectron 64:247–254
94. Kmellár B, Fodor P, Pareja L, Ferrer C, Martínez-Uroz M, Valverde A (2008) Validation and uncertainty study of a comprehensive list of 160 pesticide residues in multi-class vegetables by liquid chromatography–tandem mass spectrometry. J Chromatogr A 1215:37–50
95. Liu ZM, Zang XH, Liu WH, Wang C, Wang Z (2009) Novel method for the determination of five carbamate pesticides in water samples by dispersive liquid–liquid microextraction combined with high performance liquid chromatography. Chinese Chem Lett 20:213–216
96. Liu C, Fang G, Deng Q, Zhang Y, Feng J, Wang S (2012) Determination of metolcarb in food by capillary electrophoresis immunoassay with a laser-induced fluorescence detector. Electrophoresis 33:1471–1476
97. Sun J, Dong T, Zhang Y, Wang S (2010) Development of enzyme-linked immunoassay for the simultaneous detection of carbaryl and metolcarb in different agricultural products. Anal Chim Acta 666:76–82
98. Pan MF, Fang GZ, Liu B, Qian K, Wang S (2011) Novel amperometric sensor using metolcarb-imprinted film as the recognition element on a gold electrode and its application. Anal Chim Acta 690:175–181
99. Fang G, Yang Y, Zhu H, Qi Y, Liu J, Liu H, Wang S (2017) Development and application of molecularly imprinted quartz crystal microbalance sensor for rapid detection of metolcarb in foods. Sens Actuator B Chem 251:720–728
100. Sauerbrey G (1959) Verwendung von SchwingquarzenzurWägungdünnerSchichten und zurMikrowägung. Z Med Phys 155:206–222
101. Tuna S, Duman O, Soylu I, Kancı Bozoğlan B (2014) Spectroscopic investigation of the interactions of carbofuran and amitrol herbicides with human serum albumin. J Lumin 151:22–28
102. Çakır O, Yılmaz F, Baysal Z, Denizli A (2018) Preparation of a new quartz crystal microbalance sensor based on molecularly imprinted nanofilms for amitrole detection. Biointer Res In App Chem 8:3435–3440

Chapter 10
Controlled Release Herbicides and Allelopathy as Sustainable Alternatives in Crop Production

S. Shahena, Maya Rajan, Vinaya Chandran and Linu Mathew

Abstract Herbicides are specially designed chemicals to kill unwanted plants that adversely affect the crop growth and thereby increase the nutrient requirements. Directly applied herbicide residues finally accumulate in the soil. They are found in all environmental spheres including air, water, and soil. Most of the herbicide formulations contain active ingredients and additives so as to fulfill regulatory standards without affecting their effectiveness. Nevertheless, their excessive use leads to herbicide resistance in weed plants prompting increase in the dosages. Hence, many modified, controlled release systems with reduced hazards and environmental toxicity are developed. These methods minimize the environmental impacts and aid in the sustainable development. Also, plant derived bioformulations with weedicide activities are considered as environmentally safe alternatives. Allelochemicals are plant compounds which give a survival advantage to the producers. Allelopathy has an important role in weed control and crop productivity. Since the allelochemicals are biologically active compounds, they are less disruptive to the ecosystem than synthetic herbicides. Hence, controlled release of herbicides and allelopathy are discussed as sustainable alternatives in farming practice.

Keywords Allelopathy · Herbicides · Controlled release · Sustainable development

10.1 Introduction

Plants are the primary food source for the World. Around 10,000 B.C, human race started cultivating crops and they found weeds as their major problem in crop cultivation [1]. Out of total annual loss of agricultural production in India, weeds contribute almost 37% and rest by insects, diseases, and other pests [2–5]. Weeds can be defined as plants which are detrimental, persistent, and harmful. It affects the

S. Shahena · M. Rajan · V. Chandran · L. Mathew (✉)
School of Biosciences, Mahatma Gandhi University, Kottayam, Kerala, India
e-mail: linumathew@mgu.ac.in

Rakhimol K. R. et al. (eds.), *Controlled Release of Pesticides for Sustainable Agriculture*, https://doi.org/10.1007/978-3-030-23396-9_10

growth of other crop plants and thus affects human agricultural practices and economy of the country [3]. Herbicide application, mechanical weeding, and hand weeding are the commonly used methods of weed control and are found to be effective in agricultural production [6].

Herbicides are specially designed chemicals to kill unwanted plant tissues that affect the crop plant's growth and thereby increase the nutrient requirements. The mode of action of each herbicide is unique and targets a particular biosynthetic pathway. Some of them will prevent the sugar metabolism and the others may inhibit an essential hormone development. Herbicides control the biological and adaptive advantages of the weeds partially and totally [7].

10.2 Classification of Herbicides

The herbicides can be classified according to their mode of action, chemical family and activity, method of application, site of action, and timing of application [3].

10.2.1 Based on the Mode of Action or Translocation

According to the mode of action, the herbicides can be grouped into contact herbicides or non-systemic and systemic or translocated. The contact herbicides will kill the plant parts that it contacts, especially the upper leafy parts of the plants. The systemic herbicides will be absorbed by the plants through the leaves or stems and transport it internally throughout the plant. The systemic herbicides move along with the sap or vascular system along with the nutrients, water, and other materials, and it does not have any quick effect. The systemic herbicides will slowly kill the entire weed plant. In the case of contact herbicides, most of the weed plants will regenerate the affected plant tissues and thus cause weed resurgence. Mostly, the systemic and contact herbicides are applied together for a better and faster effect. Glyphosate and glufosinate are non-selective systemic herbicides, and 2, 4-D and dicamba are the examples of selective systemic herbicides. Bromoxynil and bentazon belong to systemic or translocated group [3].

10.2.2 Based on the Time of Application

Based on the time of application, herbicides can be classified into preplant herbicides, preemergence herbicides and postemergence herbicides. Preplant herbicides are volatile herbicides and applied to the soil before the cultivation. They are incorporated into the soil through mechanical methods to avoid photodecomposition. They are non-selective herbicides which inhibit the weed seedlings growth.

Allelocompounds can be used as preplant bioherbicides because of their allelopathic property [8, 9]. In the case of preemergence herbicides, they are applied before the weed seedling growth and their action is by affecting cell division. Dithopyr and Pendimethalin are preemergence herbicides. Postemergence herbicides are applied after the weed seedlings emerge. Postemergence herbicidal practices require multiple applications for adequate control of the weeds. They can be selective or non-selective, contact or systemic. 2, 4-D is a selective, systemic, foliar absorbed, and postemergence herbicide [3, 10].

10.2.3 Based on the Method of Application

According to the method of application, the herbicides can be grouped into soil applied and foliar applied. The soil-applied herbicides are preplant or preemergence herbicides, and they can be directly applied to the soil. These are absorbed by the root or shoot of the emerging seedlings. Thiocarbamates (e.g., EPTC) and dinitroanilines (e.g., trifluralin) are the examples of soil herbicides. The foliar herbicides are applied to the upper portion of the plants. They belong to postemergence, systemic herbicides and are absorbed by the exposed tissues of the weed seedlings. Glyphosphate and 2, 4-D are foliar-applied herbicides [3].

10.2.4 Based on the Specificity

Selective herbicides control or suppress the weeds without affecting the crop plants. 2, 4-D, mecoprop and dicamba are selective herbicides. Non-specific herbicides are not specific in their selection. They are used to clear industrial sites, waste ground, railways, etc. Paraquat, glufosinate, glyphosphate are non-selective herbicides [3].

10.3 Environmental Effects of Herbicides

There has been a tendency to maximize the crop production through increasing productivity by developing new-high yield crops, heavy use of fertilizers and agrochemicals and also by the use of improved agricultural machinery. The environment gets damaged due to these innovative practices by causing soil erosion, salinization, flooding of heavily irrigated soils, aquifer depletion, deforestation and environmental contamination which affect human health and natural ecosystems [7].

According to *Codex Alimentarius* (Commission created by the food and Agricultural Organization [FAQ]) and the World Health Organization (WHO: UNO, 1963), "a plaguicide is any substance aimed to prevent, destroy, attract, repel or fight any plague including unwelcome species of plants or animals; during the production or storage, transportation, distribution, and elaboration of food; agricultural production or food for animals; or that may be administered to animals to ectoparasites which include herbicides and compounds used as growth regulators, insecticides, defoliants, desiccants and inhibitors of fruit thinning and germination" [7].

Largely, herbicides are applied either directly to the foliage or to the soil. Approximately, 50% of the herbicidal particles finally deposits in the soil through the direct application and also a lesser amount through plant stubbles left post-harvesting. The herbicidal particles drift by the wind and rain and they finally get suspended in the atmosphere. So, they found in all environmental compartments including air, water, and soil [11–14].

The herbicidal particles reach the soil surface and undergo a series of chemical and biological reactions such as photodecomposition. The residual herbicidal particles get bioaccumulated in all kinds of organisms from plankton to human beings and have been biomagnified through trophic nets [15, 16]. Since the herbicides protect and sustain crops, they are essential and valuable tools for the modern agricultural practices. But there is an urgent need to optimize the use of herbicides to avoid the environmental harm [17].

Herbicides possess more dangerous effects as compared to insecticides and bactericides [18–21]. Due to their high mobility, they probably affect non-target organisms and affect the ecosystem adversely. Therefore, it is very important to develop an efficient technology to promote the correct dosage of herbicides and improve their utilization efficiency [22].

10.4 Herbicide Resistance in Weeds

The excessive use of herbicides without a proper release formulation leads to the transition of normal weeds into herbicide-resistant weeds [3, 23–26]. Some weeds have an inherent ability to survive adverse effect of herbicides that results in the susceptibility of the original species to that particular herbicide. Resistant weeds survive the application of herbicides and gradually their population increases causing damage to crop plants and their yield [3]. Herbicide resistance was observed in 1957 against 2, 4-D in Hawaii [27]. The first confirmed report was against triazine herbicide in *Senecio vulgaris* (common groundsel) in 1970 [28–30].

Pline et al. reported the effects of ammonium sulfate and pelargonic acid in the weed control of soybean [31]. The herbicides were applied along with glufosinate and glyphosphate. They found glufosinate-resistant and glyphosphate-resistant soybean in the greenhouse and fields. The efficacy of glufosinate can be increased by the addition of ammonium sulfate. In the case of perennial crops, its efficacy can be improved by the addition of glufosinate without affecting the soybean [32].

10.5 Mode of Applications and Herbicide Formulations

Most of the herbicides are applied along with the active ingredient and additives to fulfill regulatory standards without affecting their effectiveness. The active ingredient is combined with adjuvants such as solvents, moisturizers, and adherents. There are many modified, controlled release systems for reducing the hazardous effect of herbicides and also to reduce the environmental toxicity. These methods minimize the environmental impacts and aid in the sustainable crop production [33].

The objectives of the herbicide formulations are (a) the effectiveness should retain throughout the time required to control the harmful weeds; (b) the management and application should be safe for workers and users; (c) the use of toxic or dangerous solvents must be reduced; (d) it should not affect or minimize the effects on organisms except weeds; and (e) to extend the range of herbicide application [34, 35].

The herbicides, compatible with other formulations such as adjuvants or co-adjuvants which is already present in the formulations or during mixing them in the tank of application system is found to be physically and chemically stable throughout the time. They are able to favorably modify the physical and chemical properties of the active ingredients [36].

The traditional methods of controlled release of herbicides include:

5.1 Slow release of herbicide formulations.
5.2 Herbicide collars.
5.3 Herbicide-treated papers.
5.4 Herbicide-treated mulch.
5.5 Herbicides applied with pruning shears.

These techniques can reduce the off-site movement and other environmental concerns about herbicidal usage in container production, field production and landscape maintenance [37].

10.5.1 Slow Release Herbicide Formulations

Weed control must be maintained for a longer period of time in nursery industry, annual crops, plants, crops in landscapes, etc. In the case of annual crops, herbicides that possess long soil residual activity have a tendency to exist for a long time in the soil [38]. Multiple applications of hazardous herbicides per year are required for field production and landscape maintenance. Either herbicides with long soil residual activity or increasing the herbicide rate are the two methods which serve this purpose. But this could increase crop damage as well as cause concerns about environmental hazards [37].

10.5.1.1 Herbicide Tablets

Slow release tablets are a porous pellet containing an inert material such as Plaster of Paris or dicalcium phosphate along with herbicide. These tablets when wetted by irrigated water, low amounts of herbicides are gradually released over an extended period of time. The release of herbicides can be increased by increasing herbicide concentrations and by changing the size of porosity of the tablets [39–43]. Alachlor and 4% metolachlor in dicalcium phosphate are the examples of herbicide tablets.

Some types of slow release formulations used starch xanthide and pine craft lignin [44–49] instead of Plaster of Paris and dicalcium phosphate: e.g., ester formulations of picloram in starch xanthide and chloramben and metribuzin in pine craft lignin [39–41].

The major advantage of the herbicide tablets is the concentration of the herbicide used in this formulation is less when compared to granular or sprayable form. As a result, the herbicide leaching will be reduced. The threshold concentration and labor requirement are also less in the case of herbicide tablets [39–41].

10.5.2 Herbicide Collar

The herbicide collar formulation is a standard and better herbicide application formula. The use of physical barriers for weed control in container production is referred to as herbicide collar [50, 51]. Solid polyethylene (black plastic) is an example of herbicide collar that controls weeds in established plantings. Herbicide collar is a multiple chemical delivery system because fertilizers and other chemicals can also be added to the collar. The labor requirement is also less in the case of herbicide collar formulation. The cost-effectiveness and the difficulty in the disposal of the fabric collar are the major disadvantages of the usage of herbicide collar in the field [37].

Herbicide-impregnated cloth [52] and herbicide-impregnated string [53] are found to be more efficient than physical barriers composed of synthetic or natural fibers. Simazine, atrazine, and diuron were efficiently controlled weeds in peach planting by using these methods [37, 53].

10.5.3 Herbicide-Treated Papers

In landscape plantings, water soluble papers or herbisheets are used to deliver herbicides (American Cyanamid and Princeton N. J). Appropriate rate of herbicide pre-treated papers was cut into small pieces and fit into landscape bed and covered with a thin layer of mulch. When water droplets fall into the sheets by irrigation or through rainfall, it dissolves and releases the herbicides into the soil or mulch surface. Herbicide-treated paper sheets containing Oxyflourfen marketed under the trade names Neal and Senesac were able to control horse weeds. No calibration or spraying is required for herbicide-treated papers and it also eliminates the problem of drift through wind, water and air.

Exposure of applicator to the herbicide during the cutting and installation process makes this method difficult for the common adoption. Preinstallation will be a time-consuming process in herbicide-treated papers. The result may be varied due to irrigation practices and rainfall [37].

10.5.4 Herbicide-Treated Mulch

The controlled release of herbicides through herbicide-treated mulch is an improved method. In landscape plantings, herbicide-treated mulch is used to control weeds [50]. For example, equal or better weed control was observed when dichlobenil was incorporated into mulch [54]. No chemical mixing or spraying is required for herbicide-treated mulch. But the herbicide application rate is excessive. Exposure of applicant during mixing results in health hazards.

10.5.5 Herbicide/Hand Pruners

Weed stem is cut with a sterile blade. Herbicide containing solution is applied to the cutting edge of the weed stem. Weed stem absorbed the herbicide solution and it translocates into various plant parts resulting in the death of the weed plants. This is known as Klipkleen system (Klipkleen Pruning system, Bloomland Special Products, Atlanta, G.).Undiluted 2,4-D and 25 and 50% solutions of Clopyralid completely controlled poison ivy when applied with the hand-pruner system. Minimal herbicide usage is required for herbicide or hand pruners. Drift through spraying can be avoided with maximum elimination of desired plant. It is a labor consuming process and exposure of the applicant during application time results in health hazards [37].

10.6 Release Systems for Herbicides

Materials composed of silica, clays (bentonites and sepiolites), polymers (alginate), and synthetic polymers (polyhydroxyalkonates—PHAs) are the most commonly used bioactive substances for the transportation of herbicides.

The methods used for controlling loss of herbicide such as microcapsule [55], tablets [56, 57], organosilicone [58, 59], nano-additives [60–62] are limited due to their disadvantages.

Microcapsule and tablet methods are less applicable because of their high cost [55, 56]. They also lack sensitivity to physical and chemical factors which makes it difficult for them to control the release of herbicides [57]. The adhesion and penetration performance of the organosilicone can be increased to a certain extent. But this method has poor stability, high cost and also has low ability to control herbicide release. Nano-additives showed higher stability than organosilicone. They ensure the adhesion ability of herbicides through network-structured nano-clay [60–62]. The pH sensitive gels display low application values because of their cost [60, 63].

10.6.1 Microencapsulation

Microencapsulation is the process of coating particles of active substances of micrometric size. The active substances include coating molecules, solid particles, or liquid globules consisting of various materials. According to system particularities with regard to morphology and internal structure, the encapsulated products are denoted as microparticles, microcapsules (MICs), or microspheres (MISs), and all are <1 mm [64]. Nanocapsules or nanospheres are smaller than 1 mm in size. Most of these particles are not spherical, but they are considered as MIC or MIS due to their size and composition. The terms microparticles and nanoparticles are used in a general way to address particles presenting either small irregular or spherical shape [65, 66].

10.6.2 Polymer Micro and Nanoparticles

Polymer microparticles and nanoparticles can be used for the transport of bioactive substances. They have the ability to alter the physio-chemical properties of the substances they incorporate and are completely biodegradable [33]. These microparticles can be easily produced by the fermentation of varieties of bacteria and are easily degraded during natural biological process. They are inexpensive also. The most widely used polymers for the transport of herbicides are PHAs which includes polyhydroxybutyrate (PHB) and polyhydroxybutyrate-valerate (PHBV). The ametryn encapsulated within the PHB or PHBV microparticles will be reduced the impacts caused by the herbicide and is environmentally safe.

Tebuthiuron (TBH) is the most commonly used herbicide for the weed control in sugarcane and cotton crops and are applied as preemergent herbicide. The calcium alginate microparticles were prepared from the cross-linking of sodium alginate by Ca^{++} containing varied amounts TBH supplied in $CaCl_2$ aqueous solution. This encapsulated Tebuthiuron is found to be more effective than the normal method of application [67].

The controlled release of herbicides should be a simple procedure with a low cost. According to Chi et al. (22), attapulgite is an ideal carrier of herbicide because of their low cost and high stability [22]. Attapulgite is an eco-friendly herbicide carrier. This is a temperature-responsive controlled release herbicide particle method (TCHP) with a core-shell structure developed using a nano-composite.

The starch encapsulated agrochemicals can be dispersed as an aqueous paste of gelatinized starch through retrogradation or cross-linking [68]. The technique has many advantages. This method is highly accurate with extended activity and an economic alternative. The matrix used for the encapsulation is biodegradable and less harmful to the environment. The application time and dose of the weedicide can be reduced. This method also increases safety of handling of the plants and the applicants. It has effective control over the targets and reduced levels of losses through evaporation, degradation, leaching, etc. [69].

The starch xanthate was an effective substrate for encapsulating agrochemicals [70, 71]. The active ingredient or herbicide could be encapsulated within starch–calcium–adducts [72, 73] or starch–borate complexes [69, 74]. The natural starch is also a good material for agrochemical encapsulation [75, 76].

Zhu and Zhou suggested that starch-g-polybutyl acrylate (starch-g-PBA) is an effective material for controlling the release of carboxylic containing herbicides [68].

Teffet and Friend prepared polymeric microspheres (MS) based on ethylcellulose (EC) and polyarylsulfone (PS) and tested them for the controlled release of herbicide dicamba [77, 78].

Junior et al. proposed a new delivery system for sulfentrazone herbicide encapsulated into calcium alginate (Ca-ALG) microparticles for the weed control and the field trial were conducted using *Cucumis Sativus* or cucumber as indicator species [79]. They studied the phytotoxicity evaluation caused by Ca-ALG microparticles at different concentrations and the relationship with leaching depth. They concluded that Ca-ALG encapsulated microparticles exhibited high efficiency.

Unmodified, cyanomethylated and cross-linked corn starches were reacted with 2,4-dichlorophenoxy acetyl chloride in pyridine to produce the 2,4-D esters of high acyl content. Hydrolysis studies at pH 6 and pH 8 at ambient temperature showed that, these insoluble compositions liberated varying amounts of 2,4-D and soluble 2,4-D esters by ester link and glucoside bond cleavage, proving their possible use as controlled release herbicides [80].

Zhila et al. developed a formulation of metribuzin and tribenuron-methyl embedded in the degradable matrix of natural poly-3-hydroxy butyrate [P (3HB/MET and P (3-HB/TBM)] which were constructed in the form of films and microgranules [81]. This formulation was tested against the weeds such as white sweet clover (*Melilotos albus)* and lamb's quarters (*Chenopodium album)* in the presence of soft spring wheat (*Triticum aestivum*).

Li et al. formulated controlled release formulations of herbicide acetochlor [82]. The herbicide acetochlor prepared by using carboxy methyl cellulose (CMC) gel and different types of clay which obtained by acid activation, pillared with metal hydroxides or saturated with organic cations. The effect of formulation parameters on the acetochlor release rate from different formulations was evaluated by water-release studies.

The polymeric encapsulated system has the following advantages: The usage of minimum amount of chemical substance required for weed control in polymeric encapsulated system. It reduced the level of environmental toxicity and contamination. Minimized the energy consumption compared to the conventional methods and more safe to the persons that handled into the field [37].

Allelopathy can be integrated with agriculture that will reduce the dependence of the agriculturists to the synthetic herbicides and other agrochemicals [83].

10.7 Allelochemicals: The Biological Alternatives to Synthetic Herbicides

Allelopathy is a natural ecological phenomenon [84]. According to International Allelopathic Society, the definition of allelopathy is "any process involving secondary metabolites produced by plants, algae, bacteria and fungi that influence the growth and development of agricultural and biological systems" [8, 85]. Macias et al. defined allelochemicals as "biocommunicators," suggesting the possibility of active mixtures, because of increasing number of findings in which single compounds are not active or are not as active as a mixture [86].

Allelopathy has an important part in weed control and crop productivity. It has been used in agriculture since ancient times [84]. The allelopathic plants have the weed suppressing ability [83]. Since the allelochemicals are biologically active compounds, they are less disruptive to the ecosystem than synthetic agrochemicals. They act on the weeds without causing any harmful effects to the ecosystem. They can be used as natural herbicides [8, 9]. Allelochemicals, therefore, reduce the problems such as environmental contamination, use of hazardous agricultural products and unhealthy effects on human health [83]. Allelochemicals can enhance or suppress germination and growth of the plant, and thereby the development of crops with low phytotoxic residue amounts in water and soil. It also facilitates treatment and recycling of the wastewater [84, 87].

10.8 Natural Phytotoxic Compounds

Plants produce hundreds of compounds that are not involved in primary metabolism of plants. These compounds involved in interspecific chemical interactions or allelopathy with higher plants. Such compounds are often phytotoxic or herbicidal to other species or to the species producing them [88].

The first attempt to synthesize botanical herbicides by the isolation and identification of active principles of allelochemicals were made in the USA and Asia [89].

Four percent out of 10,000 rice germplasm line possess various degrees of allelopathic activity against certain riceweeds [90]. The plants released a great variety of organic compounds and they have an active lifespan in the soil. The active life of the organic compounds in the soil is determined by the factors such as volatility, leaching, adsorption, and microbial action [8]. The degradation of the fresh crop residues of 3–4 weeks of aerobic decomposition at 15 °C results retarded germination in the fields [91]. Many complex interactions, transformations, and synthesis occur during the decomposition of plant parts. They have an important role in the control and regulation of each developmental stage of other plants [8].

The allelopathy integrated agricultural weed management practices includes various techniques such as the use of allelopathy in crop-rotation, preceding, and cover crops, green manure, mulch, intercropping, and also the incorporation of allelochemicals or allelopathic plants in the soil [83, 92].

10.8.1 Allelochemicals as Botanical Herbicides

Since allelochemicals are phototoxic compounds, they can be used as natural herbicides and there are several examples of allelochemicals from plants that are used in the production of new herbicides [93, 94]. The light activated compounds or photosensitizes are potentially useful in agriculture as herbicides [95]. Examples are naphtha and anthraquinone of fungi and higher plants, polyines of *Asteraceae porphyrines*, extended quinones, and isoquinoline alkaloids. The terpenoids, monoterpenes, sesquiterpenes, sesquiterpene lactones, triterpenes and fatty acids with activity range of 0.25–10.5 ppb are the more potential herbicidal allelochemicals than phenolics, quinines, and alkaloids [96].

Nie et al. studied the allelopathic potential of *Wedelia trilobata* L. on growth, germination and physiological parameters of rice [97]. Wedelia was treaded into paddy fields in many areas such as Guangdong, Fujian, Taiwan, and Liaonin province. They used this mulch as organic fertilizers or as green manure [97–99].

Application of brassinolide, a steroid obtained from rape pollen (*Brassica napus*), triacontanol from alfalfa (*Medicago sativa L*), agrostemin from seeds of corncockle (*Agrostemma githago*) and several natural products isolated from the chinaberry (*Melia azedarach*) have enhanced yield of several crops and are recommended as useful materials in agricultural practice [100–102].

Worsham reported that the effective utilization of allelochemicals for weed control is through crop residues that are selectively toxic, mainly in conservation and no-tillage crop production [103]. Rye (*Secale cereale*), wheat (*Triticum aestivum*), sorghum (*Sorghum bicolor*), and certain winter legumes can be used as mulch or cover crops.

Examples of already identified allelochemicals are: (1) cyanobacterin produced by the cyanobacterium *Scytonema hofmanni* inhibit electron transport and cause extensive damage to the thylakoid membrane of the chloroplasts. The cyanobacterin has the same effect similar to that of the synthetic herbicide Diuron [104].

(2) Tentoxin produced by *Alternaria alternata* is effective against *Sorghum halepense* in maize and soybean [8]. (3) The antimalarial drug Artemisinin from *Artemisia annua L* is phytotoxic to weeds [105]. (4) Bialaphos or bilanafos is the only biosynthetically produced commercial herbicides that degrade phosphinothricin in target plants [106].

10.9 Conclusion

To conclude, the controlled release of the herbicides need only single application reducing toxicity and manpower requirement [107]. As compared to the normal practices, the controlled release of herbicides carries maximum concentration of herbicides. The reduction in costs and pollution due to the efficient utilization of herbicides is possible through improved release methods. This increases the sustainability of the practice. Utilization of environmentally safe alternatives like plant-derived bioherbicides or allelochemicals enhances the crop production by avoiding the negative and harmful impacts [108].

References

1. Hay JR (1974) Gains to the grower from weed science. Weed Sci 22(5):439–442
2. Yaduraju NT (2006) Herbicide resistant crops in weed management. In: The extended summaries, golden jubilee national symposium on conservation agriculture and environment. Banaras Hindu University Varanasi, pp 26–28
3. Vats S (2015) Herbicides: history, classification and genetic manipulation of plants for herbicide resistance. Sustainable agriculture reviews. Springer, Cham, pp 153–192
4. Bifani P (1987) Socio economic aspects of technological innovation in food production systems. Developments in Agricultural and Managed-Forest Ecology (Netherlands)
5. Yadav A, Malik RK (2005) Herbicide resistant Phalaris minor in wheat–a sustainability issue. In: Resource book. Department of Agronomy and Directorate of Extension Education, CCSHAU, Hisar, India, 152
6. Farooq M, Jabran K, Cheema ZA, Wahid A, Siddique KH (2011) The role of allelopathy in agricultural pest management. Pest Manage Sci 67(5):493–506
7. Sopena Vazquez F, Maqueda Porras C, Morillo Gonzalez E (2009) Controlled release formulations of herbicides based on micro-encapsulation. Ciencia e investigation agraria 36:27–42
8. Khalid S, Ahmad T, Shad RA (2002) Use of allelopathy in agriculture. Asian J Plant Sci 1 (3):292–297
9. Bhadoria PBS (2011) Allelopathy: a natural way towards weed management. Am J Exp Agric 1(1):7
10. Rao VS (2000) Principles of weed science. CRC Press, Florida
11. Berrada H, Font G, Molto JC (2003) Determination of urea pesticide residues in vegetable, soil, and water samples. Crit Rev Anal Chem 33(1):19–41
12. Boparai HK, Shea PJ, Comfort SD, Snow DD (2006) Dechlorinating chloroacetanilide herbicides by dithionite-treated aquifer sediment and surface soil. Environ Sci Technol 40 (9):3043–3049

13. Konstantinou IK, Hela DG, Albanis TA (2006) The status of pesticide pollution in surface waters (rivers and lakes) of Greece. Part I. Review on occurrence and levels. Environ Pollution 141(3):555–570
14. Ribeiro ML, Lourencetti C, Pereira SY, Marchi MRRD (2007) Groundwater contamination by pesticides: preliminary evaluation. Quim Nova 30(3):688–694
15. Barr DB, Needham LL (2002) Analytical methods for biological monitoring of exposure to pesticides: a review. J Chromatogr B 778(1–2):5–29
16. Albert LA (1998) Los plaguicidas persistentes y sus efectos a largo plazo. In II Simposio Internacional sobre agricultura sostenible, México DF
17. Kudsk P, Streibig JC (2003) Herbicides–a two-edged sword. Weed Res 43(2):90–102
18. Graham JH, Myers ME (2016) Evaluation of soil applied systemic acquired resistance inducers integrated with copper bactericide sprays for control of citrus canker on bearing grapefruit trees. Crop Prot 90:157–162
19. Papiernik SK, Yates SR, Koskinen WC, Barber B (2007) Processes affecting the dissipation of the herbicide isoxaflutole and its diketonitrile metabolite in agricultural soils under field conditions. J Agric Food Chem 55(21):8630–8639
20. Shrader-Frechette K, ChoGlueck C (2017) Pesticides, neurodevelopmental disagreement, and Bradford Hill's guidelines. Accountability Res 24(1):30–42
21. Vitek P, Novotna K, Hodaňová P, Rapantova B, Klem K (2017) Detection of herbicide effects on pigment composition and PSII photochemistry in Helianthus annuus by Raman spectroscopy and chlorophyll a fluorescence. Spectrochim Acta Part A Mol Biomol Spectrosc 170:234–241
22. Chi Y, Zhang G, Xiang Y, Cai D, Wu Z (2017) Fabrication of a temperature-controlled-release herbicide using a nanocomposite. ACS Sustain Chem Eng 5(6):4969–4975
23. Devine MD, Eberlein CV (1997) Physiological, biochemical and molecular aspects of herbicide resistance based on altered target sites. Herbicide activity: toxicology, biochemistry and molecular biology. Springer, Berlin, pp 159–185
24. Devine MD, Preston C (2000) The molecular basis of herbicide resistance. 72–104; Cobb AH, RC Kirkwood. Herbicides and their mechanisms of action. England Sheffield Academic Press, Sheffield. (Google Scholar)
25. Preston C, Mallory-Smith CA, Powles SB, Shaner DL (2001) Biochemical mechanisms, inheritance, and molecular genetics of herbicide resistance in weeds. Herbicide resistance and world grains. CRC Press, Boca Raton, Florida, pp 23–60
26. Smeda RJ, Vaughn KC (1997) Mechanisms of resistance to herbicides. Molecular mechanisms of resistance to agrochemicals. Springer, Berlin, Heidelberg, pp 79–123
27. Hilton HW (1957) Herbicide tolerant strains of weeds. Hawaiian sugar planters association annual report. Hawaiian Sugar Planters Association, Honolulu, HI, pp 69–72
28. Le Baron HM, Gressel J (eds) (1982) Herbicide resistance in plants. Wiley, London
29. Ryan GF (1970) Resistance of common groundsel to simazine and atrazine. Weed Sci 18 (5):614–616
30. Radosevich SR, Appleby AP (1973) Relative susceptibility of two common groundsel (senecio vulgaris L.) biotypes to six s-triazines 1. Agron J 65(4):553–555
31. Pline WA, Hatzios KK, Hagood ES (2000) Weed and herbicide-resistant soybean (Glycine max) response to glufosinate and glyphosate plus ammonium sulfate and pelargonic acid. Weed Technol 14(4):667–674
32. Pollegioni L, Schonbrunn E, Siehl D (2011) Molecular basis of glyphosate resistance–different approaches through protein engineering. FEBS J 278(16):2753–2766
33. Grillo R, Santo Pereira ADE, de Melo NFS, Porto RM, Feitosa LO, Tonello PS, … Fraceto LF (2011) Controlled release system for amteryn using polymer microspheres: preparation, characterization and release kinetics in water. J Hazard Mater 186(2–3): 1645–1651
34. Tsuji K (2001) Microencapsulation of pesticides and their improved handling safety. J Microencapsul 18(2):137–147

35. Mulqueen P (2003) Recent advances in agrochemical formulation. Adv Coll Interface Sci 106(1–3):83–107
36. Zabkiewicz JA (2000) Adjuvants and herbicidal efficacy-present status and future prospects. Weed Res Oxford 40(1):139–149
37. Derr JF (1994) Innovative herbicide application methods and their potential for use in the nursery and landscape industries. Hort Technol 4(4):345–350
38. Hagood ES, Swann CW, Wilson HP, Ritter RL, Webb FJ (1992) Weed control in forage crops. Pest management guide for field crops. Virginia Cooperative Extension, Serv Publication, pp 456–016
39. Smith AE, Verma BP (1977) Weed control in nursery stock by controlled release of alachlor. Weed Sci 25(2):175–178
40. Verma BP, Smith AE (1978) Slow release herbicide tablets for container nursery. Trans ASAE 21(6):1054–1059
41. Verma BP, Smith AE (1981) Dry-pressed slow release herbicide tablets. Trans ASAE 24(6): 1400–1403; Danielson LL (1967) Evaluation of herbicide-impregnated cloth. Weeds 15(1): 60–62
42. Koncal JJ, Gorske SF, Fretz TA (1981) Leaching of EPTC, alachlor, and metolachlor through a nursery medium as influenced by herbicide formulations. HortScience
43. Ruizzo MA, Smith EM, Gorske SF (1983) Evaluations of herbicides in slow-release formulations for container-grown landscape crops. J Am Soc Hortic Sci
44. Baur JR (1980) Release characteristics of starch xanthide herbicide formulations 1. J Environ Qual 9(3):379–382
45. Raboy V, Hopen HJ (1982) Effectiveness of starch xanthide formulations of chloramben for weed control in pumpkin (Cucurbita moschata). Weed Sci 30(2):169–174
46. Riggle BD, Penner D (1987) Evaluation of pine kraft lignins for controlled release of alachlor and metribuzin. Weed Sci 35(2):243–246
47. Riggle BD, Penner D (1988) Controlled release of three herbicides by the kraft lignin PC940C. Weed Sci 36(2):131–136
48. Schreiber MM, White MD, Wing RE, Trimnell D, Shasha BS (1988) Bioactivity of controlled release formulations of starch-encapsulated EPTC. J Controlled Release 7 (3):237–242
49. White MD, Schreiber MM (1984) Herbicidal activity of starch encapsulated trifluralin. Weed Sci 32(3):387–394
50. Appleton B, Derr J (1990) A multiple chemical delivery system for container grown nursery stock, vol 35. In: Proceedings Southern Nurserymen's association research conference, pp 68–70
51. Wells DW, Constantin RJ, Brown WL (1987) Weed control innovations for large, container grown ornamentals. In: Proceedings—southern weed science society 40: 137
52. Danielson LL (1967) Evaluation of herbicide-impregnated cloth. Weeds 15(1):60–62
53. Hamill AS, Layne REC, Von Stryk FG (1975) Weed control in a fruit tree nursery with herbicide-impregnated string [Peaches, apricots]. HortScience (USA)
54. Lanphear FO (1968) Incorporation of dichlobenil in mulches. Weed Sci 16(2):230–231
55. Vasilakoglou IB, Eleftherohorinos IG (2003) Persistence, efficacy, and selectivity of amide herbicides in corn. Weed Technol 17(2):381–388
56. Gorski SF (1993) Slow-release delivery system for herbicides in container-grown stock. Weed Technol 7(4):894–899
57. Johnson RM, Pepperman AB (1996) Leaching of alachlor from alginate-encapsulated controlled-release formulations. Pestic Sci 48(2):157–164
58. Mullin CA, Fine JD, Reynolds RD, Frazier MT (2016) Toxicological risks of agrochemical spray adjuvants: organosilicone surfactants may not be safe. Frontiers Publ Health 4:92
59. Kaiser H (2014) Stomatal uptake of mineral particles from a sprayed suspension containing an organosilicone surfactant. J Plant Nutr Soil Sci 177(6):869–874
60. Xiang Y, Wang M, Sun X, Cai D, Wu Z (2014) Controlling pesticide loss through nanonetworks. ACS Sustain Chem Eng 2(4):918–924

61. Wang M, Sun X, Zhong N, Cai D, Wu Z (2015) Promising approach for improving adhesion capacity of foliar nitrogen fertilizer. ACS Sustain Chem Eng 3(3):499–506
62. Cai D, Wang L, Zhang G, Zhang X, Wu Z (2013) Controlling pesticide loss by natural porous micro/nano composites: straw ash-based biochar and biosilica. ACS Appl Mater Interfaces 5(18):9212–9216
63. Rudzinski WE, Dave AM, Vaishnav UH, Kumbar SG, Kulkarni AR, Aminabhavi TM (2002) Hydrogels as controlled release devices in agriculture. Des Monomers Polym 5 (1):39–65
64. Remuñán-López C. y Alonso-Fernández MJ (1997) Microencapsulación de medicamentos. Tecnología farmacéutica. Formas farmacéuticas, 1, pp 577–609
65. Arshady R (eds). (1999a) Microspheres microcapsules & liposomes: preparation & chemical applications. Citus Books
66. Arshady R (1999b) Manufacturing methodology of microcapsules. MML SERIES, 1: 279
67. Faria DM, Junior D, Macias S, Nascimento JPLD, Nunes EDS, Marques RP, … Moreto JA (2017) Development and evaluation of a controlled release system of TBH herbicide using alginate microparticles. Mater Res 20(1): 225–235
68. Zhu Z, Zhuo R (2001) Controlled release of carboxylic-containing herbicides by starch-g-poly (butyl acrylate). J Appl Polym Sci 81(6):1535–1543
69. Shasha BS, Trimnell D, Otey FH (1984) Starch–borate complexes for EPTC encapsulation. J Appl Polym Sci 29(1):67–73
70. Wing RE, Otey FH (1983) Determination of reaction variables for the starch xanthide encapsulation of pesticides. J Polym Sci Polym Chem Ed 21(1):121–140
71. Wing RE, Shasha BS (1983) Encapsulation of organic chemicals within starch matrix: an undergraduate laboratory experiment. J Chem Educ 60(3):247
72. Shasha BS, Trimnell D, Otey FH (1981) Encapsulation of pesticides in a starch- calcium adduct. J Polym Sci Polym Chem Ed 19(8):1891–1899
73. Wing RE, Maiti S, Doane WM (1987) Factors affecting release of butylate from calcium ion-modified starch-borate matrices. J Controlled Release 5(1):79–89
74. Trimnell D, Shasha BS, Wing RE, Otey FH (1982) Pesticide encapsulation using a starch–borate complex as wall material. J Appl Polym Sci 27(10):3919–3928
75. Wing RE, Carr ME, Doane WM, Schreiber MM (1992) Starch encapsulated herbicide formulations: scale-up and laboratory evaluations. In: Pesticide formulations and application systems, vol 11. ASTM International, Pennsylvania
76. Trimnell D, Shasha BS (1988) Entrapment of herbicides in starch for spray applications. J Controlled Release 7(3):263–268
77. Tefft J, Friend DR (1993) Controlled release herbicide formulations based on polymeric microspheres. J Controlled Release 27(1):27–35
78. Campos EVR, de Oliveira JL, Fraceto LF (2014) Applications of controlled release systems for fungicides, herbicides, acaricides, nutrients, and plant growth hormones: a review. Adv Sci Eng Med 6(4):373–387
79. Junior SD, Nunes ES, Marques RP, Rossino LS, Quites FJ, Siqueira JR, Moreto JA (2017) Controlled release behavior of sulfentrazone herbicide encapsulated in Ca-ALG microparticles: preparation, characterization, mathematical modeling and release tests in field trial weed control. J Mater Sci 52(16):9491–9507
80. Mehltretter CL, Roth WB, Weakley FB, McGuire TA, Russell CR (1974) Potential controlled-release herbicides from 2, 4-D esters of starches. Weed Sci 22(5):415–418
81. Zhila N, Murueva A, Shershneva A, Shishatskaya E, Volova T (2017) Herbicidal activity of slow-release herbicide formulations in wheat stands infested by weeds. J Environ Sci Health Part B 52(10):729–735
82. Li J, Li Y, Dong H (2008) Controlled release of herbicide acetochlor from clay/carboxyl methylcellulose gel formulations. J Agric Food Chem 56(4):1336–1342
83. Khanh TD, Chung IM, Tawata S, Xuan TD (2007) Allelopathy for weed management in sustainable agriculture. Health 1(2)
84. Zeng RS (2014) Allelopathy-the solution is indirect. J Chem Ecol 40(6):515

85. Anonymous (1996) International allelopathy society. First World Congress on allelopathy: a science for future, Cadiz, Spain
86. Macias FA, Oliva RM, Simonet AM, Galindo JCG (1998) What are allelochemicals. In: Allelopathy in rice. Proceedings of the workshop on allelopathy in rice, 25–27 Nov 1996, pp 69–79
87. Macías FA, Marín D, Oliveros-Bastidas A, Varela RM, Simonet AM, Carrera C, Molinillo JM (2003) Allelopathy as a new strategy for sustainable ecosystems development. Biol Sci Space 17(1):18–23
88. Duke SO (1986) Naturally occurring chemical compounds as herbicides. Rev Weed Sci 2:17–65
89. Kim KU (1993) Integrated management of paddy weeds in Korea, with an emphasis on allelopathy. Allelopathy in the control of paddy weeds. Technical Bulletin 134:8–23
90. Smith RJ (1993) Biological controls as components of integrated weed management for rice in the United States. ASPAC, Food & Fertilizer Technology Center
91. Breland TA (1996) Phytotoxic effects of fresh and decomposing cover crop residues. Norwegian J Agric Sci Norway 10:355–362
92. O'connell PF (1992) Sustainable agriculture-a valid alternative. Outlook Agric 21(1):5–12
93. Anaya-Lang AL (1989) The role the allelochemicals in the management of natural resources. Botanical Sci 49:85–98
94. Duke SO, Lydon J (1987) Herbicides from natural compounds. Weed Technol 1(2):122–128
95. Towers GHN, Arnason JT (1988) Photodynamic herbicides. Weed Technol 2(4):545–549
96. Macias FA (1995) Allelopathy in the search for natural herbicide models
97. Nie C, Luo S, Zeng R, Mo M, Li H, Lin C (2005) Allelopathic potential of Wedelia trilobata L.: effects on germination, growth and physiological parameters of rice. In: 4th World Congress on allelopathy
98. Rensen Z, Xianglian L, Shiming L (1994) Studies on the allelopathic effects of Wedelia Chinensis aqueous extraction. J South China Agric Univ 4: 26–30
99. Luo SM, Lin XL, Zeng RS, Kong CH, Cao PR, Wei Q, Deng LG (1995) Allelopathy of typical plants in agroecosystem of South China. Ecol Sci 2:114–128
100. Laughlin RG, Munyon RL, Ries SK, Wert VF (1983) Growth enhancement of plants by femtomole doses of colloidally dispersed triacontanol. Science 219(4589):1219–1221
101. Maugh TH (1981) The natural occurring brassionoide in the plant species. Science 212:33–34
102. Rizvi SJH, Rizvi V (1992) Exploitation of allelochemicals in improving crop productivity. Allelopathy. Springer, Dordrecht, pp 443–472
103. Worsham AD (1989) Current and potential technique using allelopathy as an aid in weed management. Phytochemical ecology: allelochemicals, mycotoxins, insect pheromones and allomones
104. Gleason FK (1990) The natural herbicide, Cyanobacterin, specifically disrupts thylakoid membrane structure in Euglena gracilis strain Z. FEMS Microbiol Lett 68(1–2):77–81
105. Duke SO, Vaughn KC, Croom EM, Elsohly HN (1987) Artemisinin, a constituent of annual wormwood (Artemisia annua), is a selective phytotoxin. Weed Sci 35(4):499–505
106. Einhellig FA (1996) Interactions involving allelopathy in cropping systems. Agron J 88 (6):886–893
107. Collins RL, Doglia S, Mazak RA, Samulski ET (1973) Controlled release of herbicides—theory. Weed Sci 21(1):1–5
108. Einhellig FA (2018) Allelopathy—a natural protection, allelochemicals. Handbook of natural pesticides: methods. CRC Press, Boca Raton, FL, pp 161–200

Chapter 11
Controlled Release of Plant Hormones for Modifying Crop Yield

Vinaya Chandran, S. Shahena, Maya Rajan and Linu Mathew

Abstract Plant hormones have been an important component in crop improvement because they can in some way manipulate or modify plant development. Controlled application of phytohormones to crop plants alters their life processes or growth pattern to enhance yield, improve quality and facilitate better post-harvest life. Synthetic hormones those mimic natural plant hormones were used significantly and they are a major component in modern agriculture. Also, they are extensively used in plant tissue culture, as weedicides, for initiating uniform flowering parthenocarpic fruit development and even fruit set. They also have role in stress alleviation and plant's adaptation to stress.

Keywords Auxins · Cytokinins · Gibberellins · Ethylene · Abscisic acid

11.1 Introduction

Production of quality agricultural and horticultural products at quantities sufficient to the ever-increasing need is the basic purpose of farming practices. However, agricultural production not always matches the increasing demand made by human and livestock. Thus, there is always a need to improve agricultural productivity. Many approaches are tried by farmers and scientists to increase crop yield. One such approach is the use of plant hormones to augment productivity. Plant hormones affect many facets of plant life including growth, flowering, fruiting, etc. They are termed as plant growth regulators because they act by controlling or modifying plant growth processes.

V. Chandran · S. Shahena · M. Rajan · L. Mathew (✉)
School of Biosciences, Mahatma Gandhi University, Kottayam, Kerala 686-560, India
e-mail: linumathew@mgu.ac.in

Rakhimol K. R. et al. (eds.), *Controlled Release of Pesticides for Sustainable Agriculture*, https://doi.org/10.1007/978-3-030-23396-9_11

11.2 Plant Hormones

The German Botanist Julius von Sachs (1832–1897) described that certain chemical messengers are responsible for the growth and development of different plant organs. These signalling molecules were described as plant hormones [1]. They have profound effect on regulation and co-ordination of plant growth, morphogenesis and metabolism. They are defined as organic substances produced naturally in plants and controls growth and other physiological functions [2]. The exact location of synthesis of plant hormones is uncertain but actively growing tissues, leaves and developing seeds are capable of hormone production. However, it depends on plant species and climatic factors and varies according to the season [3]. Phytohormones regulate or control all aspect of growth and development from embryogenesis to reproduction. They act at genetic level [4, 5] to stimulate or inhibit specific enzyme or enzyme system that regulate plant metabolism. They act through the modification of transcription, translation and/or differential sensitivity of the tissue [6]. Generally, phytohormones required for growth and development are synthesized endogenously in plants [7]. Nevertheless, they could be applied exogenously for exploiting the full genetic potential of plants for crop production. There are five major classes of hormones that occur within the plants at extremely low concentration. These are auxins, cytokinins, ethylene, gibberellic acids and abscisic acid [8]. Some steroid hormones are also present in plants; for example, brassinosteroids have a wide range of morphological effect on plant growth and development [9].

11.2.1 Auxins

Auxins are a class of plant hormones which control plant growth and development through cell cycling and formation of vascular tissues [10] pollen [11] and embryo [12]. Auxins can enhance the root development of plant cuttings [13]. Stefen et al. (2009) noted that the differential distribution of auxin within plant tissue triggers an impressive variety of developmental processes [14]. Endogenous auxin biosynthesis takes place locally in response to both developmental and environmental signals. De novo auxin biosynthesis plays an essential role in virtually every aspect of developmental processes in plants [15]. Indole-3-acetic acid is the most abundant naturally occurring active auxin in plants [16]. However, it is unstable in heat and light and aqueous solutions are chemically labile. Therefore, its commercial use is limited as a plant growth regulator. Other naturally occurring auxins include 4-chloroindole acetic acid [17], phenylacetic acid [18] and indole-3-butyric acid. IBA is more effective than IAA and efficiently induces lateral roots at concentrations that minimally inhibit root elongation [19]. Hence, it is commonly used as a rooting hormone for accelerating rooting in plant cuttings. Applications of synthetic auxin analogs such as indole acetyl-l-alanine and indole acetyl-l-glycine also alleviate the problem associated with IAA. The uses of such stable chemical analogs are extremely effective in most cases. They cause profound changes in plant growth and development, if supplied externally.

11.2.2 Cytokinins

Cytokinins are specific plant hormones that have the ability to play a central role in regulating cell cycle and numerous other developmental pathways [20]. Cytokinins together with auxins have great influence in plant morphogenesis. Skoog and Miller demonstrated that auxin together with cytokinin play essential role in plant morphogenesis and have profound influence in the formation of root and shoot and their relative growth [21]. Cytokinins are an important regulatory factor of plant meristem activity with opposing role on shoot and root meristems [20]. Cytokinins are initially thought as a substance that stimulates cell division in cultured tobacco cells. However, externally applied cytokinins have broader actions. These include shoot initiation from callus cultures, promotion of axillary bud growth, stimulation of pigment synthesis, direct transport of nutrients and inhibition of root growth and delay of senescence.

11.2.3 Ethylene

Ethylene is a simple naturally occurring gaseous plant hormone having numerous effect on plant life, i.e. growth, development and storage life of many fruits, vegetables and many ornamental crops [22]. Ripened and diseased plant tissues are the main natural sources of ethylene. The hormone has autocatalytic activity, i.e. ethylene synthesis is promoted by itself in mature reproductive tissue and inhibited in immature tissues. Ethylene is biologically effective at very low concentration, i.e. ppm and ppb range. Most plants are continuously synthesizing small amount of ethylene to maintain biologically active levels in tissues. Because it is an organic gas, it readily diffuses from the site of production.

11.2.4 Gibberellic Acid

Gibberellic acids are large group of diterpenoid carboxylic acids functioning as potent growth regulators promoting organ expansion and developmental switches [6]. At present, about 125 gibberellins are known from all sources of plants. They differ from each other by the presence or of the location configuration (internal ester) within the ring A. Among them GA3 is universally accepted as more potent than other derivatives due to the presence of an additional ethylene double bond in ring A [23]. It is more unsaturated and more active than other gibberellins [24–26].

11.2.5 Abscisic Acid

The role of abscisic acid (ABA) as plant stress hormone is well established. ABA control a broad range of processes in plants including seed maturation, maintenance, dormancy, secondary root formation and leaf size [27]. ABA is a well-known stress hormone which improves drought tolerance in plants through osmotic adjustment and improving water relations. Under drought condition, it is synthesized in tissue and sent to the guard cell as a stress signal [28]. It induces the closure of leaf stomata to reduce water loss through transpiration [29]. Exogenous application of abscisic acid will help in conserving water within plants which is necessary for successful crop production under water deficit conditions.

11.3 Effects of Plant Growth Hormone on Plant Physiology

11.3.1 Growth and Development

Micro-application to the apical meristem demonstrates that local increase in auxin is sufficient for initiation and formation of leaves and flowers [30]. It participates in phototropism [31], geotropism and hydrotropism and other developmental changes. When plant competes for sunlight, shade induced changes in light quality triggers auxin biosynthesis that accelerates stem elongation [32]. Benkova et al. (2003) demonstrate that differential auxin activities modulate plant organ formation [33]. In early embryonic stages, the auxin accumulation site will specify the apical cell and late establishment of root pole and cotyledons [34]. Auxins accumulate at the site of organ initiation [35]. Auxin induces root formation by breaking root apical dominance induced by cytokinins. High concentration of auxin inhibits root elongation and promotes adventitious root formation [36]. Auxins may act as both bio-stimulators and bio-inhibitors of plant growth. For example, the low concentration of synthetic plant hormone 2, 4-D stimulates cell enlargement whereas its higher concentration act as a bio-inhibitor. Exogenous auxin stimulates ethylene production [37] through induction of a gene encoding enzyme in ethylene biosynthesis [38]. Auxins inhibit ethylene production in low concentration; however, its high doses stimulate ethylene production which inhibits cell elongation and leaf abscission. Therefore, high concentration of auxin is used for inducing femaleness in flowers. Exogenous application of auxins delays fruit senescence and stimulates the growth in seed removed fruits. In combination with ethylene, auxin elicits gibberellic acid (GA) production [39]. For example, auxin is necessary for the increased production of the active gibberellins GA1 and GA3 in barley [40].

Cytokinins are believed to play regulatory role in plant growth and differentiation [41]. They promote cell division, nutrient mobilization and leaf longevity [42–47]. Cytokinins are a plant growth regulator found in tissue where intense cell division is occurring in seeds, root extremities, leaves and developing fruits [8].

In horticulture, auxins—NAA and IBA—are commonly applied to stimulate root growth. In tissue culture, auxins promote rooting from undifferentiated callus [21]. The auxin and cytokinin combination in vitro, induce root and shoot development, respectively, and are inversely correlated in vivo [48]. The auxin treatment can rapidly inhibit cytokinin biosynthesis [49]. Cytokinins are an important regulatory factor of plant meristem activity with opposing role of shoot and root [20]. Numerous reports ascribe a stimulatory or inhibitory function to cytokinins in different developmental processes such as root growth and branching control of apical dominance in the shoot, chloroplast development and senescence [50]. Application of benzyladenine and a cytokinin has been reported to stimulate the growth of axillary buds in different ornamental bulbous plants including tulips [51], lilies [52] and calla bulbs [53]. Cytokinin also increases grain yield, by activating inflorescence meristem activity [54, 55]. External application of kinetin stimulates tiller bud growth in wheat [56] and rice [57].

The influence of gibberellins includes promotion of cell growth, elongation and growth of whole plant, parthenocarpic fruit development, senescence and increase cell wall plasticity [24, 5]. The obvious GA-regulated growth responses in vegetative tissues is the elongation growth of shoots, particularly stem internodes [58]. Other cell growth responses to GA occur only during specific stages of growth of the plant. For example, the transition from juvenile to adult leaf shape is regulated by GA in a range of plant species, such as sweet pea and tomato. Cell growth polarity in root tips and shoot apical meristems is also influenced by GAs. During fruit set [59], GA may stimulate pollen germination and tube growth. It delays post-harvest senescence in various crops, including roses [60] and citrus [61]. Gibberellins promote degradation of DELLA repressors [62], disrupting auxin transport [63].

11.3.2 Crop Yield

Ethylene has key regulatory role in the ripening of fruits and its associated events such as altered sugar metabolism, texture, synthesis of aroma and softening of fruits [64]. The field application ethylene became practical with the development of ethylene releasing chemicals like ethephon (2-chloroethyl-phosphonic acid). Ethephon has been used on cereals, fruits (apples, cherries, citrus, guava, cucumbers, pineapple and tomatoes), cotton, peppers, walnut, coffee, olives and sweet potatoes [65]. It has beneficial effect on colour development in fresh fruits and vegetables. For example, accumulation of red colour is enhanced through stimulation of anthocyanin synthesis in apples and tomatoes and yellowing or degreening in bananas and citrus through chlorophyll destruction. Ethylene has been used in liquid forms to effect seed germination, bulb sprouting and reduce apical dominance, to promote or delay flowering, to alter sex expression, to control insects and pests [66]. One of the major uses of gibberellins is to manage fruit crops to increasing sugar yield and fruit production in some seedless varieties [67].

Application of gibberellic acid can reduce the time of flowering and enhance the yield of flowering stem in *Limonium x* 'misty blue' and may partially replace vernalization treatment this hybrid [68]. The efficiency of this type of growth promoter depends on application timing [69]. It also has been used to increase the number of flowers [70] and promote flowering in a variety of plant species under non-inductive conditions [71, 72]. The exogenous application of gibberellic acid has been reported numerous times in various ornamental plants [73] such as tulip [74], dahlia [75], lily [76] and iris [77].

11.3.3 Seed Dormancy and Germination

Abscisic acid is the hormone that plays a prominent role in dormancy and germination control [78, 79]. It enhances dormancy [80]. The function of ABA during seed development is the induction of primary dormancy and inhibition of precocious germination [81]. Several plant hormones are necessary and sufficient to control all aspects of dormancy and germination [82]. However, the ABA/ gibberellin (GA) combination is involved in these regulatory mechanisms [83]. Certain types of dormancy including physiological dormancy, photodormancy and thermodormancy are arrested by the exogenous application of GA3 [84] and are used to increase the length or height of plants.

11.3.4 Stress Tolerance

Abscisic acid is associated with genetic difference in drought tolerant and drought sensitive plants [85]. Drought tolerance of maize [86], chilling tolerance of rice seedlings [87] and salt tolerance of rice [88] may be associated with increase of ABA in response to various environmental stresses. In sunflower, stomatal control depends on the concentration of abscisic acid in xylem sap [89]. Abscisic acid has pronounced effects on proline content and water status of different genotypes of sunflower. Unyayar et al. (2004) studied the effect of exogenous application of ABA on sunflower genotypes. Leaf area and relative leaf water content decreased under drought stress but in ABA-treated leaves relative water content had lower values than in non-treated leaves [90]. Moreover, cytokinin accumulation of *Nicotina tobacum* leads to tolerance against extreme drought stress [91]. The GA is also important phytohormones that are actively involved in regulating plant response to various stresses in many crop plants. Exogenous application of GA may increase crop growth under stressful environment. For example, Maggio et al. reported that treatment with GA decreased stomatal resistance and improved efficiency of plant water use at lower salinity level in tomato plants [92].

11.3.5 Herbicidal Action

Synthetic auxins are used not only as plant growth regulators but also as herbicides for controlling weeds [93, 94]. 2, 4-D alone is widely used as an herbicide. The infamous defoliant herbicide 'Agent Orange' was a mixture of 2, 4-D and 2, 4, 5-trichlorophenoxy butyric acid [95]. Recently, various synthetic auxin herbicides, with different weed spectra and types of selectivity, have been commercially introduced. These classes include phenoxy carboxylic acids, pyridine carboxylic acids benzoic acids, aromatic carboxymethyl derivatives and quinolone-carboxylic acids [96].

11.4 Factors Affecting the Efficiency of Plant Hormones

Effect of plant hormone depends on various factors including the application method, time of application, concentration of hormones, plant species and environmental conditions [97, 98]. The intensity of application is also an important factor affecting the efficacy of plant hormones. Some plants respond to a single hormone application, but in most cases, multiple applications are beneficial to attain good results [99]. There are various reported methods of commercial application of plant hormone, mostly including foliar application [73, 100], drenching [43], pre-plant sowing [76], pasting [101], capillary string [102] and injection [103]. Generally, pre-sowing seed treatment or through foliar application as dilute solutions at crucial stages is more effective [104–106]. This method is more promising due to many factors, including the small amount of the hormone required and low operation cost involved. Pre-plant soaking of corms inaccurate dosage of phytohormones is an efficient method for obtaining good results and also has advantages over other methods in terms of time and labour saving [107].

The effect of plant hormone is cultivar dependent which might be due to the genotype factor of the cultivar and the endogenous hormone level [108]. The application of NAA at 50 ppm on ridge guard shows higher germination percentage among all other hormone treatment during summer and kharif season [109]. In Okra, IAA/NAA at 20 ppm levels enhanced seed germination. Soaking treatment in ethephon at 480 mg/l for 24 h improved seed emergence in bottle guard, squash melon and watermelon at low temperature [110]. Controlled application of plant hormone regulates physiological processes of crop plant which are found to be beneficial for overall crop yield. It was reported that NAA sprayed at 100 ppm and GA3 at 80 ppm gave maximum yield of tomato [111]. NAA at 75 ppm were used after 30 and 60 days of transplanting gave maximum yield per chilli plant [112]. Singh et al. (2016) reported that GA3 at 200 ppm gave increased height of plant, number of branches and leaves, length of pod and seed weight on garden pea. The experiment also suggested that foliar application of hormone was fruitful to garden pea for obtaining better yield [113]. Foliar application of Ethrel at 250 ppm on potato was effective in changing total yield of the tuber [114]. Patel et al. recorded that root dipping treatment of NAA at 100 ppm on onion reduced the loss of spoilage [115]. GA3 at 50 ppm gave significant germination in asparagus seeds but as the concentration increased the germination decreased rapidly and vigour index also decreased during light and dark period [116]. The application of NAA at 15 ppm [117] significantly improved the growth yield and quality of strawberry. Hilli et al. (2008) indicated that 2, 4-D at 2 ppm gave better yield in ridge guard among other hormone treatment [109]. A combined dose of GA3 20 ppm + NAA 100 ppm showed significant increment in morphological and yield characters on cucumber [118].

The application of exogenous hormone may alter the concentration of some endogenous plant hormone. Application of ethephon an ethylene releasing compound at 100 mg/l at turning stage of earliest fruits induced fruit ripening thus increasing the early fruit yield by 30–35% [110]. Hilli et al. (2008) reported that plant growth regulators were ineffective in promoting flowering and fruiting in winter season [109]. However, the controlled application of phytohormone provides quantifiable advantages over crop improvement. For this, they must be specific in their action, less time-consuming and environmentally safe. For example, exogenous applications at low concentrations of Brassinosteroids are more effective under stress by enhancing physiological processes and improving tolerance of rice [119] and barley [120].

11.5 Conclusion

Plant hormones have been an important component in agricultural productivity by influencing the rate of plant growth and development. Their use is encouraged in many areas in agriculture, horticulture, pomiculture and moriculture for successful cultivation and greater yield. They are also helpful in altering various growth characteristics. The usage of ecologically safe formulation of synthetic plant hormones will enhance their acceptability by the consumers. Hormones like NAA and IBA are very effective in inducing roots in vegetative propagation stem cuttings. These are also be used for in grafting propagation. This way most of the plants can be vegetatively propagated in large numbers in a short time. Many horticultural plants are propagated by this way. The overall growth of plants, number of tillers, branches and leaf area of every plant in the field contributes to the total yield. Use of GA, NAA, IAA, etc. increases the percentage of germination and also greatly enhances the total growth of plants. Gibberellins ensure fruit setting and prevent premature falling of fruits. Fruits that develop from such treatments are seedless, larger in size and sweeter in content. GA is also used in inducing non-seasonal flowering. Specific synthetic hormones such as 2, 4-D, 2, 4, 5-T are available in the market to destroy the weeds selectively. Let us summarize that all these phytohormones are directly or indirectly ameliorate plant physiological responses in different crops. In many cases, seed pretreatment or exogenous applications at low concentration are more effective in modulating overall crop improvement.

References

1. Hopkins WG (1999) Introduction to plant physiology, 2nd edn. Wiley, London
2. Thimann KV, Pincus G (eds) (1948) The hormones, physiology, chemistry and applications. Academic Press
3. Weiler EW, Ziegler H (1981) Determination of phytohormones in phloem exudate from tree species by radioimmunoassay. Planta 152(2):168–170
4. Bajguz A (2000) Effect of brassinosteroids on nucleic acids and protein content in cultured cells of Chlorella vulgaris. Plant Physiol Biochem 38(3):209–215
5. Shahbaz M, Ashraf M (2008) Does exogenous application of 24-epibrassinolide ameliorate salt induced growth inhibition in wheat (Triticum aestivum L.)? Plant Growth Regul 55 (1):51–64
6. Khan F, Mazid M, Khan TA, Quddasi S, Roychowdhury R, Naqvi N (2013) Application of sesquiterpene (GA3) to spermology: a contradictory report. Res J Biol 1:45–51
7. Berg G (2009) Plant–microbe interactions promoting plant growth and health: perspectives for controlled use of microorganisms in agriculture. Appl Microbiol Biotechnol 84(1):11–18
8. Campos EVR, de Oliveira JL, Fraceto LF (2014) Applications of controlled release systems for fungicides, herbicides, acaricides, nutrients, and plant growth hormones: a review. Adv Sci Eng Med 6(4):373–387
9. Dewitte W, Van Onckelen H (2001) Probing the distribution of plant hormones by immunocytochemistry. Plant Growth Regul 33(1):67–74
10. Davies PJ (2010) The plant hormones: their nature, occurrence, and functions. In: Plant hormones. Springer, Dordrecht, pp 1–15
11. Ni DA, Yu XH, Wang LJ, Xu ZH (2002) Aberrant development of pollen in transgenic tobacco expressing bacterial iaaM gene driven by pollen-and tapetum-specific promoters. Shi Yan Sheng Wu Xue Bao 35(1):1–6
12. Benjamins R, Scheres B (2008) Auxin: the looping star in plant development. Annu Rev Plant Biol 59:443–465
13. He Z, Wang ZY, Li J, Zhu Q, Lamb C, Ronald P, Chory J (2000) Perception of brassinosteroids by the extracellular domain of the receptor kinase BRI1. Science 288 (5475):2360–2363
14. Vanneste S, Friml J (2009) Auxin: a trigger for change in plant development. Cell 136 (6):1005–1016
15. Zhao Y (2010) Auxin biosynthesis and its role in plant development. Annu Rev Plant Biol 61:49–64
16. Simon S, Petrášek J (2011) Why plants need more than one type of auxin. Plant Sci 180 (3):454–460
17. Slovin JP, Bandurski RS, Cohen JD (1999) Auxin. In: New Comprehensive Biochemistry, vol 33. Elsevier, Amsterdam, pp 115–140

18. Wightman F (1977) Gas chromatographic identification and quantitative estimation of natural auxins in developing plant organs. In: Plant growth regulation. Springer, Berlin, Heidelberg, pp 77–90
19. Zolman BK, Yoder A, Bartel B (2000) Genetic analysis of indole-3-butyric acid responses in Arabidopsis thaliana reveals four mutant classes. Genetics 156(3):1323–1337
20. Werner T, Motyka V, Strnad M, Schmülling T (2001) Regulation of plant growth by cytokinin. Proc Natl Acad Sci 98(18):10487–10492
21. Skoog F, Miller C (1957) Chemical regulation of growth and organ formation in plant tissues cultured in vitro. Symp Soc Exp Biol (11)
22. Saltveit ME (1999) Effect of ethylene on quality of fresh fruits and vegetables. Postharvest Biol Technol 15(3):279–292
23. Moore TC (2012) Biochemistry and physiology of plant hormones. Springer, New York
24. Salisbury FB, Ross CW (1992) Plant Physiology, 4th edn. Wadsworth Inc., Belmont, California, pp 27–65
25. Kumar A, Purohit SS (2014) Plant Physiology: fundamentals and applications. Student Edition
26. Camara MC, Vandenberghe LP, Rodrigues C, De Oliveira J, Faulds C, Bertrand E, Soccol CR (2018) Current advances in gibberellic acid (GA 3) production, patented technologies and potential applications. Planta 248(5):1049–1062
27. Kepka M, Benson CL, Gonugunta VK, Nelson KM, Christmann A, Grill E, Abrams SR (2011) Action of natural abscisic acid precursors and catabolites on abscisic acid receptor complexes. Plant Physiol 157(4):2108–2119
28. Hussain S, Ali A, Ibrahim M, Saleem MF, Bukhsh MAAHA (2012) Exogenous application of abscisic acid for drought tolerance in sunflower (Helianthus annuus L.): a review. J Anim Plant Sci 22(3):806–826
29. Wasilewska A, Vlad F, Sirichandra C, Redko Y, Jammes F, Valon C, dit Frey NF, Leung J (2008) An update on abscisic acid signaling in plants and more…. Mol Plant 1(2):198–217
30. Reinhardt D, Pesce ER, Stieger P, Mandel T, Baltensperger K, Bennett M, Traas J, Friml J, Kuhlemeier C (2003) Regulation of phyllotaxis by polar auxin transport. Nature 426 (6964):255
31. Muday GK (2001) Auxins and tropisms. J Plant Grow Regul 20(3):226–243
32. Tao Y, Ferrer JL, Ljung K, Pojer F, Hong F, Long JA, Li L, Moreno JE, Bowman ME, Ivans LJ, Cheng Y (2008) Rapid synthesis of auxin via a new tryptophan-dependent pathway is required for shade avoidance in plants. Cell 133(1):164–176
33. Benková E, Michniewicz M, Sauer M, Teichmann T, Seifertová D, Jürgens G, Friml J (2003) Local, efflux-dependent auxin gradients as a common module for plant organ formation. Cell 115(5):591–602
34. Friml J, Vieten A, Sauer M, Weijers D, Schwarz H, Hamann T, Offringa R, Jürgens G (2003) Efflux-dependent auxin gradients establish the apical–basal axis of Arabidopsis. Nature 426(6963):147
35. Dubrovsky JG, Sauer M, Napsucialy-Mendivil S, Ivanchenko MG, Friml J, Shishkova S, … Benková E (2008) Auxin acts as a local morphogenetic trigger to specify lateral root founder cells. Procee Nat Acad Sci 105(25):8790–8794
36. Zimmerman PW, Hitchcock AE (1942) Substituted phenoxy and benzoic acid growth substances and the relation of structure to physiological activity. Contrib Boyce Thompson Inst 12(5):321–344
37. Morgan PW, Hall WC (1962) Effect of 2, 4-dichlorophenoxyacetic acid on the production of ethylene by cotton and grain sorghum. Physiol Plant 15(3):420–427
38. Abel S, Nguyen MD, Chow W, Theologis A (1995) ASC4, a primary indoleacetic acid-responsive gene encoding 1-aminocyclopropane-1-carboxylate synthase in Arabidopsis thaliana structural characterization, expression in Escherichia coli, and expression characteristics in response to auxin. J Biol Chem 270(32):19093–19099
39. Ross JJ, O'neill DP, Smith JJ, Kerckhoffs LHJ, Elliott RC (2000) Evidence that auxin promotes gibberellin A_1 biosynthesis in pea. Plant J 21(6):547–552

40. Wolbang CM, Chandler PM, Smith JJ, Ross JJ (2004) Auxin from the developing inflorescence is required for the biosynthesis of active gibberellins in barley stems. Plant Physiol 134(2):769–776
41. Finkelstein RR (2004) The role of hormones during seed development and germination. In: Davies PJ (ed) Plant hormones: biosynthesis, signal transduction, action, pp 513–537
42. Choi J, Hwang I (2007) Cytokinin: perception, signal transduction, and role in plant growth and development. J Plant Biol 50(2):98–108
43. Kim HJ, Ryu H, Hong SH, Woo HR, Lim PO, Lee IC, Sheen J, Nam HG, Hwang I (2006) Cytokinin-mediated control of leaf longevity by AHK3 through phosphorylation of ARR2 in Arabidopsis. Proc Natl Acad Sci 103(3):814–819
44. Matsumoto TK (2006) Gibberellic acid and benzyladenine promote early flowering and vegetative growth of Miltoniopsis orchid hybrids. HortScience 41(1):131–135
45. Sakakibara H (2006) Cytokinins: activity, biosynthesis, and translocation. Annu Rev Plant Biol 57:431–449
46. Leibfried A, To JP, Busch W, Stehling S, Kehle A, Demar M, Kieber JJ, Lohmann JU (2005) WUSCHEL controls meristem function by direct regulation of cytokinin-inducible response regulators. Nature 438(7071):1172
47. Lara MEB, Garcia MCG, Fatima T, Ehneß R, Lee TK, Proels R, Tanner W, Roitsch T (2004) Extracellular invertase is an essential component of cytokinin-mediated delay of senescence. Plant Cell 16(5):1276–1287
48. Eklöf S, Åstot C, Sitbon F, Moritz T, Olsson O, Sandberg G (2000) Transgenic tobacco plants co-expressing *Agrobacterium iaa* and *ipt* genes have wild-type hormone levels but display both auxin-and cytokinin-overproducing phenotypes. Plant J 23(2):279–284
49. Nordström A, Tarkowski P, Tarkowska D, Norbaek R, Åstot C, Dolezal K, Sandberg G (2004) Auxin regulation of cytokinin biosynthesis in Arabidopsis thaliana: a factor of potential importance for auxin–cytokinin-regulated development. Proc Natl Acad Sci 101 (21):8039–8044
50. Mok MC (1994) Cytokinins and plant development. Cytokinins: chemistry, activity, and function. CRC Press, Boca Raton, 155–166
51. De Munk WJ, Gijzenberg J (1977) Flower-bud blasting in tulip plants mediated by the hormonal status of the plant. Sci Hortic 7(3):255–268
52. Ranwala AP, Miller WB (1998) Gibberellin4 + 7, benzyladenine, and supplemental light improve postharvest leaf and flower quality of cold-storedstargazer'hybrid lilies. J Am Soc Hortic Sci 123(4):563–568
53. Naor V, Kigel J, Ziv M (2004, April) The effect of gibberellin and cytokinin on floral development in Zantedeschia spp. in vivo and in vitro. In: IX International Symposium on Flower Bulbs 673, pp 255–263
54. Kurakawa T, Ueda N, Maekawa M, Kobayashi K, Kojima M, Nagato Y, Sakakibara H, Kyozuka J (2007) Direct control of shoot meristem activity by a cytokinin-activating enzyme. Nature 445(7128):652
55. Ashikari M, Sakakibara H, Lin S, Yamamoto T, Takashi T, Nishimura A, Angeles ER, Qian Q, Kitano H, Matsuoka M (2005) Cytokinin oxidase regulates rice grain production. Science 309(5735):741–745
56. Langer RHM, Prasad PC, Laude HM (1973) Effects of kinetin on tiller bud elongation in wheat (Triticum aestivum L.). Ann Bot 37(3):565–571
57. Liu Y, Gu D, Ding Y, Wang Q, Li G, Wang S (2011) The relationship between nitrogen, auxin and cytokinin in the growth regulation of rice ('Oryza sativa'l.) Tiller buds. Aust J Crop Sci 5(8):1019
58. Hooley R (1994) Gibberellins: perception, transduction and responses. In: Signals and signal transduction pathways in plants. Springer, Dordrecht, pp 293–319
59. Taylor LP, Hepler PK (1997) Pollen germination and tube growth. Annu Rev Plant Biol 48 (1):461–491
60. Goszczynska DM, Zieslin N, Mor Y, Halevy AH (1990) Improvement of postharvest keeping quality of 'Mercedes' roses by gibberellin. Plant Growth Regul 9(4):293–303

61. McDonald RE, Greany PD, Shaw PE, McCollum TG (1997) Preharvest applications of gibberellic acid delay senescence of florida grapefruit. J Hortic Sci 72(3):461–468
62. Silverstone AL, Jung HS, Dill A, Kawaide H, Kamiya Y, Sun TP (2001) Repressing a repressor: gibberellin-induced rapid reduction of the RGA protein in Arabidopsis. Plant Cell 13(7):1555–1566
63. Fu X, Harberd NP (2003) Auxin promotes Arabidopsis root growth by modulating gibberellin response. Nature 421(6924):740
64. Barry CS, Giovannoni JJ (2007) Ethylene and fruit ripening. J Plant Growth Regul 26 (2):143
65. Watada AE (1986) Effects of ethylene on the quality of fruits and vegetables. Food Technol 40(5):82–85
66. Abeles FB (1966) Auxin stimulation of ethylene evolution. Plant Physiol 41(4):585–588
67. Pandolfini T (2009) Seedless fruit production by hormonal regulation of fruit set. Nutrients 1 (2):168–177
68. Garner JM, Armitage AM (1996) Gibberellin applications influence the scheduling and flowering of Limonium × Misty Blue'. HortScience 31(2):247–248
69. Schmidt CM, Bellé RA, Nardi C, Toledo KDA (2003) The gibberelic acid (GA3) in the cut chrysanthemum (DedranthemagrandifloraTzevelev.)'viking': planting summer/autumn. Ciência Rural 33(2):267–274
70. Medina EO, Saavedra AL (2005) El uso de regulador de crecimiento en la floricultura mexicana. Ciência y Desarrollo 148–26
71. Zeevaart JAD (1983) The biochemistry and physiology of gibberellins. Gibberellins and flowering. New York, 333–374
72. Harkess RL, Lyons RE (1994) Gibberellin-and cytokinin-induced growth and flowering responses in Rudbeckia hirta l. HortScience 29(3):141–142
73. Sajjad Y, Jaskani MJ, Ashraf MY, Qasim M, Ahmad R (2014) Response of morphological and physiological growth attributes to foliar application of plant growth regulators in gladiolus'white prosperity'. Pak J Agric Sci 51(1)
74. Ramzan F, Younis A, Riaz A, Ali S, Siddique MI, Lim KB (2014) Pre-planting exogenous application of gibberellic acid influences sprouting, vegetative growth, flowering, and subsequent bulb characteristics of 'Ad-Rem'tulip. Hortic Environ Biotechnol 55(6):479–488
75. Mahgoub MH, El-Aziz NA, Mazhar AMA (2011) Response of Dahlia pinnata L. plant to foliar spray with putrescine and thiamine on growth, flowering and photosynthetic pigments. Am-Eurasian J Agric Environ Sci 10(5):769–775
76. Currey CJ, Lopez RG (2010) Paclobutrazol pre-plant bulb dips effectively control height of 'Nellie White'easter lily. HortTechnology 20(2):357–360
77. Leason T, Harkess RL (2006) Influence of Cytokinins on lateral branching of Iris germanica rhizomes, vol 50. In: Proceedings Southern Nursery Association Research Conferences, pp 300–301
78. Hilhorst HW (1995) A critical update on seed dormancy. Seed Sci Res 5(2):61–73
79. Kermode AR (2005) Role of abscisic acid in seed dormancy. J Plant Growth Regul 24 (4):319–344
80. Thompson AJ, Andrews J, Mulholland BJ, McKee JM, Hilton HW, Horridge JS, Farquhar GD, Smeeton RC, Smillie IR, Black CR, Taylor IB (2007) Overproduction of abscisic acid in tomato increases transpiration efficiency and root hydraulic conductivity and influences leaf expansion. Plant Physiol 143(4):1905–1917
81. Nambara E, Okamoto M, Tatematsu K, Yano R, Seo M, Kamiya Y (2010) Abscisic acid and the control of seed dormancy and germination. Seed Sci Res 20(2):55–67
82. Kucera B, Cohn MA, Leubner-Metzger G (2005) Plant hormone interactions during seed dormancy release and germination. Seed Sci Res 15(4):281–307
83. Karssen CM, Brinkhorst-Van der Swan DLC, Breekland AE, Koornneef M (1983) Induction of dormancy during seed development by endogenous abscisic acid: studies on abscisic acid deficient genotypes of Arabidopsis thaliana (L.) Heynh. Planta 157(2):158–165

84. Hartmann HT (1990) Anatomical and physiological basis of propagation by cuttings. Plant Propag Princ Pract 199–255
85. Ouvrard O, Cellier F, Ferrare K, Tousch D, Lamaze T, Dupuis JM, Casse-Delbart F (1996) Identification and expression of water stress-and abscisic acid-regulated genes in a drought-tolerant sunflower genotype. Plant Mol Biol 31(4):819–829
86. Pekić S, Quarrie SA (1987) Abscisic acid accumulation in lines of maize differing in drought resistance: a comparison of intact and detached leaves. J Plant Physiol 127(3–4):203–217
87. Lee TM, Lur HS, Chu C (1997) Role of abscisic acid in chilling tolerance of rice (Oryza sativa L.) seedlings.: II. Modulation of free polyamine levels. Plant Sci 126(1):1–10
88. Moons A, Bauw G, Prinsen E, Van Montagu M, Van Der Straeten D (1995) Molecular and physiological responses to abscisic acid and salts in roots of salt-sensitive and salt-tolerant Indica rice varieties. Plant Physiol 107(1):177–186
89. Tardieu F, Lafarge T, Simonneau TH (1996) Stomatal control by fed or endogenous xylem ABA in sunflower: interpretation of correlations between leaf water potential and stomatal conductance in anisohydric species. Plant Cell Environ 19(1):75–84
90. Ünyayar S, Keleþ Y, Ünal E (2004) Proline and ABA levels in two sunflower genotypes subjected to water stress. Bulg J Plant Physiol
91. Rivero RM, Kojima M, Gepstein A, Sakakibara H, Mittler R, Gepstein S, Blumwald E (2007) Delayed leaf senescence induces extreme drought tolerance in a flowering plant. Proc Natl Acad Sci 104(49):19631–19636
92. Maggio A, Barbieri G, Raimondi G, De Pascale S (2010) Contrasting effects of GA 3 treatments on tomato plants exposed to increasing salinity. J Plant Growth Regul 29(1):63–72
93. Sterling TM, Hal JC (1997) Mechanism of action of natural auxins and the auxinic herbicides. Rev Toxicol 1:111–142
94. Grossmann K (2003) Mediation of herbicide effects by hormone interactions. J Plant Growth Regul 22(1):109–122
95. Fallon H, Tollerud D, Breslow N (1994) Veterans and agent orange: health effects of herbicides used in Vietnam. In: Committee to review the health effects in Vietnam veterans of exposure to herbicides, Division of Health Promotion and Disease Prevention, Institute of Medicine, 26
96. Grossmann K (2010) Auxin herbicides: current status of mechanism and mode of action. Pest Manag Sci 66(2):113–120
97. Grzesik M (1988, September). Factors influencing the effectiveness of growth regulators in nursery production. In: III International Symposium on Growth Regulators in Ornamental Horticulture 251, pp 371–376
98. Wróblewska K, Debicz R (2013) Influnce of time of benzyladenine application on rooting of cuttings and subsequent development of portulaca umbraticola kunth. Acta Sci Pol-Hortoru Cultus 12(1):89–99
99. Carey D, Whipker B, McCall I, Buhler W (2007) Cytokinin based PGR affects growth of vegetative petunia. ASG/PGRSA, p 94
100. Saniewski M, Góraj J, Węgrzynowicz-Lesiak E, Okubo H, Miyamoto K, Ueda J (2010) Different growth of excised and intact fourth internode after removal of the flower bud in growing tulips: focus and auxin action. J Fruit Ornam Plant Res 18:297–308
101. Weaver RJ (1958) Effect of gibberellic acid on fruit set and berry enlargement in seedless grapes of Vitis vinifera. Nature 181(4612):851
102. Carswell FE, Day JS, Gould KS, Jameson PE (1996) Cytokinins and the regulation of plant form in three species of Sophora. NZ J Bot 34(1):123–130
103. De Vries DP, Dubois LA (1988) The effect of BAP and IBA on sprouting and adventitious root formation of 'Amanda'rose single-node softwood cuttings. Sci Hortic 34(1–2):115–121
104. Ahmad A, Hayat S, Fariduddin Q, Ahmad I (2001) Photosynthetic efficiency of plants of Brassica juncea, treated with chlorosubstituted auxins. Photosynthetica 39(4):565–568

105. Hayat S, Ahmad A, Mobin M, Fariduddin Q, Azam ZM (2001) Carbonic anhydrase, photosynthesis, and seed yield in mustard plants treated with phytohormones. Photosynthetica 39(1):111–114
106. Afroz S, Mohammad F, Hayat S, Siddiqui MH (2006) Exogenous application of gibberellic acid counteracts the ill effect of sodium chloride in mustard. Turkish J Biol 29(4):233–236
107. Ranwala AP, Legnani G, Reitmeier M, Stewart BB, Miller WB (2002) Efficacy of plant growth retardants as preplant bulb dips for height control in LA and oriental hybrid lilies. HortTechnology 12(3):426–431
108. Gray WM (2004) Hormonal regulation of plant growth and development. PLoS Biol 2(9): e311
109. Hilli JS, Vyakarnahal BS, Biradar DP, Ravi H (2010) Effect of growth regulators and stages of spray on growth, fruit set and seed yield of ridgegourd (Luffa acutangula L. Roxb). Karnataka J Agric Sci 23(2):239–242
110. Prajapati S, Jamkar T, Singh O, Raypuriya N, Mandloi R, Jain P (2015) Plant growth regulators in vegetable production: an overview. Plant Arch 15:619–626
111. Prasad RN, Singh SK, Yadava RB, Chaurasia SNS (2013) Effect of GA3 and NAA on growth and yield of tomato. Veg Sci 40(2):195–197
112. Balraj R, Kurdikeri MB (2002) Effect of growth regulators on growth and yield of chilli (Capsicum annuum) at different pickings. Indian J Hortic 59(1):84–88
113. Singh M, John SA, Rout S, Patra SS (2015) Effect of GA3 and NAA on growth and quality of garden pea (Pisum sativum L.) cv. Arkel Bioscan 10(3):381–383
114. Ravindra A, Anjanabha B, Bharat C (2016) Effect of foliar application of plant growth regulators on growth and yield of potato seed tubers propagated from micro plantlets on soilless solid media in greenhouse. Adv Res J Crop Improv 7(2):234–239
115. Patel MJ, Patel HC, Chavda JC (2010) Influence of plant growth regulators and their application methods on yield and quality of onion (Allium cepa L.). Asian J Hortic 5 (2):263–265
116. Dhoran VS, Gudadhe SP (2012) Effect of plant growth regulators on seed germination and seedling vigour in Asparagus sprengeri Regelin. Int Res J Biol Sci 1(7):6–10
117. Mir MM, Barche S, Singh DB (2004) Effect of plant growth regulators on growth, yield and quality of strawberry (Fragaria ananassa Duch) cv.'Sweet charley'. Appl Biol Res 6(1/2):48–51
118. Dalai S, Singh MK, Singh KV, Kumar M, Malik S, Kumar V (2015) Effect of foliar application of GA3 and NAA on growth, flower-ing yield and yield attributes of cucumber [Cucumis Sativus L.]. Ann Hortic 8(2):181–194
119. Anuradha S, Rao SSR (2001) Effect of brassinosteroids on salinity stress induced inhibition of seed germination and seedling growth of rice (Oryza sativa L.). Plant Growth Regul 33 (2):151–153
120. Krishna P (2003) Brassinosteroid-mediated stress responses. J Plant Growth Regul 22 (4):289–297

Printed by Printforce, the Netherlands